T0213689

# Lecture Notes in Computer Science 10196

Commenced Publication in 1973
Founding and Former Series Editors:
Gerhard Goos, Juris Hartmanis, and Jan van Leeuwen

James McDermott · Mauro Castelli
Lukas Sekanina · Evert Haasdijk
Pablo García-Sánchez (Eds.)

# Genetic Programming

20th European Conference, EuroGP 2017
Amsterdam, The Netherlands, April 19–21, 2017
Proceedings

 Springer

*Editors*

James McDermott
University College Dublin
Dublin
Ireland

Mauro Castelli
Universidade Nova de Lisboa
Lisbon
Portugal

Lukas Sekanina
Brno University of Technology
Brno
Czech Republic

Evert Haasdijk
Vrije Universiteit Amsterdam
Amsterdam
The Netherlands

Pablo García-Sánchez
University of Cádiz
Cádiz
Spain

ISSN 0302-9743          ISSN 1611-3349   (electronic)
Lecture Notes in Computer Science
ISBN 978-3-319-55695-6          ISBN 978-3-319-55696-3   (eBook)
DOI 10.1007/978-3-319-55696-3

Library of Congress Control Number: 2017934458

LNCS Sublibrary: SL1 – Theoretical Computer Science and General Issues

Printed on acid-free paper

This Springer imprint is published by Springer Nature
The registered company is Springer International Publishing AG
The registered company address is: Gewerbestrasse 11, 6330 Cham, Switzerland

# Preface

The 20th European Conference on Genetic Programming (EuroGP) took place in Amsterdam, The Netherlands, during April 19–21, 2017.

Genetic programming (GP) is a unique field of research. It uses the principles of Darwinian evolution, already well-known in genetic algorithms and other areas of evolutionary computation, to approach problems in the synthesis, improvement, and repair of computer programs. The universality of computer programs, and their importance in so many areas of our lives, means that the automation of these tasks is an exceptionally ambitious challenge with far-reaching implications. It has attracted a very large number of researchers: over 12,000 articles now appear in the online GP bibliography[1]. GP researchers have demonstrated some extraordinary successes, many presented at EuroGP itself.

Since the first EuroGP event, in Paris in 1998, EuroGP has been the only conference exclusively devoted to the evolutionary generation of computer programs. Indeed, EuroGP represents the single largest venue at which GP results are published. It plays an important role in the success of the field, by serving as a forum for expressing new ideas, meeting fellow researchers, and starting collaborations. It attracts scholars from all over the world. The 20th anniversary edition of EuroGP was characterized by the same friendly and welcoming atmosphere for which it has always been known.

EuroGP 2017 received 34 submissions from around the world. The papers underwent a rigorous double-blind peer-review process, each being reviewed by at least three members of the international Program Committee from 24 countries.

The members of the Program Committee encountered an exceptionally high standard this year, and the chairs decided to recognize this by allowing an increase compared with the typical EuroGP acceptance rate. The selection process resulted in the papers presented in this volume, with 14 accepted for full-length oral presentation (41.2% acceptance rate) and eight for short talks (64.7% acceptance rate for both categories of papers combined). Authors of both categories of papers also had the opportunity to present their work in poster sessions.

The wide range of topics in this volume reflects the current state of research in the field. Thus, we see topics and applications including program synthesis, genetic improvement, grammatical representations, self-adaptation, multi-objective optimization, program semantics, search landscapes, mathematical programming, games, operations research, networks, evolvable hardware, and program synthesis benchmarks.

---

[1] http://liinwww.ira.uka.de/bibliography/Ai/genetic.programming.html.

Together with three other co-located evolutionary computation conferences (EvoCOP 2017, EvoMusArt 2017, and EvoApplications 2017), EuroGP 2017 was part of the Evo* 2017 event. This meeting could not have taken place without the help of many people. The EuroGP Organizing Committee is particularly grateful to the following.

- SPECIES – the Society for the Promotion of Evolutionary Computation in Europe and its Surroundings, and its board:

  Marc Schoenauer, President
  Anna Esparcia Alcázar, Secretary and Vice-President
  Wolfgang Banzhaf, Treasurer

- The high-quality and diverse EuroGP Program Committee. Each year the members give freely of their time and expertise, in order to maintain high standards in EuroGP, and provide constructive feedback to help authors improve their papers.
- Marc Schoenauer of Inria-Saclay, France, for his continued hosting and maintaining of the MyReview conference management system.
- The local organizing team, Evert Haasdijk and Jacqueline Heinerman of VU Amsterdam.
- Pablo García-Sánchez (University of Cádiz, Spain) for the Evo* 2017 publicity and website.
- Our invited speakers, Kenneth De Jong and Arthur Kordon, who gave inspiring, enlightening, and entertaining keynote talks.
- Jennifer Willies, Evo* coordinator, whose dedicated and continued involvement in Evo* since 1998 remains essential for building the image, status, and unique atmosphere of this series of events.

April 2017

James McDermott
Mauro Castelli
Lukas Sekanina
Evert Haasdijk
Pablo García-Sánchez

# Organization

## Organizing Committee

### Program Co-chairs

James McDermott          University College Dublin, Ireland
Mauro Castelli           Universidade Nova de Lisboa, Portugal

### Publication Chair

Lukas Sekanina           Brno University of Technology, Czech Republic

### Local Chair

Evert Haasdijk           Vrije Universiteit Amsterdam, The Netherlands

### Publicity Chair

Pablo García-Sánchez     University of Cádiz, Spain

### Conference Administration

Jennifer Willies         EvoStar Coordinator

## Program Committee

Alexandros Agapitos         University College Dublin, Ireland
R. Muhammad Atif Azad       University of Limerick, Ireland
Ignacio Arnaldo             MIT, USA
Douglas Augusto             LNCC/UFJF, Brazil
Wolfgang Banzhaf            Michigan State University, USA
Mohamed Bahy Bader          University of Portsmouth, UK
Helio Barbosa               LNCC/UFJF, Brazil
Heder Bernardino            LNCC/UFJF, Brazil
Anthony Brabazon            University College Dublin, Ireland
Nicolas Bredeche            Université Pierre et Marie Curie, France
Stefano Cagnoni             University of Parma, Italy
Ernesto Costa               University of Coimbra, Portugal
Antonio Della Cioppa        University of Salerno, Italy
Grant Dick                  University of Otago, New Zealand
Federico Divina             Pablo de Olavide University, Spain
Marc Ebner                  Ernst-Moritz-Arndt Universität Greifswald, Germany
Aniko Ekart                 Aston University, UK
Francisco Fernandez         Universidad de Extremadura, Spain
  de Vega

# Contents

**Posters**

# Oral Presentations

# Evolutionary Program Sketching

Iwo Błądek[(✉)] and Krzysztof Krawiec

Institute of Computing Science, Poznan University of Technology,
Piotrowo 2, 60-965 Poznań, Poland
{ibladek,krawiec}@cs.put.poznan.pl

**Abstract.** Program synthesis can be posed as a satisfiability problem and approached with generic SAT solvers. Only short programs can be however synthesized in this way. Program sketching by Solar-Lezama assumes that a human provides a partial program (*sketch*), and that synthesis takes place only within the uncompleted parts of that program. This allows synthesizing programs that are overall longer, while maintaining manageable computational effort. In this paper, we propose Evolutionary Program Sketching (EPS), in which the role of sketch provider is handed over to genetic programming (GP). A GP algorithm evolves a population of partial programs, which are being completed by a solver while evaluated. We consider several variants of EPS, which vary in program terminals used for completion (constants, variables, or both) and in the way the completion outcomes are propagated to future generations. When applied to a range of benchmarks, EPS outperforms the conventional GP, also when the latter is given similar time budget.

**Keywords:** Program synthesis · Satisfiability modulo theory · Program sketching · Genetic programming

## 1 Introduction

Program synthesis (PS), as many other search tasks, can be posed as a *satisfiability problem*: given a *contract*, i.e., a logical predicate that describes the desired input-output behavior, and an encoding of program's structure parametrized with Boolean variables, determine whether a valuation of those variables exists that makes the program satisfy the contract. To obtain this valuation, called in propositional logic a *model*, the synthesis formula is passed to a SAT *solver*, which produces a feasible variable assignment, and thus a program that is guaranteed to meet the contract, or otherwise states that the sought program does not exist. In practice, the solver is equipped with an additional abstraction layer, a *theory* that enables reasoning in terms of, for instance, integer arithmetic. This leads to the concept of *satisfiability modulo theories* (SMT) used in a range of past works on program synthesis [1–3].

SMT solvers implement exact algorithms supported by heuristics which are guaranteed to find the sought program in the prescribed search space (unless they

© Springer International Publishing AG 2017
J. McDermott et al. (Eds.): EuroGP 2017, LNCS 10196, pp. 3–18, 2017.
DOI: 10.1007/978-3-319-55696-3_1

time out). Nevertheless, the above approach to PS suffers from poor scalability: the assumed maximum length of a synthesized program determines the number of involved variables, and search space grows exponentially with that number. Additionally, solving an SMT problem involves solving a SAT problem as a subtask, which is known to be NP-complete. Even with contemporary sophisticated SMT solvers, only short programs can be synthesized with this approach in a reasonable time.

To address this problem, Solar-Lezama proposed *program sketching* [4]. Therein, one assumes that a partial program (*sketch*) is provided, with one or more *holes* marking the locations of missing code pieces. The synthesis takes place only in the holes, while the sketch remains intact. In this way, the total length of the program (sketch length plus the length of the synthesized code pieces) is increased, while the task remains manageable for the solver.

In the original sketching, it is assumed that a human provides the sketch. Indeed, there are plausible scenarios in which a programmer may come up with an overall program structure, yet fails to implement all the details. This endows sketching with certain interactive flavor, which is often desirable in software development. On the other hand, human-provided sketch can be suboptimal for completion, or in an extreme case incorrect, i.e., such that cannot be completed to satisfy the contract.

*Evolutionary Program Sketching* (EPS) we propose in this paper substitutes the human sketcher with an evolutionary process. A GP algorithm evolves partial programs with holes. When evaluating a program, the solver attempts to complete the holes with code pieces; in this preliminary study, we fill the holes with constants and input variables. Fitness measures the extent to which the holes can be completed to meet the contract. We consider several variants of EPS, which vary in the way they handle code completions, in particular in whether the holes persist after evaluation. Experimental verification proves EPS feasible and points to interesting potential extensions of this approach.

## 2    Program Sketching

Presentation of program sketching requires brief introduction to SMT-based program synthesis. The synthesis task is given by a *contract*, typically a pair of logical formulas: a *precondition $Pre$* – the constraint imposed on program input, and a *postcondition $Post$* – a logical clause that should hold upon program completion. The content (code) of a candidate program is controlled by a vector of variables $\mathbf{b}$. For instance, when synthesizing sequential programs $n$ instructions long, with each instruction taken from an instruction set of cardinality $k$, one would assume $\mathbf{b} \in [1, k]^n$, i.e., that each program corresponds one-to-one to a vector of $n$ variables controlling the choices of instructions on particular positions.

Let $p_{\mathbf{b}}$ denote a program determined by a specific vector $\mathbf{b}$, and let $p_{\mathbf{b}}(in)$ denote the output produced by $p_{\mathbf{b}}$ when applied to input $in$. Solving a synthesis task $(Pre, Post)$ is equivalent to proving that

$$\exists_{\mathbf{b}} \forall_{in} Pre(in) \implies Post(in, p_{\mathbf{b}}(in)),$$ (1)

where $Pre(in)$ is the precondition valuated for the input $in$, and $Post(in, p_{\mathbf{b}}(in))$ is the postcondition valuated for the input $in$ and the output produced by $p_{\mathbf{b}}$ for $in$.

For illustration, consider synthesizing a program that calculates the maximum of two integers $(x, y)$. For this synthesis task, the contract is defined as follows:

$$Pre((x, y)) \iff (x, y) \in \mathbb{Z}^2$$
$$Post((x, y), o) \iff o \in \mathbb{Z} \wedge o \geq x \wedge o \geq y \wedge (o = x \vee o = y)$$ (2)

This is an example of *a complete specification*, which defines the desired behavior of the sought program for *all possible inputs*, the number of which happens to be infinite here. If programs are expressed in terms of the theory known to the solver, the solver can prove (1), and so determine $\mathbf{b}$ and the sought program $p_{\mathbf{b}}$. The solver achieves this without actually running any program, because the properties of the output can be logically, *modulo the theory*, inferred from the properties of the input and the properties of program code. For the above problem to be solved, it would be sufficient to provide the solver with the Linear Integer Arithmetic (LIA) theory [5]. The resulting synthesized program $p_{\mathbf{b}}$ is guaranteed to adhere to the contract or, in other words, it is correct by construction.

This approach to synthesis, as elegant as it seems, is nevertheless feasible only when the sought program is short. As the above example of sequential programs shows, the cardinality of the search space grows exponentially with program length $n$, and even the modern SMT solvers, equipped with sophisticated heuristics for prioritizing search, become quickly computationally inefficient.

Program sketching [4,6] extends the effective program length while keeping the computational expense of synthesis at bay. This is achieved by assuming that the sought program is to some extent fixed, and the fixed part forms a *sketch*, a template that should be completed by a solver. For the max$(x, y)$ synthesis problem mentioned above, a template could have the following form:

$$\text{if } (h_1) \text{ then } h_2 \text{ else } h_3,$$ (3)

where $h_1, h_2$ and $h_3$ are *holes*, i.e. program parts not specified by the sketch. Crucially, only the instructions in the holes can be varied by manipulating the control variables in $\mathbf{b}$. The number and domains of these control variables depend on assumed structure for the missing code. This structure is usually defined by a grammar, and variables in $\mathbf{b}$ determine the traversal of production rules of that grammar.

*Example 1.* Consider the max$(x, y)$ problem presented above. The user starts by constructing a sketch of the solution as in (3). She assumes that $h_1$ should be filled with a Boolean expression of the form *var op var*, where *op* is an arithmetic operator ($\{>, =, <\}$) and *vars* are either $x$ or $y$, and that both $h_2$ and $h_3$ should be also filled with *var*. All such hole completions can be enumerated (encoded) with five variables: four binary variables that control whether $x$ or $y$ should be be filled in for $h_2$, $h_3$, and for both sides of *op*, and one ternary variable that controls the choice of *op*. These five variables together determine the search space for sketching (of size $2^4 \cdot 3 = 48$) and form the vector **b** that is controlled by the solver.                                                                                    □

We present here only the aspects that are relevant for this paper; the original program sketching involves more mechanisms, among them maintaining a set of test cases and augmenting them with counterexamples produced by the solver. Overall, sketching has multiple merits: it not only increases the effective program length, but delegates some control on the synthesis process to a human. Allowing a user to express her *intent* in this way is often desirable in the practice of software development. Nevertheless, there are limitations too. A human might find it difficult to come up with a sketch featuring a number of holes small enough for a solver to find a solution in an acceptable time. The provided sketch can be suboptimal in not forming the most elegant (or the shortest) solution to a given problem, or not enabling the solver to find the solution fast enough. In the worst scenario, a human may propose a wrong sketch, which cannot be completed so as to satisfy the contract. Evolutionary Program Sketching detailed in the next section addresses some of these issues.

# 3   Evolutionary Program Sketching

EPS evolves partial programs (sketches) and evaluates them based on the substitution for the missing parts determined by an SMT solver. The workflow of the method is presented in Fig. 1, and in the following we detail its key components.

## 3.1   Problem Specification

As the conventional GP, EPS assumes that a synthesis problem is given by a set of instructions $I$ of which the programs are to be built, and a set of examples $T$ (tests) on which the programs are evaluated. Each example is a pair $(in, out)$ of an input $in$ and the corresponding desired output $out$. Such specification is *partial*: the universal quantifier in (2) is bound to $T$ ($\forall_{(in, out) \in T}$), the precondition is always *true*, and the postcondition simply checks whether $p_b(in) = out$. This stands in contrast to the complete formal specification of the $\max(x, y)$ example in the previous section, and places GP and EPS in the realm of inductive approaches to PS, where the behavior of the synthesized program beyond the set of examples cannot be in general predicted.

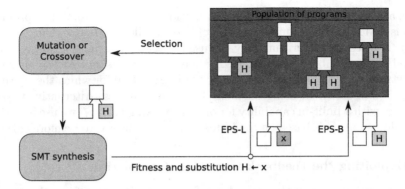

**Fig. 1.** The flow of candidate programs in EPS. $H$ represents a hole, and $x$ the content assigned to the hole. The EPS-L and EPS-B variants are described in Sect. 3.4.

## 3.2   Instruction Set

As in sketching (Sect. 2), we allow for incomplete (partial) programs. To this aim, we extend the instruction set $I$ with a set of terminal instructions $H$ containing a symbol for each kind of hole allowed in a program. The kind of a hole determines the content it may be filled with, e.g. an integer or Boolean expression, or linear or nonlinear arithmetic expression. In conclusion, the GP process works with the set of instructions $I \cup H$, where each $h \in H$ is treated like other terminal symbols (i.e., it is subject to search operators, and in the case of strongly-typed GP it has an assigned type).

## 3.3   Fitness Function

Partial programs are incomplete and thus cannot be evaluated in the common GP fashion, i.e. by executing them on examples in $T$. This is, however, not a problem if evaluation is to be based on a query to an SMT solver which can substitute the missing code pieces, as in the original sketching. But the outcome of the completion process described in Sect. 2 is only twofold: either a perfect completion (and thus a correct program) is found and the search terminates, or there is no feasible completion[1]. There is no obvious way of eliciting a fine-grained fitness from this binary outcome.

To address this issue, we reformulate the synthesis problem, originally posed as a search problem in (1), as an *optimization problem*, asking the solver to determine the hole completion that maximizes the number of tests passed by the evaluated program. The number of passed tests becomes the (maximized) fitness of the program:

$$f(p) = \max_{\mathbf{b}} |(in, out) \in T : Pre(in) \implies Post(in, p_{\mathbf{b}}(in))|. \qquad (4)$$

---

[1] Technically, the solver may also time-out, which we interpret as lack of feasible completion too.

Optimization is beyond the original formulation of SMT satisfiability problem, and thus SMT solvers cannot be expected to handle it. However, in the case of $f(p)$, a bisection algorithm may be used, because we have a discrete and bounded set of possible fitness values. By halving intervals and adding appropriate constraints, it is possible to determine the largest $f(p)$ for which the synthesis formula is still satisfied, using only $\log_2 |T|$ solver queries. Alternatively, there are solvers with a built-in capability for optimization, like Z3 [7] we use in Sect. 5. Our implementation is based on the latter, because it proved to be more efficient.

### 3.4  Exploiting the Feedback from Hole Completion

The optimization process that calculates fitness in (4), apart from the number of passed tests, produces also the optimal completion of holes (the model). Let $\mathbf{b}^*$ be the associated optimal assignment of variables found in (4). In the default scenario, we discard it. However, one may argue that the assignment defined by $\mathbf{b}^*$ contains useful knowledge that can be leveraged. Thus, we consider an alternative variant in which the completion defined by $\mathbf{b}^*$ is incorporated in the evaluated program, and the modified program replaces the original candidate solution in the population (in other words, $f$ has a *side effect* consisting in $p$ being modified).

In both variants, the process of hole completion can be seen as a local search, or in evolutionary terms as an adaptation that takes place during individual's lifetime. It seems thus justified to liken the former variant to *Baldwinian evolution*, in which such adaptations impact individual's fitness, but do not get explicitly inherited, and the latter to *Lamarckian evolution*, in which the acquired adaptations do get inherited directly. We will refer to these variants in this way and use the respective acronyms EPS-B and EPS-L.

## 4  Related Work

The work most directly related to EPS is obviously program sketching [4,6], presented in Sect. 2. There are, however, other studies that involve the two distinguishing features of EPS:

– its formal (and unusual in GP) approach to program evaluation,
– evolution of partial programs.

We group them according to these characteristics.

Concerning the **use of formal techniques** in GP, Johnson [8] was probably the first to use *model checking* for calculating fitness in GP. Model checking is a specific approach to *formal verification* of programs and systems, which essentially consists in determining whether a given program $p$ meets the contract $(Pre, Post)$:

$$\forall_{in} Pre(in) \implies Post(in, p(in)). \tag{5}$$

Verification applies to an existing program and is thus computationally less demanding than synthesis (1). In [8], Johnson used temporal logic to express formal specifications that describe the desired time-wise behavior of finite state machines. A fairly conventional GP algorithm was employed to evolve candidate state machines, with fitness defined as the number of fulfilled constituent clauses in the contract. The work demonstrated successful application of this approach to synthesis of control programs for a vending machine.

Temporal logic and GP are also the underlying mechanisms in other related work by Katz and Peled [9]. Similarly to [8] – and in contrast to this study – the authors evolve complete programs. When evaluating them, they distinguish four levels of program correctness: first, in which no scenario of program execution can satisfy the contract; second, in which some program executions satisfy the contract; third, in which all terminating executions meet the contract, and the highest level, in which all program executions meet the contract. Given this distinction, the fitness measure counts the satisfied postconditions. The authors apply the methods to examples from [10] and to synthesis of mutual exclusion algorithms and correction of erroneous programs. To verify programs, they consider both model checking and an SMT solver. Their use of SMT solver is, however, different than in this paper. In EPS, SMT solver is used to synthesize the content of a hole, while in [9] it is used solely for verification and producing counterexamples.

Concerning **evolving and completing partial programs**, the latter EPS feature identified at the beginning of this section, the related work is very limited. Though partial programs are occasionally considered in GP (cf., e.g., program contexts in semantic GP [11]), no GP approach known to us explicitly maintains them in population. However, program completion in EPS is limited to single-node terminals, and as such can be likened to optimizing constants in programs, which attracted significant attention in GP research. Among past contributions, Sarafopoulos [12] hybridized GP with evolutionary strategies (ES), where the ES component was responsible solely for fine-tuning the constants in candidate programs. Azad and Ryan [13] extended GP with a simple local search that tunes the instructions of individuals (including the internal nodes of program trees), and implemented a caching mechanism to reduce the computational overhead of tuning. When evaluated on a range of benchmarks, their approach synthesizes fitter and smaller programs than standard GP. The cited work features comprehensive review of analogous techniques, which we redirect an interested reader to.

In a broader perspective, EPS capability to improve candidate programs pertains also to *memetic* approaches, researched in numerous studies in the past. For instance, semantic backpropagation [14] and memetic semantic genetic programming [15] offer search operators that improve programs locally, i.e. at the level of particular instructions/subprograms.

## 5    Experimental Evaluation

**Objectives.** We compare the Baldwinian (EPS-B) and Lamarckian (EPS-L) variants of EPS on a range of problems, in a few configurations detailed in the

$$I ::= I + I \mid I - I \mid I * I \mid I / I \mid \text{ite}(B,I,I) \mid c$$
$$\mid v_1 \mid v_2 \mid \ldots \mid v_k$$
$$\mid h_1 \mid h_2 \mid \ldots \mid h_l$$
$$B ::= I < I \mid I <= I \mid I = I \mid B = B \mid I >= I \mid I > I$$

**Fig. 2.** The NIA grammar defining the set of considered programs. $c$ stands for an integer constant, $v_i$ for the $i$th input variable, and $h_j$ for the $j$th hole allowed in the program. ite stands for *if-then-else*, the conventional conditional statement.

following, within the conventional tree-based GP. Our goal is to find out which of EPS variants and configurations fair the best and how its performance compares to that of standard GP.

**Domain.** As follows from Sect. 2, applicability of EPS is conditioned on a theory that supports the reasoning conducted by the SMT solver. Past research led to elaboration of several popular theories and associated logics, now systematized by the SMT-LIB standard [5,16]. The theories vary in the data types they support (e.g., Booleans, bit vectors, integers, floating point, reals) and in logics that constrain the form of expressions/formulas (e.g., linear, nonlinear)[2]. Wider logics offer more expressibility but typically require higher computational effort from the solver. In this preliminary study, we settle on a mid-way compromise in that trade-off, the Nonlinear Integer Arithmetic (NIA) logic. This choice determines:

1. The types that can be used in expressions: integer (I) and Boolean (B),
2. The set of expressions that can be passed to the solver, which, by the design of EPS, becomes also the instruction set to be used by the evolutionary process.

A solver equipped with NIA can prove theorems that obey the grammar shown in Fig. 2. For the sake of synthesizing programs that are $k$-ary integer functions ($I^k \rightarrow I$), we assume that the starting symbol of the grammar is I, even though the top-level type of the predicates passed to the solver is naturally B, as follows from the synthesis formula (1). The grammar diverges from the conventional NIA in two ways:

– It features additional terminal symbols $h_i$, which implement the holes to be substituted by the solver in EPS.
– It does not contain some of the less common nonterminals, e.g. mod and abs.

**Configurations.** In sketching as introduced by Solar-Lezama [4], holes can be filled by arbitrary code pieces (of the compatible type). However, the larger the code pieces one considers to substitute for holes, the larger the search space and the more expensive synthesis becomes. In this study, we consider the simplest approach, i.e., we allow the holes to be substituted only with single-instruction code pieces, more precisely the integer-valued terminals available in the NIA

---

[2] http://smtlib.cs.uiowa.edu/logics.shtml.

**Table 1.** Compared configurations.

| Configuration | Terminals that can be substituted for holes | |
|---|---|---|
| | Constants $c$ | Input variables $v_i$ |
| GP | | |
| $EPS_c$ | ✓ | |
| $EPS_v$ | | ✓ |
| $EPS_{cv}$ | ✓ | ✓ |

**Table 2.** Program synthesis benchmarks.

| Benchmark | #vars. | Formula | Tests | #tests |
|---|---|---|---|---|
| Keijzer12 | 2 | $x_1^4 - x_1^3 + x_2^2/2 - x_2$ | $x_1, x_2 \in \{-3, \ldots, 0, \ldots, 3\}$ | 49 |
| Koza1 | 1 | $x^4 + x^3 + x^2 + x$ | $x \in \{-5, -4, \ldots, 0, \ldots, 4, 5\}$ | 11 |
| Koza1-p | | $3x^4 - 2x^3 + 6x^2 + 3x - 4$ | | |
| Koza1-2D | 2 | $x_1^4 + x_2^3 + x_1^2 + x_2$ | $x_1, x_2 \in \{-3, \ldots, 0, \ldots, 3\}$ | 49 |
| Koza1-p-2D | | $3x_1^4 - 2x_2^3 + 6x_1^2 + 3x_2 - 4$ | | |

grammar. Nevertheless, even this simple design choice leads to several configurations summarized in Table 1, which vary in the terminals that are substituted for holes: constants only ($EPS_c$), variables only ($EPS_v$), or both ($EPS_{cv}$). These three configurations together with two EPS variants (EPS-B, EPS-L) lead to six setups. Naturally, standard GP cannot handle holes, so the hole terminals are removed from the grammar for this method.

**Benchmarks.** NIA allows us to use benchmarks that are similar in spirit to symbolic regression, albeit dwell in the integer domain. We employ the benchmarks presented in Table 2; these are based on their real-valued counterparts from the GP benchmarks suite [17], but by necessity use integer inputs, typically from a wider interval than in the original benchmark.

**Search Operators.** The presence of two types (I and B) implies that the GP part of EPS implementation has to be typed. We impose the correct typing by means of a grammar in Fig. 2 and constrain the actions of initialization, subtree mutation and tree-swapping crossover operators, so that they guarantee producing programs that follow the grammar. When generating a random program for the initial population, we traverse the grammar rules starting from the I symbol, reducing the probability of nonterminals when closing to the allowable tree height (Table 3); if the resulting program exceeds that limit, we scrap it and initialize a new program. For subprograms to be inserted by mutation we proceed similarly, however starting either from the I or from the B symbol, depending on the type of the instruction being replaced. The crossover operator picks a random location from the first parent, draws a random location of the same type in the second parent, and swaps the subtrees rooted at those locations. Should it fail to find a type-compatible location in the second parent, it discards both

**Table 3.** Parameters of the evolutionary algorithm.

| Parameter | Value |
|---|---|
| Population size | 250 |
| Maximum height of initial programs | 4 |
| Maximum height of subprograms inserted by mutation | 4 |
| Constant terminals drawn from interval | $[0, 5]$ |
| Maximum number of generations | 100 |
| Probability of mutation | 0.5 |
| Probability of crossover | 0.5 |
| Tournament size | 7 |

parents and starts anew with another selected pair of parents (this may happen only for the B type, as at least one instruction of type I is guaranteed to exist in every program).

**Solver Budget.** In EPS, the solver is given the computational budget of 1.5 s for a single query. If it fails to find an optimal assignment in this time, the evaluated program receives the worst possible fitness of zero. Handling timeouts is essential, because it is very hard in general to estimate the upper bound on solver's computation time.

**Implementation.** EPS has been implemented in authors' PySV (Python Synthesis and Verification) framework (responsible for constructing queries to Z3 solver) and SMTGP Scala framework (responsible for running evolution with holes). Sources of both of these frameworks are accessible on Github[3]. The latter framework is based on two Scala libraries: *Functional Evolutionary Algorithms* (FUEL) and *Synthesis with Metaheuristics* (SWIM), both originating in [18] and also available on GitHub[4]. The SMTGP framework implements the conventional GP workflow, with the exception of fitness function that passes the individuals with holes to the PySV framework, which in turn handles the call to the SMT solver. The communication with the solver is realized using the SMT-LIB standard [16]. We employ the Z3 SMT solver by Microsoft [7], one of the most efficient and powerful non-commercial solvers.

**Results.** Table 4 presents the success rate of particular configurations on individual benchmarks, and Table 5 and Fig. 3 the average fitness of the best-of-run programs. Applying the configurations that substitute holes with variables only (EPS-$L_v$ and EPS-$B_v$) to univariate benchmarks is pointless, so such cases are excluded from presentation. The figure reveals clear, repetitive pattern of relative performances of individual configurations. EPS-$B_c$ fares the best: it tops the other configurations in terms of success rate, and reliably produces an optimal

---

[3] https://github.com/iwob.
[4] https://github.com/kkrawiec.

**Table 4.** The number of optimal solutions found (maximum: 100).

| | GP | | | EPS | | | | | |
|---|---|---|---|---|---|---|---|---|---|
| | GP | $GP_T$ | $GP_{5000}$ | $L_c$ | $L_v$ | $L_{cv}$ | $B_c$ | $B_v$ | $B_{cv}$ |
| Keijzer12 | 0 | 0 | 5 | 0 | 0 | 1 | 39 | 1 | 0 |
| Koza1 | 19 | 68 | 96 | 33 | - | 32 | 100 | - | 100 |
| Koza1-p | 0 | 0 | 0 | 5 | - | 3 | 100 | - | 100 |
| Koza1-2D | 1 | 12 | 20 | 2 | 0 | 11 | 80 | 21 | 23 |
| Koza1-p-2D | 0 | 0 | 0 | 0 | 0 | 1 | 75 | 0 | 0 |

**Table 5.** Average end-of-run fitness.

| | GP | | | EPS | | | | | |
|---|---|---|---|---|---|---|---|---|---|
| | GP | $GP_T$ | $GP_{5000}$ | $L_c$ | $L_v$ | $L_{cv}$ | $B_c$ | $B_v$ | $B_{cv}$ |
| Keijzer12 | 15.85 | 23.02 | 25.06 | 23.92 | 18.05 | 27.77 | 39.05 | 20.45 | 17.47 |
| Koza1 | 5.89 | 9.74 | 10.87 | 9.93 | - | 9.83 | 11.00 | - | 11.00 |
| Koza1-p | 2.59 | 4.45 | 3.98 | 9.05 | - | 8.78 | 11.00 | - | 11.00 |
| Koza1-2D | 16.54 | 29.73 | 33.18 | 23.39 | 19.47 | 31.29 | 45.42 | 27.36 | 23.70 |
| Koza1-p-2D | 9.29 | 17.18 | 14.60 | 22.60 | 10.66 | 29.47 | 46.23 | 12.56 | 15.41 |

program in each run for Koza1 and Koza1-p. The figure suggests that EPS-$L_{cv}$ and EPS-$L_c$ are the two competing runners-up; however, Table 4 leaves no doubts that they are much less likely to synthesize a correct program.

Overall, the configurations that complete the holes with variables only (EPS-$L_v$ and EPS-$B_v$) fare the worst. This suggests that substituting with constants, available in the other configurations of EPS, is essential. This capability is particularly important for the benchmarks considered here, which feature at most two variables, and manipulating them does not leave much space for improvement.

Table 6 presents the average runtimes of configurations on particular benchmarks, which reveals that engaging the SMT solver comes at a price: EPS runs take up to four orders of magnitude longer than standard GP. One may question thus whether comparing EPS with short-timed GP is entirely fair. To address this issue, we devise another configuration, $GP_T$, in which genetic programming uses the same parameters as previously (Table 3), except for the maximum number of generations, which is replaced by the time limit, equal to the average runtime of the EPS configurations on a given benchmark. For instance for the Keijzer12 benchmark, $GP_T$ is allowed to run for 11,300 s.

By definition, $GP_T$ should not be worse than GP, which is confirmed in Fig. 3, where the non-overlapping inter-quartile boxes suggest superiority of the former to the latter. $GP_T$ also manages to produce more fit best-of-run individuals than EPS-$L_v$ and EPS-$B_v$. However, it seems incapable to catch up with the other EPS configurations, in particular with the leading EPS-$B_c$.

We also include configuration $GP_{5000}$, which is identical to GP except forpopulation holding 5000 programs. It proves to be much better than $GP_T$ on all the tested benchmarks and often outperforms all EPS-L configurations. However, in terms of success rate it still fares worse than EPS-$B_C$.

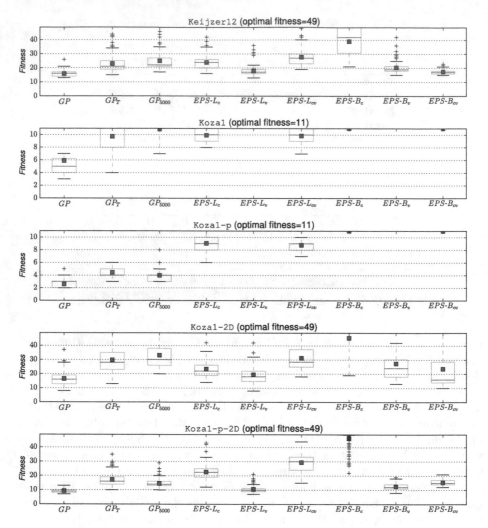

**Fig. 3.** Box-and-whiskers plots of the (maximized) fitness of the final solutions across all configurations and benchmarks. Boxes mark lower and upper quartiles, red line – median, red square – mean, whiskers – 1.5 of inter-quartile range below/above the corresponding quartile, and crosses – the outliers. The missing plots for EPS-L$_v$ and EPS-B$_v$ applied to Koza1 and Koza1-p are due to those benchmarks being univariate, which makes variable substitution pointless. (Color figure online)

**Table 6.** Average runtime in seconds.

|            |      | GP        |           |       |       | EPS     |         |         |         |
|------------|------|-----------|-----------|-------|-------|---------|---------|---------|---------|
|            | GP   | GP$_T$    | GP$_{5000}$ | L$_c$ | L$_v$ | L$_{cv}$ | B$_c$   | B$_v$   | B$_{cv}$ |
| Keijzer12  | 14.8 | 11330.7   | 493.0     | 772.3 | 488.0 | 1578.6  | 15439.8 | 21172.6 | 28354.0 |
| Koza1      | 4.8  | 291.0     | 46.3      | 699.8 | –     | 801.4   | 652.0   | –       | 695.8   |
| Koza1-p    | 4.5  | 962.9     | 344.0     | 892.3 | –     | 971.6   | 978.2   | –       | 982.0   |
| Koza1-2D   | 16.0 | 7635.8    | 431.9     | 793.1 | 478.7 | 1790.6  | 9076.9  | 16280.5 | 23033.8 |
| Koza1-p-2D | 15.4 | 9206.1    | 515.9     | 750.4 | 511.3 | 1725.7  | 11986.4 | 12390.8 | 27875.4 |

# 6    Discussion

Overall, the EPS-B configurations perform better than or at least as good as the corresponding EPS-L configurations in terms of average end-of-run fitness (Table 5). This holds for 10 out of 13 pairs of corresponding EPS-B and EPS-L configurations. It seems thus that EPS favors the Baldwinian approach, in which local, within-lifetime modifications (hole completions) affect individual's fitness but do not propagate to its offspring. Our working explanation is that, by propagating to offspring, the unfilled holes in EPS-B prospectively make it possible to find even better completions. In the Lamarckian variant, to the contrary, fitness evaluation leaves no holes in the evaluated individuals. The only supply of 'fresh' holes is the mutation operator, which affects on average only half of the offspring (Table 3), but even in them it is not guaranteed to introduce any new holes. Apparently, those holes are not sufficient in numbers to permit completions that would make EPS-L outperform EPS-B.

The above pattern is however reversed for the configurations that permit completion with both constants and variables, when applied to bivariate benchmarks, i.e., EPS-B$_{cv}$ vs. EPS-L$_{cv}$ on Keijzer12, Koza1-2d and Koza-1-p-2d. This may result from the overall worse performance of configurations that complete holes with variables: when juxtaposing such configurations with their counterparts that do not involve variables (i.e., EPS-B$_v$ vs. EPS-B$_c$, EPS-L$_v$ vs. EPS-L$_c$, EPS-B$_{cv}$ vs. EPS-B$_c$, and EPS-L$_{cv}$ vs. EPS-L$_c$), the former almost always fare worse. The only exception is the last pair, EPS-L$_{cv}$ vs. EPS-L$_c$, where the former may occasionally perform better (for Koza-1-2d and Koza-p-2d), but the differences do not seem to be statistically significant. As signaled in the previous section, this could be to some extent explained with the very low number of variables in considered benchmarks: with only two variables at its disposal, the solver has limited chance to complete the holes in a way that leads to high fitness. However, this argument is unconvincing for the mixed configurations ($cv$), where both variables and constants can be substituted.

Explanation for this phenomenon turns out to be of a different nature: in EPS-B$_{cv}$, with the possibility of completing with both constants and variables, and with relatively many holes to complete (due to following the Baldwinian process), the search space of possible completions is on average the largest compared to the other configurations. As a consequence, the solver faced with such large problems is more likely to fail to return a definitive answer within the prescribed computational budget of 1.5 s. This results with assigning the worst possible fitness to an evaluated program and likely loss of potentially useful code it may contain. This is confirmed by the statistics on solver behavior we gathered: in EPS-B$_{cv}$, the solver fails to provide optimal completion on time in roughly 35compared to only 13this does not turn out to be problematic for EPS-B$_c$, which also suffers from quite high incidence of such cases (around 25yet performs the best. Overall, these relatively high numbers suggest that solver timeouts may have significant impact on search dynamics. However, this phenomenon does not need to be pathological. To the contrary, it can serve as a natural parsimony pressure: in EPS-L as well as in EPS-B, large programs tend to have higher number of holes than small programs, and large number of holes makes solver timeout more likely.

Interestingly, program evaluation in EPS-B can be said in resulting in *prospective fitness*: the fitness that the partial program being evaluated *could* achieve in the future, given the optimal completion of its holes. This concept bears some resemblance to potential fitness considered in past work [19].

Concerning the runtime, it is not surprising that the Baldwinian configurations are, by a huge margin, more time consuming (except for the simpler Kozal and Kozal-p benchmarks) than the corresponding Lamarckian variants – after all they contain more holes. It is also interesting that the Lamarckian configurations generally achieve end-of-run fitness (and, to a lesser extent, success rate) that is comparable to $GP_T$, which was granted much larger time budgets. This however may be a result of bloat, which could have lessened the effectiveness of GP.

## 7    Conclusion

This paper presented Evolutionary Program Sketching, a novel approach to program synthesis that combines selected elements of genetic programming and formal synthesis methods. EPS evolves partial programs and uses an SMT solver to complete them so as to maximize the number of passed test cases. The experiments have shown that EPS has the potential to be more efficient than standard GP in some scenarios. Nevertheless, empirical evaluation conducted here was rather constrained, which makes approaching a larger and more diverse benchmark suite our priority in further work on this topic.

As in all SMT-based approaches to program synthesis, the fact that candidate programs are never executed opens interesting possibilities. In principle, EPS can be used synthesize programs written in programming languages for which no interpreter exists or program execution is particularly costly (albeit the theory that backs up the language has to be known).

EPS as presented in this paper is an inductive synthesis method. However, engaging a solver for evaluation opens up an interesting possibility of using GP to synthesize programs from a complete formal specification (like the $\max(x, y)$ specification in Formula (2)). Apart from this, future work may include filling holes with more complex content than just constants and variables, and optimizing the mechanism of querying the solver.

**Acknowledgments.** This work was supported by grant 2014/15/B/ST6/05205 funded by the National Science Centre, Poland.

## References

1. Gulwani, S., Jha, S., Tiwari, A., Venkatesan, R.: Component based synthesis applied to bitvector programs. Technical report, MSR-TR-2010-12, February 2010
2. Jha, S., Gulwani, S., Seshia, S.A., Tiwari, A.: Oracle-guided component-based program synthesis. In: 29th International Conference on Software Engineering (ICSE 2010), pp. 215–224, May 2010

3. Srivastava, S., Gulwani, S., Foster, J.S.: From program verification to program synthesis. In: Proceedings of the 37th Annual ACM SIGPLAN-SIGACT Symposium on Principles of Programming Languages, POPL 2010, pp. 313–326. ACM, New York (2010)

4. Solar-Lezama, A.: Program synthesis by sketching. Ph.D. thesis, Electrical Engineering and Computer Science, University of California, Berkeley, USA (fall 2008)

5. Barrett, C., Fontaine, P., Tinelli, C.: The Satisfiability Modulo Theories Library (SMT-LIB) (2016). www.SMT-LIB.org

6. Solar-Lezama, A.: Program sketching. Int. J. Softw. Tools Technol. Transfer **15**(5), 475–495 (2013)

7. Moura, L., Bjørner, N.: Z3: an efficient SMT solver. In: Ramakrishnan, C.R., Rehof, J. (eds.) TACAS 2008. LNCS, vol. 4963, pp. 337–340. Springer, Heidelberg (2008). doi:10.1007/978-3-540-78800-3_24

8. Johnson, C.G.: Genetic programming with fitness based on model checking. In: Ebner, M., O'Neill, M., Ekárt, A., Vanneschi, L., Esparcia-Alcázar, A.I. (eds.) EuroGP 2007. LNCS, vol. 4445, pp. 114–124. Springer, Heidelberg (2007). doi:10. 1007/978-3-540-71605-1_11

9. Katz, G., Peled, D.: Synthesis of parametric programs using genetic programming and model checking. In: Clemente, L., Holik, L. (eds.) Proceedings 15th International Workshop on Verification of Infinite-State Systems, vol. 140, EPTCS, Hanoi, Vietnam, 14 October 2013, pp. 70–84 (2013). Invited talk

10. Warren, H.S.: Hacker's Delight. Addison Wesley, Boston (2002)

11. McPhee, N.F., Ohs, B., Hutchison, T.: Semantic building blocks in genetic programming. In: O'Neill, M., Vanneschi, L., Gustafson, S., Esparcia Alcázar, A.I., Falco, I., Cioppa, A., Tarantino, E. (eds.) EuroGP 2008. LNCS, vol. 4971, pp. 134–145. Springer, Heidelberg (2008). doi:10.1007/978-3-540-78671-9_12

12. Sarafopoulos, A.: Evolution of affine transformations and iterated function systems using hierarchical evolution strategy. In: Miller, J., Tomassini, M., Lanzi, P.L., Ryan, C., Tettamanzi, A.G.B., Langdon, W.B. (eds.) EuroGP 2001. LNCS, vol. 2038, pp. 176–191. Springer, Heidelberg (2001). doi:10.1007/3-540-45355-5_14

13. Azad, R.M.A., Ryan, C.: A simple approach to lifetime learning in genetic programming based symbolic regression. Evolut. Comput. **22**(2), 287–317 (2014)

14. Pawlak, T.P., Wieloch, B., Krawiec, K.: Semantic backpropagation for designing search operators in genetic programming. IEEE Trans. Evol. Comput. **19**(3), 326–340 (2015)

15. Ffrancon, R., Schoenauer, M.: Memetic semantic genetic programming. In: Silva, S., Esparcia-Alcazar, A.I., Lopez-Ibanez, M., Mostaghim, S., Timmis, J., Zarges, C., Correia, L., Soule, T., Giacobini, M., Urbanowicz, R., Akimoto, Y., Glasmachers, T., Fernandez de Vega, F., Hoover, A., Larranaga, P., Soto, M., Cotta, C., Pereira, F.B., Handl, J., Koutnik, J., Gaspar-Cunha, A., Trautmann, H., Mouret, J.B., Risi, S., Costa, E., Schuetze, O., Krawiec, K., Moraglio, A., Miller, J.F., Widera, P., Cagnoni, S., Merelo, J., Hart, E., Trujillo, L., Kessentini, M., Ochoa, G., Chicano, F., Doerr, C. (eds.) GECCO 2015; Proceedings of the 2015 on Genetic and Evolutionary Computation Conference, Madrid, Spain, 11–15 July 2015, pp. 1023–1030. ACM (2015). GP Track best paper

16. Barrett, C., Fontaine, P., Tinelli, C.: The SMT-LIB standard: version 2.5. Technical report, Department of Computer Science, The University of Iowa (2015). www.SMT-LIB.org

17. McDermott, J., White, D.R., Luke, S., Manzoni, L., Castelli, M., Vanneschi, L., Jaskowski, W., Krawiec, K., Harper, R., De Jong, K., O'Reilly, U.M.: Genetic programming needs better benchmarks. In: Soule, T., Auger, A., Moore, J., Pelta, D., Solnon, C., Preuss, M., Dorin, A., Ong, Y.S., Blum, C., Silva, D.L., Neumann, F., Yu, T., Ekart, A., Browne, W., Kovacs, T., Wong, M.L., Pizzuti, C., Rowe, J., Friedrich, T., Squillero, G., Bredeche, N., Smith, S.L., Motsinger-Reif, A., Lozano, J., Pelikan, M., Meyer-Nienberg, S., Igel, C., Hornby, G., Doursat, R., Gustafson, S., Olague, G., Yoo, S., Clark, J., Ochoa, G., Pappa, G., Lobo, F., Tauritz, D., Branke, J., Deb, K. (eds.) GECCO 2012: Proceedings of the Fourteenth International Conference on Genetic and Evolutionary Computation Conference, Philadelphia, Pennsylvania, USA, 7–11 July 2012, pp. 791–798. ACM (2012)
18. Krawiec, K.: Behavioral Program Synthesis with Genetic Programming. Studies in Computational Intelligence, vol. 618. Springer International Publishing, New York (2015)
19. Krawiec, K., Polewski, P.: Potential fitness for genetic programming. In: Ebner, M., Cattolico, M., van Hemert, J., Gustafson, S., Merkle, L.D., Moore, F.W., Congdon, C.B., Clack, C.D., Moore, F.W., Rand, W., Ficici, S.G., Riolo, R., Bacardit, J., Bernado-Mansilla, E., Butz, M.V., Smith, S.L., Cagnoni, S., Hauschild, M., Pelikan, M., Sastry, K. (eds.) GECCO 2008 Late-Breaking Papers, Atlanta, GA, USA, 12–16 July 2008, pp. 2175–2180. ACM (2008)

# Exploring Fitness and Edit Distance of Mutated Python Programs

Saemundur O. Haraldsson[⊠], John R. Woodward, Alexander E.I. Brownlee,
and David Cairns

University of Stirling, Stirling FK9 4LA, Scotland
soh@cs.stir.ac.uk

**Abstract.** Genetic Improvement (GI) is the process of using computational search techniques to improve existing software e.g. in terms of execution time, power consumption or correctness. As in most heuristic search algorithms, the search is guided by fitness with GI searching the space of program variants of the original software. The relationship between the program space and fitness is seldom simple and often quite difficult to analyse. This paper makes a preliminary analysis of GI's fitness distance measure on program repair with three small Python programs. Each program undergoes incremental mutations while the change in fitness as measured by proportion of tests passed is monitored. We conclude that the fitnesses of these programs often does not change with single mutations and we also confirm the inherent discreteness of bug fixing fitness functions. Although our findings cannot be assumed to be general for other software they provide us with interesting directions for further investigation.

**Keywords:** Search Based Software Engineering · Genetic Improvement · Genetic Programming · Automatic programming · Software repair

## 1 Introduction

In recent years work has been emerging from the Search Based Software engineering (SBSE) community called Genetic Improvement (GI) [17]. GI is where computational search techniques have been applied to already functioning software for the purpose of improvement. The improvement criteria can be various properties of the existing software such as speed, accuracy or energy efficiency. The most commonly used search method for GI is Genetic Programming (GP) although various other techniques like Genetic Algorithms [3] and Grammatical Evolution [33] have also been applied.

The examples of work that can be categorised as GI are widespread. Most of them, with few exceptions, are similar in the sense that they tackle software that can be considered large, with lines of code (LOC) numbering in the thousands. Nearly one third of the GI literature is dominated by examples of automatic bug fixing while approximately another third is concerned with improving non-functional properties [12,40]. Of those, execution time is perhaps the

© Springer International Publishing AG 2017
J. McDermott et al. (Eds.): EuroGP 2017, LNCS 10196, pp. 19–34, 2017.
DOI: 10.1007/978-3-319-55696-3_2

property that is easiest to measure and therefore the most commonly researched non-functional property. Bug fixing and execution time make very different fitness functions. The bug fixing fitness function is a *discrete* integer function, counting the number of test cases passed and many non-functional fitness functions, like the execution time and memory consumption, are *continuous* measurements. The main emphasis to date has been on results, rather than an understanding of the GI search space. There is a need for empirical and theoretical analysis of the GI process. Is it easy or difficult to traverse the search space of programs and what can possibly be done to increase the chance that the programs improves? We begin to answer these questions in the space of smaller programs. Using small programs to begin with allows us to easily analyse the full impact of changes introduced to the code. It also keeps uncertainties introduced by the rest of the program to a minimum.

This paper explores the relationship between fitness and number of accumulated incremental changes while using GI to break programs and gain an understanding of the program repair process. We chose three small Python programs to make a preliminary study on empirical properties of GI's fitness and edit lists. The following questions were addressed:

- Is it feasible to apply GI to fix multiple bugs at a time? Will fixing one bug introduce another?
- If a fix needs multiple edits, will GI be able to find it within reasonable time/number of iterations?
- Can we identify any similarities or patterns in fitness distance relationship that might be worth exploring in more detail on larger sets of programs? I.e. is there some rule there that has not been discovered yet?

Although the experiments with only three programs are limiting for generalization of the answers we find, they help us narrow down the most promising route of research for larger and more resource consuming experimentation. Choosing to operate on Python programs also serves two purposes:

- There are very few examples of GI being applied to Python programs [1] and given that it is a very popular language, there is a large gap in the literature.
- Because of dynamic typing of Python programs, the search space is possibly less restricted than for statically typed languages. Therefore changes to the source code might have more possibilities than a static typed language.

The remainder of the paper is structured as follows. Section 2 gives an overview of related work. Section 3 describes the implementation of GI that was used for this paper while Sect. 4 gives more details on the configuration and data collection. Section 5 summarizes the results with discussion followed by the conclusions in Sect. 6.

## 2   Related Work

A search through publications in software engineering and computer science suggests that approximately one third of GI related material is on bug fixing.

GenProg [25, 26, 38] is one of the better known automatic bug fixing frameworks and uses GP to evolve patches for a fraction of the price it would cost manually [24]. GenProg itself is derived from earlier work by Weimar *et al.* [11, 37, 39]. Smith *et al.* is an example of a more recent use of this framework [34]. Although GenProg is perhaps the most commonly known framework, there is also a large body of GI literature dedicated to automatic bug fixing [1, 2, 4, 28, 31] that use alternative approaches.

Automatic program repair can be considered an improvement of a functional property, like improving the quality of hash code functions in Hadoop [15] or grafting new features to existing software [14, 27]. However, optimizing attributes like execution time, memory consumption and power consumption is generally considered an improvement of a non-functional property which spans another big part of the GI literature. Of those attributes, execution time seems to be very popular, with Langdon's work on the 50k line DNA sequencing tool Bowtie [17, 19] possibly the best known. Langdon has also reported 100 fold speed-up of another DNA sequencing tool BarraCUDA [18, 20–23] and the GI improvements have now been included in the official release. Langdon's GI implementation has furthermore been used by others for specializing and optimizing the execution time of MiniSAT [30], a boolean satisfiability solver and for optimizing power consumption of that same solver [5, 6]. Many others have applied or suggested GI for improving non-functional properties such as execution time [9, 10, 35, 41], energy consumption [7, 8, 13, 42, 43] and memory usage [32, 44].

The literature of GI counts around 100 papers and there is not much work on empirical analysis of the search landscapes of GI like there is for GP [36], nor is there much on Python programs which is the topic of this paper.

## 3    Our Implementation of Genetic Improvement

There are multiple ways to implement GI. We have implemented a variation of the work of Langdon *et al.* [14, 17]. Like their GI, ours operates on the source code with no need to convert the program to a different representation like abstract syntax trees (AST) [1]. Therefore our approach is directly transferable between programming languages with minimal configuration. The source code is read as a text file and stored in a data structure ($x$) of program lines. For each line we recorded the raw text as it appears in the source file, along with information on: indentation[1]; line type; whether the line can be altered; and any variables or built-in operators that the line includes.

To manipulate the source code we evolve *edit lists* (Fig. 1). Each edit consists of: the operation of *Replace*; the source code snippet before and after the edit; and the location of where to apply this edit (line and character number).

The code snippets can be whole lines from the source code, or one of the various single operators or numerical constants listed in Table 1. For other programming languages this table would vary slightly, like the incremental operator ++ which exist in C and Java but not in Python.

---

[1] Code blocks are defined by indentation in Python and not by {} as in JAVA/C.

**Table 1.** Sets of single operators available to the GI. One member of a given set can be changed to another member of the same set.

|    |                        | Operations |
|----|------------------------|------------|
|    |                        | Operations |
| S1 | Numerical constants    | Can increment by $\pm 1$ |
| S2 | Arithmetic operators   | $+, -, *, /, //, \%, **$ |
| S3 | Arithmetic assignments | $+ =, - =, * =, / =,$ |
| S4 | Relational operators   | $<, >, <=, >=, ==, ! =,$ $is, is\ not, not$ |
| S5 | Logical operators      | $and, or$ |
| S6 | Logical constants      | $True, False$ |

An individual genome consists of a list of edit operations sampled from the set shown in Table 1 and is applied in sequence from the first edit to the last item in the list.

### 3.1 Fitness Function

In our experiments the fitness function counts the number of test cases for which the program passes. It is inherently discrete and possibly provides no obvious gradient for the search to follow, which poses a difficulty for many search methods, although evolutionary algorithms have been applied successfully [30]. We want to analyse this discrete nature and in addition, explore ways to report on GI problem difficulty. The relationship between the fitness landscape and genotype/phenotype is far from trivial. However some work has been done on introducing a partial evaluation of programs and hence some kind of guidance to the search process in cases where the fitness landscape was largely flat [16].

### 3.2 Search Algorithm

Generally, the GI's search algorithm is guided by a fitness function, applying selection pressure towards better programs. GI methods that evolve edit lists also have to produce a new generation of edit lists from a previous generation. Our implementation uses the base type of mutation illustrated in Fig. 1 that is applied to parents to produce offspring: Append randomly generated edits to the parent (*Grow*).

For our experiments, analysing the relationship between edit list size and fitness, we start from an assumed correct implementation of the program and apply a single edit. Then incrementing by using *Grow* with single edit and without any pressure to search, resetting the edit list to a single edit when all conditions are met; maximum size of edit list and minimum fitness.

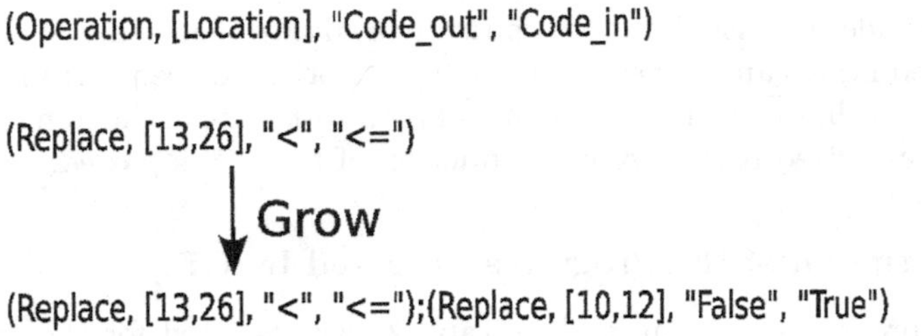

(Operation, [Location], "Code_out", "Code_in")

(Replace, [13,26], "<", "<=")

↓ Grow

(Replace, [13,26], "<", "<=");(Replace, [10,12], "False", "True")

Fig. 1. An example of an edit list and how it can evolve with *Grow*.

## 4 Experimental Setup

Each program source code is subjected to experiments to assess fitness distance which is the change in fitness given a particular number of mutations. The proportion of passed test cases is the chosen measurement for the fitness function.

---

**Procedure 1.** A single experimental run

---
1: $F \leftarrow$ empty array     {list of fitnesses}
2: $x \leftarrow [random \quad edit]$     {edit list}
3: **for** $i = 0$ until $i \le 50$ **do**
4:     append $f(x)$ to $F$
5:     **if** $|x| \ge 20$ and $f(x) = 0$ **then**
6:         break
7:     **end if**
8:     append $[random \quad edit]$ to $x$
9: **end for**

---

The experiment in Procedure 1 is repeated 100 times by initiating an edit list $x$ with a single randomly chosen edit and then appending to the list incrementally. Apart from recording the fitness $f(x)$ for each added edit in every experiment, three variables are recorded for each run. These are the size of the edit list ($|x|$) when the fitness ($0 \le f(x) \le 1$) satisfies the following:

$\Delta$ decreases for the first time:

$$\text{when} \quad f(x_i) < f(x_0) \quad \text{and} \quad f(x_0) = f(x_j) \qquad \forall j \in (1, ..., i-1) \subset \mathcal{Z}$$

$\Omega$ reaches zero:

$$\text{when} \quad f(x_i) = 0 \quad \text{and} \quad f(x_j) > 0 \qquad \forall j \in (0, ..., i-1) \subset \mathcal{Z}$$

$\Psi$ starts to increase again:

$$\text{when} \quad f(x_i) > f(x_{i-1}) \quad \text{and} \quad f(x_j) \le f(x_{j-1}) \qquad \forall j \in (1, ..., i-1) \subset \mathcal{Z}$$

This provides us with data to empirically evaluate the nature of the relationship between fitness and the size of the edit list for the three programs described in Sect. 4.1. Each program is accompanied by a test suite of different sizes so the fitness is normalized to represent the fraction of test cases passed.

## 4.1   Description of the Programs Targeted by GI

The three programs[2] summarised in Table 2 were selected for the experiment. They are all comprised of 3 or fewer functions. They are not a complete Python module but are either part of a module, like **P2** or a standalone function like **P1** and **P3**. However they can be integrated into any Python module and have well defined input, output types.

**P1** is a simple text input calculator that reads text from left to right, parses a single character at a time into operator, digits bins and calculates a result using a reverse Polish notation. It is a beginners programming exercise and the only program of the three that is not a part of publicly available software. It branches out for each of the four basic arithmetic operations; addition, subtraction, multiplication and division as well as a special branch for parentheses.

**P2** is an initialization function for the K-means clustering method and is a part of the scikit-learn Python toolbox [29]. It determines the initial $k$ centres for the algorithm. **P2** does this with a random factor that can be seeded for consistency purposes.

**P3** is a string manipulation function that reads through a text replacing HTML tags with latex equivalent commands. It is a part of a larger software system, *Janus Manager* that is in commercial use by a vocational rehabilitation centre in Iceland.

Table 2 shows basic info about the programs, **P1**, **P2** and **P3**. The numbers in the second column are the number of lines of code and the number of lines that can be changed, i.e. excluding definitions, comments that do not include executable code and functions out of scope. The third and fourth columns are the count of mutable points, the number of instances in the source that fit into any of the sets defined in Table 1 and the count for each set. The fifth column describes the input and output of each program and the last column is a short description of its purpose.

**P2** and **P3** are accompanied by test modules which are used to evaluate fitness. However **P2**'s test suite, comprising of approximately 60 test cases, was expanded to 500 by sampling from an estimated distribution of inputs from the original tests and using the **P2** as an oracle. **P3** comes with 124 test cases based on HTML input from users and their verified output, so expanding that test suite is not feasible. The test suite for **P1** is 600 cases made by combining two sets; $A = \{0, 1, 2, 3, 4\}$ and $B = \{+, -, *, /\}$ into:

(a) All possible combinations of a single operator from $B$ with two digits from $A$, an example would be $2 + 2$.

---

[2] The programs and their test suites are available on https://github.com/saemundo/
Exploring-Fitness-and-Edit-Distance-of-Mutated-Python-Programs.git.

**Table 2.** Information about the programs that were used in the experiments

| Program | LOC (mLOC*) | Mutable points | Types of mutable points | Input ⇓ Output | Description |
|---------|-------------|----------------|-------------------------|----------------|-------------|
| P1 | 99 (98) | 147 | S1:   19<br>S2:   37<br>S3:   50<br>S4–6:41 | String ⇓ Float | Simple text input calculator |
| P2 | 177 (75) | 106 | S1:   11<br>S2:   14<br>S3:   57<br>S4–6:24 | Cluster data** ⇓ Matrix | Initiation of K-means cluster centers |
| P3 | 103 (63) | 59 | S1:   26<br>S2:    3<br>S3:   29<br>S4–6: 1 | HTML ⇓ Latex | Html to Latex conversion tool |

*Changeble lines of code.
**Data points, number of clusters, initialization method and 3 optional arguments.

(b) All combinations of $(Xo_1Y)o_2Z$ where $\{X, Y, Z\} \subseteq A$, $o_1 \in \{+, -\} \subseteq B$ and $o_2 \in \{*, /\} \subseteq B$. An example would be $(4 + 2) * 2$.

P1's test suite was verified with the Python built in *eval* function.

# 5   Results

We will look at the programs from three different angles; size of edit list versus a specific change in fitness, average fitness as a function of edit list size and, unique and discrete steps in fitness. When statistically comparing the mean of two variables we use Welch's t-test for two independent samples with unequal variance. For testing the likelihood of two variables coming from the same distribution, we compute a two sample Kolmogorov-Smirnoff statistic.

## 5.1   Change in Fitness

Table 3 lists the basic descriptive statistics for the edit list size ($|x|$) for the three different changes in fitness (see Sect. 4) and the total number of fitness evaluations for each program. That number varies between programs due to the termination conditions described in Sect. 4. Firstly we measure the edit distance required for the fitness to decrease for the first time (i.e. $f(x) < 1$).

For **P1** $\Delta$ occurs on average when the edit list is 9 edits long, although there is a lot of variations as shown by the standard deviation and the range. We find this a quite surprising result. It should also be noted that in 1 run out of the 100 repeated experiments the fitness did not drop at all, reaching the maximum

**Table 3.** Statistics for the variables defined in Sect. 4, edit list size $|x|$ when changes in fitness $f(x)$ are detected and total number of fitness evaluations during the experiments.

| | Variable | Mean (std) | (Min, Max) | Number of occurrences | Evaluations |
|---|---|---|---|---|---|
| P1 | $\Delta$ | 8.91 (9.83) | (1, 50) | 99 | 2213 |
| | $\Omega$ | 12.68 (10.55) | (1, 50) | 97 | |
| | $\Psi$ | 12.0 (7.53) | (3, 25) | 12 | |
| P2 | $\Delta$ | 2.56 (3.20) | (1, 19) | 100 | 2267 |
| | $\Omega$ | 10.28 (8.33) | (1, 42) | 100 | |
| | $\Psi$ | 7.64 (5.42) | (2, 26) | 34 | |
| P3 | $\Delta$ | 2.69 (3.30) | (1, 19) | 100 | 1980 |
| | $\Omega$ | 3.92 (3.74) | (1, 19) | 100 | |
| | $\Psi$ | 4.76 (3.44) | (2, 16) | 17 | |

size of 50 edits. There is a highly significant difference ($p < 0.001$) between **P1** and **P2** for the mean $\Delta$. We can also reject the null hypothesis that the samples come from the same distribution ($p < 0.001$) as is evident when comparing Figs. 2a and b. However statistically we cannot rule out the possibility that the distribution ($p = 0.99$) and mean ($p = 0.78$) are the same for those measures when comparing **P2** and **P3**. Notice that the bar plots in Figs. 2b and c are very similar.

**P1** and **P2** are closer together when comparing the number of edits it takes to reach zero fitness and until the fitness might increase again. Testing for the same distribution gives $p = 0.08$ and $p = 0.10$ for $\Omega$ and $\Psi$ respectively and we also cannot reject the hypothesis that the means are the same (0.08 and 0.09). Figures 2d and e corroborate the observation about zero fitness. However looking at Figs. 2g and h we are less sure about the number of edits it takes to increase the fitness again and might infer that we do not have enough data to be confident the test results are accurate.

**P3** has very different means and distributions than **P1** on all measured variables ($p < 0.01$) which is validated on looking at Figs. 2g–i. The lack of data for $\Psi$ is further verified by the outcome of tests comparing increased fitness between **P3** and **P2**: Rejecting that they have same mean ($p = 0.025$) but failing to reject that they come from the same distribution ($p = 0.09$).

These results indicate that: **P1** is unaffected by many edits and **P2** and **P3** are easily broken and fixed even though they are very different.

## 5.2   Average Fitness with Respect to Edit List Size

Repeating the experiments 100 times provided us with enough datapoints to construct fitness distance graphs and approximate the distribution of fitness for each increment in edit list size. Looking at the boxplots in Figs. 3a–c we see that overall, the distributions are quite different. **P1**'s first three increments (Fig. 3a) have

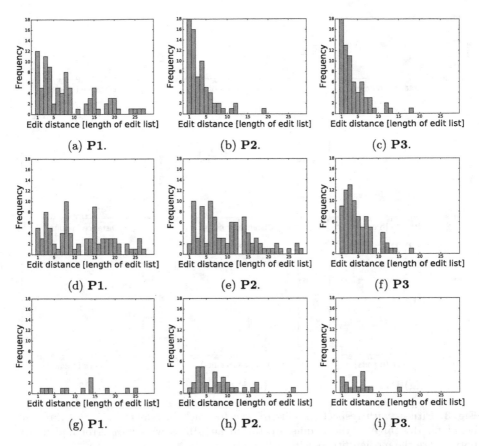

**Fig. 2.** Distributions of the three different measurements in Table 3 during the experiments for each program. Top three are $\Delta$, the middle are $\Omega$ and the bottom three are $\Psi$

very narrow distributions close to $f(x) = 100\%$ and then the distributions widen considerably until increment 15 where they start narrowing towards the bottom. For **P2** it is a smoother transition (Fig. 3b) from top to bottom, maintaining a similar rate of descent for the mean, maximum and minimum throughout. Then **P3** stands out completely with seemingly only two distributions; covering the entire range and nearly collapsed on either extreme (Fig. 3c).

Having a closer look at how the mean fitness changes in Figs. 3d–f we see that these programs are as dissimilar as initially assumed. While both **P2** (Fig. 3e) and **P3** (Fig. 3f) both follow curves that are concave upwards, **P3** seems to follow a curvature of higher magnitude. Now it is **P1** that is the outlier (Fig. 3d) following a noisy line with a negative slope until the edit list reaches size 22 when it jumps up to $f(x) = 80\%$ again. **P2** shows signs of starting to recover in increment 22 as well but on a much slower rate than **P1** and **P3** shows no such signs at all.

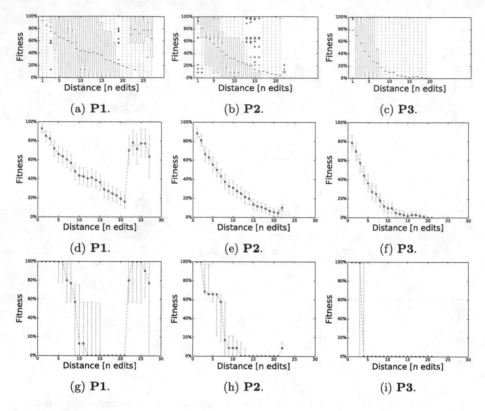

**Fig. 3.** Fitness with respect to edit list size for each increment. The top three are boxplots, the three in the middle are the mean fitness with 95% error and at the bottom are the median fitness with 95% error.

The plots for medians in Figs. 3g–i paint a completely different story, displaying no hint of smoothness to the transition from one increment to another. However there are obvious steps that highlight the discreteness of each program's fitness function. We see in Fig. 3h that **P2** has the most number of steps, while **P1** comes second (Fig. 3g) and **P3** last (Fig. 3g).

## 5.3   Discrete Steps in Fitness

Following the observation of the different steps for each program's median fitness seen in Figs. 3g–i, we counted the unique number of fitness evaluations throughout the entire experiment. As previously inferred, **P2** has by far the largest number of discrete steps, with 46 in total as seen in Fig. 4. **P2** had its fitness evaluated 2067 times, as seen in Table 3, and Fig. 4 shows how often the fitness changed from one value to another by adding a single edit to the list. The count matrix is very sparse as can be seen by the white squares that denote zero counts, for example the fitness never went from 0.655 to zero with one edit. The diagonal line of shaded boxes from $(0, 0)$ to $(1, 1)$ indicates that the majority of single edits had little to no effect on the fitness, especially when the fitness is already zero. However there is an abundance of blue squares at the bottom which tells us that a single edit can often lead to complete failure of the program.

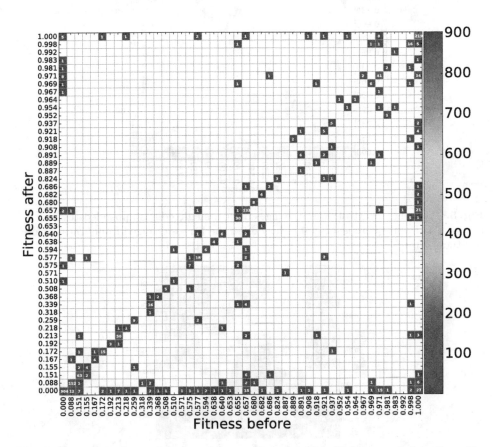

**Fig. 4.** Frequency chart of fitness changes after a single edit is appended to the edit list for **P2**. Each square is a count of how often the fitness changed from fitness before to fitness after. (Color figure online)

**P1** has the second most number of discrete steps, 15 as seen in Fig. 5, the total number of counts is 2213 (Table 3). This chart looks like a scaled version of **P2**'s Fig. 4, displaying the same dominant diagonal line as well as the number of single edits that make the program pass 0% of test cases. We also see here that there are a lot of single edits that decrease the fitness from 1, shown by the large number of non-white squares in the right most column. This is to be expected since every experimental run starts with a correct program, so it visits that state at least once each time while it is not guaranteed to visit other specific steps.

Figure 6 is the least sparse of the three, only 5 fitness steps in total. What is surprising however, is that even though there were 1980 fitness evaluations

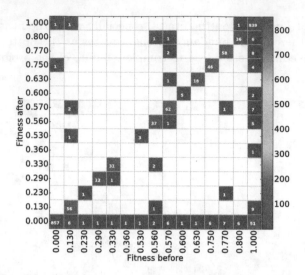

**Fig. 5.** Frequency chart of fitness changes after a single edit is appended to the edit list for **P1**. Each square is a count of how often the fitness changed from fitness before to fitness after.

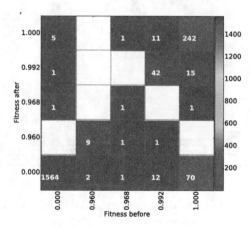

**Fig. 6.** Frequency chart of fitness changes after a single edit is appended to the edit list for **P3**. Each square is a count of how often the fitness changed from fitness before to fitness after.

there are still some zero counts. As with **P1** and **P2**, there are more of them above the diagonal line than below, meaning that adding an edit is more likely to decrease fitness than to increase. The highest counts are also on the diagonal line, so the same behaviour can be observed: most likely a single edit will leave the fitness unchanged.

# 6   Conclusions

The aim of this preliminary exploration of three Python programs' fitness distance is to provide a greater understanding of the search process encountered by GI.

We can conclude that for these three programs and the assumption that the landscape has no inaccessible areas to our GI implementation:

- It is feasible to apply GI to fix multiple bugs simultaneously. Starting from any variation of these programs and applying search pressure can result in a fix.
- If a fix needs multiple edits, GI will be able to find it within reasonable time. If it takes on average 10 edits to completely break these programs (i.e. zero test cases passed), the reverse is likely to be true.
- We have identified a similarity in fitness distance relationship that is worth exploring in more detail on larger set of programs.

Although we cannot conclude general rules or patterns from these preliminary experiments, a common denominator for our programs is that a single edit will likely have no impact on fitness. However, if a change does occur then it can be expected that the change will be large.

It would be interesting to explore the cause of the difference in sparsity of the frequency charts in Figs. 4, 5 and 6. The programs and the test suites were quite different in nature which might explain the stark contrast in discreteness of our programs. While **P1** and **P3** had a single input argument, **P2** had 6, and the test suites reflected this difference. **P2**'s test suite could be divided into multiple categories, testing various aspects and combinations of input arguments. **P1**'s test suite could also be categorised but only in 4–6 groups and it would be a stretch to try and group **P3**'s test suite.

Our next task is to apply the same analysis that we have done here to a larger set of programs with the goal of helping us to form more general and applicable rules for GI fitness distance. With larger sets we can also analyse the fitness distance correlation of GI with variety of edit list mutation operators, such as removing edits or changing individual edits.

**Acknowledgements.** The work presented in this paper is part of the DAASE project which is funded by the EPSRC. The authors would like to thank Janus Rehabilitation Centre for allowing the use of their software in the experiments and consequently making the relevant part of the source code available for others to use in their experiments.

# References

1. Ackling, T., Alexander, B., Grunert, I.: Evolving patches for software repair. In: 13th Annual Conference on Genetic and Evolutionary Computation, GECCO 2011, pp. 1427–1434. ACM, Dublin, July 2011

2. Arcuri, A.: On the automation of fixing software bugs. In: Companion of the 30th International Conference on Software Engineering, ICSE Companion 2008, pp. 1003–1006. ACM, New York (2008)
3. Arcuri, A., Yao, X.: A novel co-evolutionary approach to automatic software bug fixing. In: 2008 IEEE World Congress on Computational Intelligence, pp. 162–168. IEEE Computational Intelligence Society (2008)
4. Bradbury, J.S., Jalbert, K.: Automatic repair of concurrency bugs. In: Proceedings of the 2nd International Symposium on Search Based Software Engineering, p. 2 (2010)
5. Bruce, B.R.: Energy optimisation via genetic improvement a SBSE technique for a new era in software development. In: Proceedings of the Companion Publication of the 2015 Annual Conference on Genetic and Evolutionary Computation, GECCO Companion 2015, pp. 819–820. ACM, Madrid, July 2015
6. Bruce, B.R., Petke, J., Harman, M.: Reducing energy consumption using genetic improvement. In: Proceedings of the 2015 Annual Conference on Genetic and Evolutionary Computation, GECCO 2015, pp. 1327–1334. ACM, Madrid, July 2015
7. Burles, N., Bowles, E., Brownlee, A.E.I., Kocsis, Z.A., Swan, J., Veerapen, N.: Object-oriented genetic improvement for improved energy consumption in Google Guava. In: Barros, M., Labiche, Y. (eds.) SSBSE 2015. LNCS, vol. 9275, pp. 255–261. Springer, Cham (2015). doi:10.1007/978-3-319-22183-0_20
8. Burles, N., Swan, J., Brownlee, A.E.I., Kocsis, Z.A., Veerapen, N.: Embedded dynamic improvement. In: Proceedings of the Companion Publication of the 2015 Annual Conference on Genetic and Evolutionary Computation, GECCO Companion 2015, pp. 831–832. ACM, Madrid, July 2015
9. Cody-Kenny, B., Barrett, S.: The emergence of useful bias in self-focusing genetic programming for software optimisation. In: Ruhe, G., Zhang, Y. (eds.) SSBSE 2013. LNCS, vol. 8084, pp. 306–311. Springer, Heidelberg (2013). doi:10.1007/978-3-642-39742-4_29
10. Cody-kenny, B., Galván-López, E., Barrett, S.: locoGP: improving performance by genetic programming java source code. In: Proceedings of the Companion Publication of the 2015 Annual Conference on Genetic and Evolutionary Computation, GECCO Companion 2015, pp. 811–818. ACM, Madrid, July 2015
11. Forrest, S., Nguyen, T., Weimer, W., Goues, C.L.: A genetic programming approach to automated software repair. In: Genetic and Evolutionary Computation Conference, pp. 947–954 (2009)
12. Haraldsson, S.O., Woodward, J.R.: Automated design of algorithms and genetic improvement: contrast and commonalities. In: Proceedings of the 2014 Conference Companion on Genetic and Evolutionary Computation Companion, GECCO Comp 2014, pp. 1373–1380. ACM, Vancouver, July 2014
13. Haraldsson, S.O., Woodward, J.R.: Genetic improvement of energy usage is only as reliable as the measurements are accurate. In: Proceedings of the 2015 Conference Companion on Genetic and Evolutionary Computation Companion, pp. 831–832. ACM, Madrid (2015)
14. Harman, M., Jia, Y., Langdon, W.B.: Babel Pidgin: SBSE can grow and graft entirely new functionality into a real world system. In: Goues, C., Yoo, S. (eds.) SSBSE 2014. LNCS, vol. 8636, pp. 247–252. Springer, Heidelberg (2014). doi:10.1007/978-3-319-09940-8_20
15. Kocsis, Z.A., Neumann, G., Swan, J., Epitropakis, M.G., Brownlee, A.E.I., Haraldsson, S.O., Bowles, E.: Repairing and optimizing hadoop *hashCode* implementations. In: Goues, C., Yoo, S. (eds.) SSBSE 2014. LNCS, vol. 8636, pp. 259–264. Springer, Heidelberg (2014). doi:10.1007/978-3-319-09940-8_22

16. Krawiec, K., Swan, J.: Pattern-guided genetic programming. In: Proceedings of the 15th Annual Conference on Genetic and Evolutionary Computation, GECCO 2013, pp. 949–956. ACM, Amsterdam (2013)
17. Langdon, W.B.: Genetic improvement of programs. In: 18th International Conference on Soft Computing, MENDEL 2012 (2012)
18. Langdon, W.B.: Improved CUDA 3D medical image registration. In: UK Many-Core Developer Conference 2014, UKMAC 2014, December 2014
19. Langdon, W.B.: Performance of genetic programming optimised Bowtie2 on genome comparison and analytic testing (GCAT) benchmarks. BioData Min. **8**(1), 1 (2015)
20. Langdon, W.B., Harman, M.: Genetically improved CUDA C++ software. In: Nicolau, M., Krawiec, K., Heywood, M.I., Castelli, M., García-Sánchez, P., Merelo, J.J., Rivas Santos, V.M., Sim, K. (eds.) EuroGP 2014. LNCS, vol. 8599, pp. 87–99. Springer, Heidelberg (2014). doi:10.1007/978-3-662-44303-3_8
21. Langdon, W., Harman, M.: Genetically improved CUDA kernels for StereoCamera. Technical report, UCL Department of Computer Science, London, UK (2014)
22. Langdon, W.B., Harman, M.: Grow and Graft a better CUDA pknotsRG for RNA pseudoknot free energy calculation. In: Proceedings of the Companion Publication of the 2015 Annual Conference on Genetic and Evolutionary Computation, GECCO Companion 2015, pp. 805–810. ACM, Madrid, July 2015
23. Langdon, W.B., Lam, B.Y.H., Petke, J., Harman, M.: Improving CUDA DNA analysis software with genetic programming. In: Proceedings of the 2015 Annual Conference on Genetic and Evolutionary Computation, GECCO 2015, pp. 1063–1070. ACM, Madrid, July 2015
24. Le Goues, C., Dewey-Vogt, M., Forrest, S., Weimer, W.: A systematic study of automated program repair: fixing 55 out of 105 bugs for $8 each. In: 2012 34th International Conference on Software Engineering (ICSE). pp. 3–13. IEEE, Zurich, June 2012
25. Le Goues, C., Forrest, S., Weimer, W.: Current challenges in automatic software repair. Softw. Qual. J. **21**(3), 421–443 (2013)
26. Le Goues, C., Nguyen, T., Forrest, S., Weimer, W.: GenProg: a generic method for automatic software repair. IEEE Trans. Softw. Eng. **38**(1), 54–72 (2012)
27. Marginean, A., Barr, E.T., Harman, M., Jia, Y.: Automated transplantation of call graph and layout features into Kate. In: Barros, M., Labiche, Y. (eds.) SSBSE 2015. LNCS, vol. 9275, pp. 262–268. Springer, Heidelberg (2015). doi:10.1007/978-3-319-22183-0_21
28. Nguyen, H.D.T., Qi, D., Roychoudhury, A., Chandra, S.: SemFix: program repair via semantic analysis. In: Proceedings - International Conference on Software Engineering, pp. 772–781 (2013)
29. Pedregosa, F., Varoquaux, G., Gramfort, A., Michel, V., Thirion, B., Grisel, O., Blondel, M., Prettenhofer, P., Weiss, R., Dubourg, V., Vanderplas, J., Passos, A., Cournapeau, D., Brucher, M., Perrot, M., Duchesnay, E.: Scikit-learn: machine learning in Python. J. Mach. Learn. Res. **12**, 2825–2830 (2011)
30. Petke, J., Harman, M., Langdon, W.B., Weimer, W.: Using genetic improvement and code transplants to specialise a C++ program to a problem class. In: Nicolau, M., Krawiec, K., Heywood, M.I., Castelli, M., García-Sánchez, P., Merelo, J.J., Rivas Santos, V.M., Sim, K. (eds.) EuroGP 2014. LNCS, vol. 8599, pp. 137–149. Springer, Heidelberg (2014). doi:10.1007/978-3-662-44303-3_12
31. Qi, Z., Long, F., Achour, S., Rinard, M.: An analysis of patch plausibility and correctness for generate-and-validate patch generation systems. In: Xie, T. (ed.) ISSTA 2015, pp. 24–36. ACM, Baltimore (2015)

32. Risco-Martín, J.L., Colmenar, J.M., Hidalgo, J.I., Lanchares, J., Díaz, J.: A methodology to automatically optimize dynamic memory managers applying grammatical evolution. J. Syst. Softw. **91**, 109–123 (2014)
33. Ryan, C., Collins, J.J., Neill, M.O.: Grammatical evolution: evolving programs for an arbitrary language. In: Banzhaf, W., Poli, R., Schoenauer, M., Fogarty, T.C. (eds.) EuroGP 1998. LNCS, vol. 1391, pp. 83–96. Springer, Heidelberg (1998). doi:10.1007/BFb0055930
34. Smith, E.K., Barr, E.T., Le Goues, C., Brun, Y.: Is the cure worse than the disease? Overfitting in automated program repair. In: Proceedings of the 2015 10th Joint Meeting on Foundations of Software Engineering, ESEC/FSE 2015, pp. 532–543. ACM, Bergamo (2015)
35. Swan, J., Epitropakis, M.G., Woodward, J.R.: Gen-O-Fix: an embeddable framework for dynamic adaptive genetic improvement programming. Technical report CSM-195, Department of Computing Science and Mathematics University of Stirling, Stirling, UK (2014)
36. Tomassini, M., Vanneschi, L., Collard, P., Clergue, M.: A study of fitness distance correlation as a difficulty measure in genetic programming. Evol. Comput. **13**(2), 213–239 (2005)
37. Weimer, W., Forrest, S., Le Goues, C., Nguyen, T.: Automatic program repair with evolutionary computation. Commun. ACM **53**(5), 109 (2010)
38. Weimer, W., Fry, Z.P., Forrest, S.: Leveraging program equivalence for adaptive program repair: models and first results. In: 28th IEEE/ACM International Conference on Automated Software Engineering, Palo Alto, USA, pp. 356–366, November 2013
39. Weimer, W., Nguyen, T., Le Goues, C., Forrest, S.: Automatically finding patches using genetic programming. In: Proceedings of the 31st International Conference on Software Engineering, pp. 364–374. IEEE, Vancouver (2009)
40. White, D.R.: An unsystematic review of genetic improvement. In: 45th CREST Open Workshop on Genetic Improvement, London (2016)
41. White, D.R., Arcuri, A., Clark, J.A.: Evolutionary improvement of programs. IEEE Trans. Evol. Comput. **15**(4), 515–538 (2011)
42. White, D.R., Clark, J., Jacob, J., Poulding, S.M.: Searching for resource-efficient programs. In: Proceedings of the 10th Annual Conference on Genetic and Evolutionary Computation, GECCO 2008, vol. 1, p. 1775 (2008)
43. White, D.R.: Genetic programming for low-resource systems, December 2009
44. Wu, F., Weimer, W., Harman, M., Jia, Y., Krinke, J.: Deep parameter optimisation. In: Proceedings of the 2015 Annual Conference on Genetic and Evolutionary Computation, GECCO 2015, pp. 1375–1382. ACM, Madrid, July 2015

# Differentiable Genetic Programming

Dario Izzo[1][(✉)], Francesco Biscani[2], and Alessio Mereta[1]

[1] Advanced Concepts Team, European Space Agency, Noordwijk, The Netherlands
{dario.izzo,alessio.mereta}@esa.int
[2] Interdisciplinary Center for Scientific Computing, Heidelberg University,
Heidelberg, Germany
bluescarni@gmail.com

**Abstract.** We introduce the use of high order automatic differentiation, implemented via the algebra of truncated Taylor polynomials, in genetic programming. Using the Cartesian Genetic Programming encoding we obtain a high-order Taylor representation of the program output that is then used to back-propagate errors during learning. The resulting machine learning framework is called differentiable Cartesian Genetic Programming (dCGP). In the context of symbolic regression, dCGP offers a new approach to the long unsolved problem of constant representation in GP expressions. On several problems of increasing complexity we find that dCGP is able to find the exact form of the symbolic expression as well as the constants values. We also demonstrate the use of dCGP to solve a large class of differential equations and to find prime integrals of dynamical systems, presenting, in both cases, results that confirm the efficacy of our approach.

**Keywords:** Genetic programming · Truncated Taylor polynomials · Machine learning · Symbolic regression · Back-propagation

## 1 Introduction

Many of the most celebrated techniques in Artificial Intelligence would not be as successful if, at some level, they were not exploiting differentiable quantities. The back-propagation algorithm, at the center of learning in artificial neural networks, leverages on the first and (sometimes) second order differentials of the error to update the network weights. Gradient boosting techniques make use of the negative gradients of a loss function to iteratively improve over some initial model. More recently, differentiable memory access operations were successfully implemented [6,7] and shown to give rise to new, exciting, neural architectures. Even in the area of evolutionary computations, also sometimes referred to as *derivative-free* optimization, having the derivatives of the fitness function is immensely useful, to the extent that many derivative-free algorithms, in one way or another, seek to approximate such information (e.g. the covariance matrix in CMA-ES as an approximation of the inverse Hessian [8]). In the context of Genetic Programming too, previous works made use of the differential properties of the encoded programs [5,15,17,18], showing the potential use of such

© Springer International Publishing AG 2017
J. McDermott et al. (Eds.): EuroGP 2017, LNCS 10196, pp. 35–51, 2017.
DOI: 10.1007/978-3-319-55696-3_3

information, though a systematic use of the program differentials is still lacking in this field.

In this paper, we introduce the use of high-order automatic differentiation as a tool to obtain a complete representation of the differential properties of a program encoded by a genetic programming expression and thus of the model it represents. We make use of the truncated Taylor polynomial algebra to compute such information efficiently in one forward pass of the encoded program. The non trivial implementation of the necessary algebraic manipulations, as well as the resulting framework, is offered to the community as an open source project called AuDi [10] (a C++11 header only library and a Python module) allowing the machine learning community to gain access to a tool we believe can lead to several interesting advances. We use AuDi to evaluate Cartesian Genetic Programming expressions introducing a number of applications where the differential information is used to back-propagate the error on a number of model parameters. The resulting new machine learning tool is called dCGP and is also released as an open source project [9]. Using dCGP we study in more detail three distinct application domains where the Taylor expansion of the encoded program is used during learning: we perform symbolic regression on expressions including real constants, we search for analytical solutions to differential equations, we search prime integrals of motion in dynamical systems.

The paper is structured as follows: In Sect. 2 we describe the program encoding used, essentially a weighted version of the standard CGP encoding. In Sect. 3 we introduce the differential algebra of truncated Taylor polynomial used to obtain a high order automatic differentiation system. Some examples (Sect. 4) are also given to help the reader go through the possibly unfamiliar notation and algebra. In Sect. 5, using dCGP, we propose two different new methods to perform symbolic regression on expressions that include real constants. In Sect. 6 we use dCGP to find the analytical solution to differential equations. In the following Sect. 7, we propose a method to systematically search for prime integrals of motions in dynamical systems.

## 2 Program Encoding

To represent our functional programs we use the Cartesian Genetic Programming framework [13], but any other representation would be valid (such as tree-based GP). CGP is a form of Genetic Programming in which computations are organized using a matrix of nodes as depicted in Fig. 1. In the original formulation, nodes are arranged in $r$ rows and $c$ columns, and each is defined by a function $F_i$ with arity $a$ and by its connections $C_{ij}$, $j = 1 \ldots a$ indicating the nodes whose outputs to use to compute the node output via $F_i$. Connections can only point to nodes in previous columns and a maximum of $l$ columns back (levels back). Input and output nodes are also present. By evaluating all nodes in cascade starting from the $N_{in}$ input nodes, the $N_{out}$ outputs nodes are computed. They represent a complex combination of the inputs able to easily represent computer programs, digital circuits, mathematical relations and, in general, computational

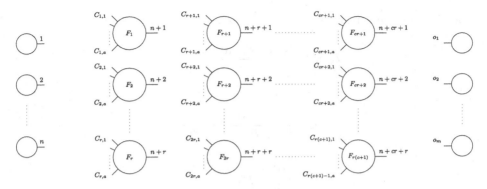

$$F_1, C_{1,1}..C_{1,a}, F_2, C_{2,1}..C_{2,a}, ..., o_1..o_m, w_{1,1}..w_{1,a}, ..., w_{2,1}..w_{2,a}$$

**Fig. 1.** The general form of a CGP as described in [13]. In our version weights are also assigned to the connections $C_{ij}$ so that the program is defined by some integer values (the connections, the functions and the outputs) and by some floating point values (the connection weights, not represented in the figure for clarity)

structures. Depending on the various $C_{ij}$, not all nodes affects the output so that only the active nodes need to be actually computed. The connections $C_{ij}$ are also here associated to multiplicative factors (weights) $w_{i,j}$. In this way, a CGP is allowed to represent highly successful models such as feed forward artificial neural networks as suggested in [11,19].

## 3    The Algebra of Truncated Polynomials

Consider the set $\mathcal{P}_n$ of all polynomials of order $\leq n$ with coefficients in $\mathbb{R}$. Under the truncated polynomial multiplication $\mathcal{P}_n$ becomes a field (*i.e.*, a ring whose nonzero elements form an abelian group under such multiplication). The meaning of this last statement is, essentially, that we may operate in $\mathcal{P}_n$ using four arithmetic operations $+, -, \cdot, /$ as we normally do in more familiar fields such as $\mathbb{R}$ or $\mathbb{C}$. In order to formally define the division operator, we indicate the generic element of $\mathcal{P}_n$ as $p = p_0 + \hat{p}$ so that the constant and the non-constant part of the polynomial are clearly separated. The multiplicative inverse of $p \in \mathcal{P}_n$ is then defined as:

$$p^{-1} = 1/p = \frac{1}{p_0}\left(1 + \sum_{k=1}^{m}(-1)^k(\hat{p}/p_0)^k\right) \tag{1}$$

As an example, to compute, in $\mathcal{P}_2$, the inverse of $p = 1 + x - y^2$ we write $p_0 = 1$ and $\hat{p} = x - y^2$. Applying the definition we get $p^{-1} = 1 - \hat{p} + \hat{p}^2 = (1 - (x - y^2) + x^2)$. It is then trivial to verify that $p \cdot p^{-1} = 1$.

### 3.1    The Link to Taylor Polynomials

We make use of the multi-index notation according to which $\alpha = (\alpha_1, \ldots, \alpha_n)$ and $\mathbf{x} = (x_1, \ldots, x_n)$ are n-tuples and the Taylor expansion around the point $\mathbf{a}$

to order $m$ of a multivariate function $f$ of $\mathbf{x}$ is written as:

$$T_f(\mathbf{x}) = \sum_{|\alpha|=0}^{m} \frac{(\mathbf{x}-\mathbf{a})^{\alpha}}{\alpha!}(\partial^{\alpha}f)(\mathbf{a}) \tag{2}$$

where:

$$\partial^{\alpha}f = \frac{\partial^{|\alpha|}f}{\partial^{\alpha_1}x_1\partial^{\alpha_2}x_2\ldots\partial^{\alpha_n}x_n}$$

$$\alpha! = \prod_{j=1}^{n}\alpha_j!$$

and

$$|\alpha| = \sum_{j=1}^{n}\alpha_j$$

The summation $\sum_{|\alpha|=0}^{n}$ must then be taken over all possible combinations of $\alpha_j \in N$ such that $\sum_{j=1}^{n}\alpha_j = |\alpha|$. The expression in Eq. (2), *i.e.* the Taylor expansion truncated at order $m$ of a generic function $f$, is a polynomial $\in \mathcal{P}_n$ in the $m$ variables $\mathbf{dx} = \mathbf{x} - \mathbf{a}$. It follows from Eq. (2) that a Taylor polynomial contains the information on all the derivatives of $f$.

The remarkable thing about the field $\mathcal{P}_n$ is that its arithmetic represents also Taylor polynomials, in the sense that if $T_f, T_g \in \mathcal{P}_n$ are truncated Taylor expansions of two functions $f, g$ then the truncated Taylor expansions of $f \pm g$, $fg, f/g$ can be found operating on the field $\mathcal{P}_n$ simply computing $T_f \pm T_g, T_f \cdot T_g, T_f/T_g$. We may thus compute high order derivatives of multivariate rational functions by computing their Taylor expansions in $\mathcal{P}_n$ and then extracting the desired coefficients.

**Example - A Division.** Consider the following rational function of two variables:

$$h = (x - y)/(x + 2xy + y^2) = f/g$$

Its Taylor expansion $T_h \in \mathcal{P}_2$ around the point $x = 0$, $y = 1$ can be computed as follows:

$$T_x = 0 + dx$$

$$T_y = 1 + dy$$

$$T_g = T_x + 2T_x \cdot T_y + T_y \cdot T_y = 1 + 3dx + 2dy + 2dxdy + dy^2$$

applying now Eq. (1), we get:

$$T_{1/g} = (1 - \hat{p} + \hat{p}^2)$$

where $\hat{p} = 3dx + 2dy + 2dxdy + dy^2$, hence:

$$T_{1/g} = 1 - 3dx - 2dy + 10dxdy + 9dx^2 + 3dy^2$$

and,

$$T_h = (-1 + dx - dy) \cdot T_{1/g} = -1 + 4dx + dy - 9dxdy - 12dx^2 - dy^2$$

which allows to compute the value of the function and all derivatives up to the second order in the point $x = 0, y = 1$:

$$h = -1, \partial_x h = 4, \partial_y h = 1, \partial_{xy} h = -9, \partial_{xx} h = -24, \partial_{yy} h = -2.$$

## 3.2   Non Rational Functions

If the function $f$ is not rational, $i.e.$ it is not the ratio between two polynomials, it is still possible, in most cases, to compute its truncated Taylor expansion operating in $\mathcal{P}_n$. This remarkable result is made possible leveraging on the nil-potency property of $\hat{p}$ in $\mathcal{P}_n$, $i.e.$ $\hat{p}^k = 0$ if $k > n$. We outline the derivation in the case of the function $f = \exp(g)$, other cases are treated in details in [2]. We write $T_f = \exp(T_g) = \exp(p_0 + \hat{p}) = \exp p_0 \exp \hat{p}$. Using the series representation of the function exp we may write $f = \exp p_0 \sum_{i=0}^{\infty} \frac{\hat{p}^i}{i!}$. As per the nil-potency property this infinite series is finite in $\mathcal{P}_n$ and we can thus write $\exp(p) = \exp p_0 \sum_{i=0}^{n} \frac{\hat{p}^i}{i!}$, or equivalently:

$$T_f = \exp p_0 \cdot (1 + \hat{p} + \hat{p}^2 + \ldots + \hat{p}^n)$$

With similar derivations it is possible to define most commonly used functions including $\exp, \log, \sin, \cos, \tan, \arctan, \arcsin, \arccos, \text{abs}$ as well as exponentiation. Operating in this algebra, rather than in the common arithmetical one, we compute the programs encoded by CGP obtaining not only the program output, but also all of its differential variations (up to order $n$) with respect to any of the parameters we are interested in. This idea is leveraged in this paper to propose new learning methods in genetic programming.

## 4   Example of a dCGP

Consider a CGP expression having $n = 3$, $m = 1$, $r = 1$, $c = 10$, $l = 3$, $a = 2$, $w_{i,j} = 1$ and the kernel functions: $+, *, /, \sigma$ where $\sigma$ is the sigmoid function. Using the same convention as in [13] for node numbering, the chromosome:

$$x = [2, 1, 1, 0, 2, 0, 1, 2, 2, 1, 2, 1, 3, 1, 2, 3, 6, 3, 3, 4, 2, 2, 8, 0, 2, 7, 2, 2, 3, 10, 10]$$

will produce, after simplifications, the following expression at the terminal node:

$$o_0 = \frac{\sigma(yz + 1)}{x}$$

In Fig. 2 the actual graph expressing the above expression is visualized. If we now consider the point $x = 1$, $y = 1$ and $z = 1$ and we evaluate classically the CGP expression (thus operating on float numbers), we get, trivially, $O_1 = 0.881$ where, for convenience, we only report the output rounded up to 3 significant

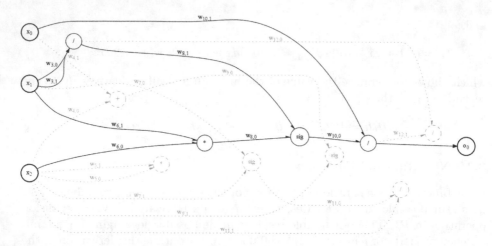

**Fig. 2.** A weighted CGP graph expressing (assuming $w_{i,j} = 1$) $o_0 = \frac{\sigma(yz+1)}{x}$ at the output node. The inactive part of the graph is greyed out.

digits. Let us, instead, use dCGP to perform such evaluation. One option could be to define $x, y, z$ as generalized dual numbers operating in $\mathcal{P}_2$. In this case, the output of the dCGP program will then be a Taylor polynomial in $x, y, z$ truncated at second order:

$$o_0 = 0.881 - 0.881dx + 0.105dy + 0.105dz + 0.881dx^2$$
$$- 0.0400dy^2 - 0.0400dz^2 - 0.105dxdz + 0.025dydz - 0.105dxdy$$

carrying information not only on the actual program output, but also on all of its derivatives (in this case to order 2) with respect to the chosen parameters (*i.e.* all inputs, in this case). Another option could be to define some of the weights as generalized dual numbers, in which case it is convenient to report the expression at the output node with the weights values explicitly appearing as parameters:

$$o_0 = w_{10,0} \frac{\sigma(w_{3,0}w_{8,1}/w_{3,1} + w_{6,0}w_{6,1}w_{8,0}yz)}{w_{10,1}x}$$

If we define $w_{3,1}$ and $w_{10,1}$ as generalized dual numbers operating in $\mathcal{P}_3$ then, evaluating the dCGP will result in:

$$o_0 = 0.881 - 0.104dw_{3,1} - 0.881dw_{10,1} + 0.104dw_{10,1}dw_{3,1} + 0.0650dw_{3,1}^2 + 0.881dw_{10,1}^2$$
$$- 0.0315dw_{3,1}^3 - 0.881dw_{10,1}^3 - 0.104dw_{10,1}^2dw_{3,1} - 0.0650dw_{10,1}dw_{3,1}^2$$

## 5    Learning Constants in Symbolic Regression

A first application of dCGP we study is the use of the derivatives of the expressed program to learn the values of constants by, essentially, back-propagating the error to the constants' value. In 1997, during a tutorial on Genetic Programming in Paolo Alto, John Koza [12] one of the fathers of Genetic Programming research noted that "finding of numeric constants is a skeleton in the GP closet ... an area of research that requires more investigation". Years later, the problem, while investigated by multiple researchers, is still open [14]. The standard approach to this issue is that of the *ephemeral constant* where a few additional input terminals containing constants are added and used by the encoded program to build more complex representations of any constant. Under this approach one would hope that to approximate, say, $\pi$ evolution would assemble, for example, the block $\frac{22}{7} = 3.1428$ from the additional terminals. In some other versions of the approach the values of the input constants are subject to genetic operators [5], or are just random. Here we first present a new approach that back-propagates the error on the ephemeral constants values, later we introduce a different approach using weighted dCGP expressions and back-propagating the error on the weights.

## 5.1   Ephemeral Constants Approach

The idea of learning ephemeral constants by back-propagating the error of a GP expression was first studied in [17] where gradient descent was used during evolution of a GP tree. The technique was not proved, though, to be able to solve the problems there considered (involving integers), but to reduce the RMSE at a fixed number of generations with respect to a standard genetic programming technique. Here we will, instead, focus on using dCGP to solve exactly symbolic regression problems involving real constants such as $\pi$ and $e$. Consider the quadratic error $\epsilon_q(\mathbf{c}) = \sum_i (y_i(\mathbf{c}) - \hat{y}_i)^2$ where $\mathbf{c}$ contains the values of the ephemeral constants, $y_i$ is the value of the dCGP evaluated on the input point $x_i$ and $\hat{y}_i$ the target value. Define the fitness (error) of a candidate dCGP expression as:

$$\epsilon = \min_{\mathbf{c}} \epsilon_q(\mathbf{c})$$

This "inner" minimization problem can be efficiently solved by a second order method. If the number of input constants is reasonably small, the classic formula for the Newton's method can be conveniently applied iteratively:

$$\mathbf{c}_{i+1} = \mathbf{c}_i - \mathbf{H}^{-1}(\mathbf{c}_i)\nabla\epsilon_q(\mathbf{c}_i)$$

starting from some initial value $\mathbf{c}_0$. The Hessian $\mathbf{H}$ and the gradient $\nabla\epsilon$ are extracted from the Taylor expansion of the error $\epsilon$ computed via the dCGP. Note that if the constants $\mathbf{c}$ appear only linearly in the candidate expression, one single step will be sufficient to get the exact solution to the inner minimization problem, as $\epsilon_q(c)$ is the quadratic error. This suggests the use of the following approximation of the error defined above, where one single learning step is taken:

$$\epsilon = \epsilon_q(\mathbf{c}_0 - \mathbf{H}^{-1}(\mathbf{c}_0)\nabla\epsilon_q(\mathbf{c}_0)) \tag{3}$$

where $\mathbf{c}_0$ is the current value for the constants. In those cases where this Newton step does not reduce the error (when $\mathbf{H}$ is not positive definite we do not have any guarantee that the search direction will lead to a smaller error) we, instead,

**Table 1.** Learning the ephemeral constants: experiments definition and results. In all cases the constants and the expressions are found exactly (final error is $\epsilon < 10^{-14}$).

| | Target | Expression found | Constants found | ERT | Bounds | $g_{max}$ |
|---|---|---|---|---|---|---|
| P1: | $x^5 - \pi x^3 + x$ | $-cx^3 + x^5 + x$ | $c = 3.1415926$ | 19902 | [1,3] | 1000 |
| P2: | $x^5 - \pi x^3 + \frac{2\pi}{x}$ | $cx^3 - 2c/x + x^5$ | $c = -3.1415926$ | 124776 | [0.1,5] | 5000 |
| P3: | $\frac{ex^5 + x^3}{x+1}$ | $\frac{-cx^5 + x^3}{x+1}$ | $c = -2.7182818$ | 279400 | [−0.9, 1] | 5000 |
| P4: | $\sin(\pi x) + \frac{1}{x}$ | $\sin(cx + x) + \frac{1}{x}$ | $c = 2.1415926$ | 105233 | [−1, 1] | 5000 |
| P5: | $ex^5 - \pi x^3 + x$ | $c_1 x^3 - c_2 x^5 + x$ | $c_1 = -3.1415926$ | 45143 | [1,3] | 2000 |
| | | | $c_2 = -2.7182818$ | | | |
| P6: | $\frac{ex^2 - 1}{\pi(x+2)}$ | $\frac{c_1 + c_2 x}{x+2}$ | $c_1 = -0.3183098$ | 78193 | [−2.1, 1] | 10000 |
| | | | $c_2 = 0.86525597$ | | | |
| P7: | $\cos(\pi x) + \sin(ex)$ | $\sin(c_1^2 c_2 x + c_1^2 x) + \cos(c_1^2 x)$ | $c_1 = 1.7724538$ | 210123 | [−1, 1] | 5000 |
| | | | $c_2 = -0.1347440$ | | | |

take a few steps of gradient descent (learning rate set to 0.05). $\mathbf{c}_0$ is initialized to some random value and, during evolution, is set to be the best found so far. Note that when one offspring happens to be the unmutated parent, its fitness will still improve on the parent's thanks to the local learning, essentially by applying one or more Newton steps. This mechanism (also referred to as Lamarckian evolution in memetic research [17]) ensures that the exact value, and not an approximation, is eventually found for the constants even if only one Newton step is taken at each iteration. Note also that the choice of the quadratic error function in connection with a Newton method will greatly favour, during evolution, expressions such as, for example, $c_1 x + c_2$ rather than the equivalent $c_1^2 c_2 x + \frac{1}{c_2}$.

**Experiments.** We consider seven different symbolic regression problems (see Table 1) of varying complexity and containing the real constants $\pi$ and $e$. Ten points $x_i$ are chosen on a uniform grid within some lower and upper bound. For all problems the error defined by Eq. (3) is used as fitness to evolve an unweighted dCGP with $r = 1$, $c = 15$, $l = 16$, $a = 2$, $N_{in} = 2$ or 3 according to the number of constants present in the target expression and $N_{out} = 1$. As Kernel functions we use: $+, -, *, /$ for problems P1, P2, P3, P5, P6 and $+, -, *, /, \sin, \cos$ for problems P4, P7. The evolution is driven by a (1+$\lambda$)-ES evolution strategy with $\lambda = 4$ where the $i - th$ mutant of the $\lambda$ offspring is created by mutating $i$ active genes in the dCGP. We run all our experiments[1] 100 times and for a maximum of $g_{max}$ generations. The 100 runs are then considered as multi starts of the same algorithm and are used to compute the Expected Run Time [1] (ERT), that is the ratio between the overall number of dCGP evaluations ($fevals$) made across all trials and the number of successful trials $n_s$: $ERT = \frac{fevals}{n_s}$. As shown in Table 3 our approach is able to solve all problems exactly (final error is $\epsilon < 10^{-14}$) and to represent the real constants with precision.

---

[1] The exact details on all these experiments can be found in two IPython notebooks available here: https://goo.gl/iH5GAR, https://goo.gl/0TFsSv.

## 5.2   Weighted dCGP Approach

While the approach we developed above works very well for the selected problems, it does have a major drawback: the number of ephemeral constants to be used as additional terminals must be pre-determined. If one were to use too few ephemeral constants the regression task would fail to find a zero error, while if one were to put too many ephemeral constants the complexity would scale up considerably and the proposed solution strategy would quickly lose its efficacy. This is particularly clear in the comparison between problems P1 and P5 where the only difference is a factor $e$ in the quintic term of the polynomial. Ideally this should not change the learning algorithm. As detailed in Sects. 2 and 4, a dCGP gives the possibility to associate weights to each edge of the acyclic graph (see Fig. 2) and then compute the differential properties of the error w.r.t. any selection of the weights. This introduces the idea of performing symbolic regression tasks using a weighted dCGP expression with no additional terminal inputs but with the additional problem of having to learn values for all the weights appearing in the represented expression. In particular we now have to define the fitness (error) of a candidate dCGP expression as:

$$\epsilon = \min_{\mathbf{w}} \epsilon_q(\mathbf{w})$$

where $\mathbf{w}$ are the weights that appear in the output terminal. Similarly to what we did previously, we need a way to solve this inner minimization problem. Applying a few Newton steps could be an option, but since the number of weights may be large we will not follow this idea. Instead, we iteratively select, randomly, $n_w$ weights $\tilde{\mathbf{w}}$ active in the current dCGP expression and we update them applying one Newton step:

$$\tilde{\mathbf{w}}_{\mathbf{i+1}} = \tilde{\mathbf{w}}_{\mathbf{i}} - \tilde{\mathbf{H}}^{-1}(\tilde{\mathbf{w}}_{\mathbf{i}})\nabla_{\tilde{\mathbf{w}}}\epsilon_q(\tilde{\mathbf{w}}_{\mathbf{i}})$$

where the tilde indicates that not all, but only part of the weights are selected. If no improvement is found we discard the step. At each iteration we select randomly new weights (no Lamarckian learning). This idea (that we call weight batch learning), is effective in our case as it avoids computing and inverting large Hessians, while also having more chances to actually improve the error at each step without the use of a learning rate for a line search. For each candidate expression we perform $N$ iterations of weight batch learning starting from normally distributed initial values for the weights.

**Experiments.** We use the same experimental set-up employed to perform symbolic regression using the ephemeral constants approach (see Table 2). For each iteration of the weight batch learning we use a randomly selected $n_w \in \{2,3\}$ and we perform $N = 100$ iterations[2]. The initial weights are drawn from a zero mean, unit standard deviation normal distribution. By assigning weights to all the edges, we end up with expressions where every term has its own constant

---

[2] The exact details on all these experiments can be found in the IPython notebook available here: https://goo.gl/8fOzYM.

**Table 2.** Learning constants using weighted dCGP: experiments definition and results. In all cases the constants and expressions are found exactly (final error is $\epsilon < 10^{-14}$)

| | Target | Expression found | Constants found | ERT | Bounds | $g_{max}$ |
|---|---|---|---|---|---|---|
| P1: | $x^5 - \pi x^3 + x$ | $c_1 x^5 + c_2 x^3 + c_3 x$ | $c_1 = 1.0$ $c_2 = -3.1415926$ $c_3 = 0.9999999$ | 106643 | [1,3] | 200 |
| P2: | $x^5 - \pi x^3 + \frac{2\pi}{x}$ | $c_1 x^5 + c_2 x^3 + \frac{c_3}{c_4 x}$ | $c_1 = 0.9999999$ $c_2 = -3.1415926$ $c_3 = 6.2831853$ $c_4 = 1.0$ | 271846 | [0.1,5] | 200 |
| P3: | $\frac{ex^5 + x^3}{x+1}$ | $\frac{c_1 x^5 + c_2 x^3}{c_3 x + c_4}$ | $c_1 = 4.3746892$ $c_2 = 1.6093582$ $c_3 = 1.6093582$ $c_4 = 1.6093582$ | 1935500 | [-0.9, 1] | 200 |
| P4: | $\sin(\pi x) + \frac{1}{x}$ | $c_1 \sin(c_2 x) + \frac{c_3}{c_4 x}$ | $c_1 = 1.0$ $c_2 = 3.1415926$ $c_3 = 1.0$ $c_4 = 1.0$ | 135107 | [-1, 1] | 200 |
| P5: | $ex^5 - \pi x^3 + x$ | $c_1 x^5 + c_2 x^3 + c_3 x$ | $c_1 = 2.7182818$ $c_2 = -3.1415926$ $c_3 = 1.0$ | 122071 | [1,3] | 200 |
| P6: | $\frac{ex^2 - 1}{\pi(x+2)}$ | $\frac{c_1 x^2 + c_2}{c_3 x + c_4}$ | $c_1 = 1.5963630$ $c_2 = -0.5872691$ $c_3 = 1.8449604$ $c_4 = 3.6899209$ | 628433 | [-2.1, 1] | 200 |
| P7: | $\cos(\pi x) + \sin(ex)$ | $c_1 \sin(c_2 x) + c_3 \cos(c_4 x)$ | $c_1 = 1.0$ $c_2 = 2.7182818$ $c_3 = 1.0$ $c_4 = 3.1415926$ | 243629 | [-1, 1] | 200 |

(*i.e.*, a combination of weights), hence no distinction is made by the learning algorithm between integer and real constants. This gives the method a far greater generality than the ephemeral constants approach as the number of real constants appearing in the target expression is not used as an information to design the encoding. It is thus not a surprise that we obtain, overall, higher ERT values. These higher values mainly come from the Newton steps applied on each weighted candidate expression and not from the number of required generations which was, instead, observed to be generally much lower across all experiments. Also note that for problems P3 and P6 the final expression found can have an infinite number of correct constants, obtained by multiplying the numerator and denominator by the same number. Overall, the new method here proposed to perform symbolic regression on expressions containing constants was able to consistently solve the problems considered and is also suitable for applications where the number of real constants to be learned is not known in advance.

# 6    Solution to Differential Equations

We show how to use dCGP to search for expressions $S(x_1, \ldots, x_n) = S(\mathbf{x})$ that solve a generic differential equation of order $m$ in the form:

$$f\left(\partial^\alpha S, \mathbf{x}\right) = 0, |\alpha| \leq m \tag{4}$$

with some boundary conditions $S(\mathbf{x}) = S_\mathbf{x}, \forall \mathbf{x} \in \mathcal{B}$. Note that we made use of the multi-index notation for high order derivatives described previously. Such formal representation includes initial value problems for ordinary differential equations and boundary value problems for partial differential equations. While this representation covers only the case of Dirichlet boundary conditions (values of $S$ are specified on a border), the system devised here can be used also for Neumann boundary conditions (values of $\partial S$ are specified on a border). Similarly, also systems of equations can be studied. Assume $S$ is encoded by a dCGP expression: one may easily compute Eq. (4) over a number of control points placed in some domain, and boundary values $S(\mathbf{x})$ may also be computed on some other control points placed over $\mathcal{B}$. It is thus natural to compute the following expression and use it as error:

$$\epsilon = \sum_i f^2\left(\partial^\alpha S, \mathbf{x}_i\right) + \alpha \sum_j \left(S(\mathbf{x}) - S_\mathbf{x}\right)^2 \tag{5}$$

which is, essentially, the sum of the quadratic error of the violation of the differential equations plus the quadratic error of the violation on the boundary values. Symbolic regression was studied already in the past as a new tool to find solutions to differential equations by Tsoulos and Lagaris [18] who used grammatical evolution to successfully find solutions to a large number of ordinary differential equations (ODEs and NLODEs), partial differential equations (PDE), and systems of ordinary differential equations (SODEs). To find the derivatives of the encoded expression Tsoulos and Lagaris add additional stacks where basic rules for differentiation are applied in a chain. Note that such system (equivalent to a basic automatic differentiation system) is not easily extended to the computation of mixed derivatives, necessary for example when Neumann boundary conditions are present. As a consequence, all of the differential test problems introduced in [18] do not involve mixed derivatives. To test the use of dCGP on this domain, we consider a few of those problems and compare our results to the ones obtained by Tsoulos and Lagaris. From the paper it is possible to derive the average number of evaluations that were necessary to find a solution to the given problem, by multiplying the population size used and the average number of generations reported. This number can be then compared to the ERT [1] obtained in our multi-start experiments.

**Experiments.** For all studied problems we use the error defined by Eq. (5) to train an unweighted dCGP with $r = 1$, $c = 15$, $l = 16$, $a = 2$, $N_{in}$ equal to the number of input variables and $N_{out} = 1$. As Kernel functions we use the same as that used by Tsoulos and Lagaris: $+, -, *, /, \sin, \cos, \log, \exp$. The

**Table 3.** Expected run-time for different test cases taken from [18]. The ERT can be seen as the average number of evaluation of the program output and its derivatives needed (on average) to reduce the error to zero (*i.e.* to find an exact solution)

| Problem | d-CGP | Tsoulos [18] |
|---------|-------|--------------|
| ODE1    | 8123  | 130600       |
| ODE2    | 35482 | 148400       |
| ODE5    | 22600 | 88200        |
| NLODE3  | 896   | 38200        |
| PDE2    | 24192 | 40600        |
| PDE6    | 327020| 797000       |

evolution is made using a $(1+\lambda)$-ES evolution strategy with $\lambda = 10$ where the $i - th$ mutant of the $\lambda$ offspring is created mutating $i$ active genes in the dCGP. We run all our experiments[3] 100 times and for a maximum of 2000 generation. We then record the successful runs (*i.e.* the runs in which the error is reduced to $\epsilon \leq 10^{-16}$) and compute the expected run-time as the ratio between the overall number of evaluation of the encoded program and its derivatives ($fevals$) made across successful and unsuccessful trials and the number of successful trials $n_s$: $ERT = \frac{fevals}{n_s}$ [1]. The results are shown in Table 3. It is clear how, in all test cases, the dCGP based search is able to find solutions very efficiently, outperforming the baseline results.

## 7   Discovery of Prime Integrals

We now show how to use dCGP to search for expressions $P$ that are prime integrals of some set of differential equations. Consider a set of ordinary differential equations (ODEs) in the form:

$$
\begin{cases}
\frac{dx_1}{dt} = f_1(x_1, \cdots, x_n) \\
\quad\vdots \\
\frac{dx_n}{dt} = f_n(x_1, \cdots, x_n)
\end{cases}
$$

Solutions to the above equations are denoted with $x_i(t)$ to highlight their time dependence. Under relatively loose conditions on the functions $f_i$, the solution to the above equations always exists unique if initial conditions $x_i(t_0)$ are specified. A prime integral for a set of ODEs is a function of its solutions in the form $P(x_1(t), \cdots, x_n(t)) = 0, \forall t$. Prime integrals, or integral of motions, often express a physical law such as energy or momentum conservation, but also the conservation of some more "hidden" quantity as its the case in the Kovalevskaya top [3] or of the Atwood's machine [4]. Each of them allows to decrease the number

---

[3] The full details on these experiments can be found in an IPython notebook available here: https://goo.gl/wnCkO9.

of degrees of freedom in a system by one. In general, finding prime integrals is a very difficult task and it is done by skillful mathematicians using their intuition. A prime integral is an implicit relation between the variables $x_i$ and, as discussed in [16], when symbolic regression is asked to find such implicit relations, the problem of driving evolution towards non trivial, informative solutions arises. We thus have to devise an experimental set-up that is able to avoid such a behaviour. Assuming $P$ to be a prime integral, we differentiate it obtaining:

$$\frac{dP}{dt} = \sum_i \frac{\partial P}{\partial x_i} \frac{dx_i}{dt} = \sum_i \frac{\partial P}{\partial x_i} f_i = 0$$

The above relation, and equivalent forms, is essentially what we use to compute the error of some candidate expression representing a prime integral $P$. In more details, we define $N$ points in the phase space and denote them with $x_i^j, j = 1 \ldots N$. Thus, we introduce the error function:

$$\epsilon = \sum_j \sum_i \left[ \frac{\partial P}{\partial x_i}(x_1^j, \ldots, x_n^j) f_i(x_1^j, \ldots, x_n^j) \right]^2 \tag{6}$$

Since the above expression is identically zero if $P$ is, trivially, a constant we introduce a "mutation suppression" method during evolution. Every time a new mutant is created, we compute $\frac{\partial P}{\partial x_i}, \forall i$ and we ignore it if $\frac{\partial P}{\partial x_i} = 0, \forall i$, that is if the symbolic expression is representing a numerical constant. Our method bears some resemblance to the approach described in [15] where physical laws are searched for to fit some observed system, but it departs in several significant points: we do not use experimental data, rather the differential description of a system, we compute our derivatives using the Taylor polynomial algebra (*i.e.* automatic differentiation) rather than estimating them numerically from observed data, we use the mutation suppression method to avoid evolving trivial solutions, we need not to introduce a *variable pairing* [15] choice and we do not assume variables as not dependent on all others. All the experiments[4] are made using a non-weighted dCGP with $N_{out} = 1$, $r = 1$, $c = 15$, $lb = 16$, $a = 2$. We use a $(1+\lambda)$-ES evolution strategy with $\lambda = 10$ where the $i - th$ mutant of the $\lambda$ offspring is created mutating $i$ active genes in the dCGP. A set of $N$ control points are then sampled uniformly in some bounds. Note how these are not belonging to any actual trajectory of the system, thus we do not need to numerically integrate the ODEs. We consider three different dynamical systems: a mass-spring system, a simple pendulum and the two body problem.

**Mass-Spring System.** Consider the following equations:

$$MSS : \begin{cases} \dot{v} = -kx \\ \dot{x} = v \end{cases}$$

---

[4] The full details on our experiments can be found in an IPython notebook available here: https://goo.gl/ATrQR5.

describing the motion of a simple, frictionless mass-spring system (MSS). We use $N = 50$ and create the points at random as follows: $x \in [2, 4]$, $v \in [2, 4]$ and $k \in [2, 4]$. For the error, rather than using directly the form in Eq. (6) our experiments indicate that a more informative (and yet mathematically equivalent) variant is, in this case, $\epsilon = \sum_j \left[ \frac{\frac{\partial P}{\partial r}}{\frac{\partial P}{\partial v}} + \frac{f_v}{f_r} \right]^2$.

**Simple Pendulum.** Consider the following equations:

$$SP : \begin{cases} \dot{\omega} = -\frac{g}{L} \sin \theta \\ \dot{\theta} = \omega \end{cases}$$

describing the motion of a simple pendulum (SP). We use $N = 50$ and create the points at random as follows: $\theta \in [-5, 5]$, $\omega \in [-5, 5]$ and $c = \frac{g}{L} \in (0, 10]$. Also in this case, we use a variant for the error expression: $\epsilon = \sum_j \left[ \frac{\frac{\partial P}{\partial \theta}}{\frac{\partial P}{\partial \omega}} + \frac{f_\omega}{f_\theta} \right]^2$.

**Two Body Problem.** Consider the following equations:

$$TBP : \begin{cases} \dot{v} = -\frac{\mu}{r^2} + r\omega^2 \\ \dot{\omega} = -2\frac{v\omega}{r} \\ \dot{r} = v \\ \dot{\theta} = \omega \end{cases}$$

describing the Newtonian motion of two bodies subject only to their own mutual gravitational interaction (TBP). We use $N = 50$ random control points sampled uniformly as follows: $r \in [0.1, 1.1]$, $v \in [2, 4]$, $\omega \in [1, 2]$ and $\theta \in [2, 4]$ and $\mu \in [1, 2]$ (note that these conditions allow for both elliptical and hyperbolic motion). The unmodified form for the error (Eq. (6)) leads to the identification of a first prime integral (the angular momentum conservation). Since the evolution quickly and consistently converges to that expression, the problem arises on how to find possibly different ones. Changing the error expression to $\epsilon = \sum_j \left[ \frac{\frac{\partial P}{\partial r}}{\frac{\partial P}{\partial v}} + \frac{f_v}{f_r} + \frac{\frac{\partial P}{\partial \theta}}{\frac{\partial P}{\partial v}} \frac{f_\theta}{f_r} + \frac{\frac{\partial P}{\partial \omega}}{\frac{\partial P}{\partial v}} \frac{f_\omega}{f_r} \right]^2$ forces evolution away from the angular conservation prime integral and thus allows for other expressions to be found.

**Experiments.** For each of the above problems we run 100 independent experiments letting the dCGP expression evolve up to a maximum of 2000 generations (brought up to 100000 for the most difficult case of the second prime integral in the two-body problem). We record the successful runs and the generation number to then evaluate the expected run time (ERT) [1]. The results are shown in Table 4 where we also report some of the expressions found and their simplified form. In all cases the algorithm is able to find all prime integrals with the notable case of the two-body problem energy conservation being very demanding in terms of the ERT. In this case we note how the angular momentum $\omega r^2$ is often present in the final expression found as it there acts as a constant. The

**Table 4.** Results of the search for prime integrals in the mass-spring system (MSS), simple pendulum (SP), and two body problem (TBP). Some prime integrals found are reported, in both the original and simplified form. The Expected Run Time (ERT), *i.e.* the average number of evaluations of the dCGP expression required, is also reported.

| MSS: Energy Conservation (ERT = 50000) | |
|---|---|
| Expression found | Simplified |
| $k*((x*x)-k)+k+(v*v)$ | $-k^2+kx^2+k+v^2$ |
| $(x*x)+(v/k)*v$ | $x^2+v^2/k$ |
| $(k*((x*x)/v)/(k*(x*x)/v)+v)/(x*x)$ | $k/(kx^2+v^2)$ |
| $(v+x)*((v+x)-x)-x*(((v+x)-x)-(x*k))$ | $kx^2+v^2$ |

| SP: Energy Conservation (ERT = 114000) | |
|---|---|
| Expression found | Simplified |
| $((((\omega*\omega)/c)-\cos(\theta))-\cos(\theta))/c$ | $-2\cos(\theta)/c+\omega^2/c^2$ |
| $(\omega/(c+c))*(\omega-(\cos(\theta)/(\omega/(c+c))))$ | $-\cos(\theta)+\omega^2/c$ |
| $((\omega*\omega)-((c+c)*\cos(\theta)))$ | $-2c\cos(\theta)+\omega^2$ |
| $(((c+c)*\cos(\theta))-(\omega*\omega))-\sin((\theta/\theta))$ | $2c\cos(\theta)-\omega^2-\sin(1)$ |

| TBP: Angular momentum Conservation (ERT = 5270) | |
|---|---|
| Expression found | Simplified |
| $(((\mu/r)/(r/\mu))/\omega)$ | $\mu^2/(\omega r^2)$ |
| $((((r*r)*\omega)+\mu)/\mu)$ | $1+\omega r^2/\mu$ |
| $((\omega*\mu)*((r*r)*\mu))$ | $\mu^2*\omega r^2$ |
| $((\mu/(\mu+\mu))-((r*\omega)*r))$ | $-\omega r^2+1/2$ |

| TBP: Energy Conservation (ERT = 6144450) | |
|---|---|
| Expression found | Simplified |
| $((v-r)*(v+r))-(((\mu-r)+(\mu-r))/r-$ $((r-(r*\omega))*(r-(r*\omega))))$ | $-2\mu/r+\omega^2r^2-2\omega r^2+v^2+2$ |
| $((\mu/r)-((v*v)-(\mu/r)))-((r*\omega)*(r*\omega))$ | $2\mu/r-\omega^2r^2-v^2$ |
| $(((\omega*r)*(r-(\omega*r)))+((\mu/r)-((v*v)-(\mu/r))))$ | $2\mu/r-\omega^2r^2+\omega r^2-v^2$ |
| $(((r*\omega)*\omega)*r)-(((\mu/r)+(\mu/r))-(v*v))$ | $-2\mu/r+\omega^2r^2+v^2$ |

systematic search for prime integrals is successful in these cases and makes use of no mathematical insight into the system studied, introducing a new computer assisted procedure that may help in future studies of dynamical systems.

# 8    Conclusions

We have introduced a novel machine learning framework called differentiable genetic programming, which makes use of a high order automatic differentiation system to compute any-order derivatives of the program outputs and errors. We test the use of our framework on three distinct open problems in symbolic regression: learning constants, solving differential equations and searching for physical laws. In all cases, we find our model able to successfully solve selected problems and, when applicable, to outperform previous approaches. Of particular interest is the novel solution proposed to the long debated problem of constant finding in genetic programming, here proved to allow to find the exact solution in a number of interesting cases. Our work is a first step towards the systematic use of differential information in learning algorithms for genetic programming.

# References

1. Auger, A., Hansen, N.: Performance evaluation of an advanced local search evolutionary algorithm. In: 2005 IEEE Congress on Evolutionary Computation, vol. 2, pp. 1777–1784. IEEE (2005)
2. Bertz, M.: Modern Map Methods in Particle Beam Physics, vol. 108. Academic Press, Cambridge (1999)
3. Borisov, A.V., Kholmskaya, A., Mamaev, I.S.: S.V. Kovalevskaya top and generalizations of integrable systems. Regul. Chaotic Dyn. **6**(1), 1–16 (2001)
4. Casasayas, J., Nunes, A., Tufillaro, N.: Swinging Atwood's machine: integrability and dynamics. J. de Phys. **51**(16), 1693–1702 (1990)
5. Cerny, B.M., Nelson, P.C., Zhou, C.: Using differential evolution for symbolic regression and numerical constant creation. In: Proceedings of the 10th Annual Conference on Genetic and Evolutionary Computation, pp. 1195–1202. ACM (2008)
6. Graves, A., Wayne, G., Danihelka, I.: Neural turing machines. arXiv preprint arXiv:1410.5401 (2014)
7. Graves, A., Wayne, G., Reynolds, M., Harley, T., Danihelka, I., Grabska-Barwińska, A., Colmenarejo, S.G., et al.: Hybrid computing using a neural network with dynamic external memory. Nature **538**(7626), 471–476 (2016)
8. Hansen, N.: The CMA evolution strategy: a comparing review. In: Lozano, J.A., Larrañaga, P., Inza, I., Bengoetxea, E. (eds.) Towards a new evolutionary computation. SFSC, vol. 192, pp. 75–102. Springer, Heidelberg (2006). doi:10.1007/3-540-32494-1_4
9. Izzo, D.: dCGP: first release, November 2016. https://doi.org/10.5281/zenodo.164627
10. Izzo, D., Biscani, F.: AuDi: first release, November 2016. https://doi.org/10.5281/zenodo.164628
11. Khan, M.M., Ahmad, A.M., Khan, G.M., Miller, J.F.: Fast learning neural networks using cartesian genetic programming. Neurocomputing **121**, 274–289 (2013)
12. Koza, J.: Tutorial on advanced genetic programming, at genetic programming 1997, Palo Alto, CA (1997)
13. Miller, J.F.: Cartesian genetic programming. In: Miller, J.F. (ed.) Cartesian Genetic Programming. NCS, pp. 17–34. Springer, Heidelberg (2011). doi:10.1007/978-3-642-17310-3_2
14. O'Neill, M., Vanneschi, L., Gustafson, S., Banzhaf, W.: Open issues in genetic programming. Genet. Program. Evolvable Mach. **11**(3–4), 339–363 (2010)
15. Schmidt, M., Lipson, H.: Distilling free-form natural laws from experimental data. Science **324**(5923), 81–85 (2009)

16. Schmidt, M., Lipson, H.: Symbolic regression of implicit equations. In: Riolo, R., O'Reilly, U.-M., McConaghy, T. (eds.) Genetic Programming Theory and Practice VII. GEC, pp. 73–85. Springer, Heidelberg (2010)
17. Topchy, A., Punch, W.F.: Faster genetic programming based on local gradient search of numeric leaf values. In: Proceedings of the Genetic and Evolutionary Computation Conference (GECCO-2001), pp. 155–162 (2001)
18. Tsoulos, I.G., Lagaris, I.E.: Solving differential equations with genetic programming. Genet. Program. Evolvable Mach. **7**(1), 33–54 (2006)
19. Turner, A.J., Miller, J.F.: Cartesian genetic programming encoded artificial neural networks: a comparison using three benchmarks. In: Proceedings of the 15th Annual Conference on Genetic and Evolutionary Computation, pp. 1005–1012. ACM (2013)

# Evolving Game State Features from Raw Pixels

Baozhu Jia[✉] and Marc Ebner

Institut für Mathematik und Informatik,
Ernst Moritz Arndt Universität Greifswald,
Walther-Rathenau-Strasse 47, 17487 Greifswald, Germany
{baozhuj,marc.ebner}@uni-greifswald.de

**Abstract.** General video game playing is the art of designing artificial intelligence programs that are capable of playing different video games with little domain knowledge. One of the great challenges is how to capture game state features from different video games in a general way. The main contribution of this paper is to apply genetic programming to evolve game state features from raw pixels. A voting method is implemented to determine the actions of the game agent. Three different video games are used to evaluate the effectiveness of the algorithm: Missile Command, Frogger, and Space Invaders. The results show that genetic programming is able to find useful game state features for all three games.

**Keywords:** Genetic programming · General video game playing · Voting method

## 1 Introduction

All types of games, including video games and board games, provide a testbed for artificial intelligence research. AI technology has developed successful game players for many games, including go and chess. In 1996, Deep Blue [12] became the first computer to win a chess game against the world champion Garry Kasparov. In 2016, AlphaGo [18] beat Lee Sedol in a five-game match. This was the first time that a computer Go program beat a 9-dan professional Go player. These examples show that a computer program can be better than a human being in certain specific areas. Some would even argue that the program has a higher intelligence for this particular area than the human player. However, usually games where computers have become good players are no longer considered to require intelligent behaviour.

The success of these AI technologies has inspired researchers to explore machines with more general-purpose intelligence. The computer program is no longer limited to play one specific game, instead knowledge can be transferred to another game. General Video Game Playing (GVGP) is the design of artificial intelligence computer programs that can play many different video games. Therefore, the computer program should be game-independent and use as little game specific knowledge as possible during the learning process.

© Springer International Publishing AG 2017
J. McDermott et al. (Eds.): EuroGP 2017, LNCS 10196, pp. 52–63, 2017.
DOI: 10.1007/978-3-319-55696-3_4

The main contribution of this paper is using genetic programming [9,10] to evolve game state features. A voting based method is used to determine the behaviour of the game agent, i.e. direction of motion and shooting behaviour. During the learning process, the only game knowledge used in this paper are game screen grabs and game scores. This knowledge is obtained from the game engine and then passed on to the learning algorithm.

In order to evaluate the efficiency of the algorithm, three different video games are used to evaluate the algorithms: Frogger, Missile Command and Space Invaders. All games run on the general video game engine [2]. We compare our results with another genetic programming algorithm [8] which uses handcrafted game state features. The results show that the algorithm which uses evolved game state features from raw pixels performs better than the algorithm which uses handcrafted features.

This paper is organised as follows. Section 2 briefly introduces previous work. Section 3 describes the game engine and games which are used to evaluate the algorithm. Section 4 presents how genetic programming is used to evolve visual features. The results are shown in Sect. 5. Section 6 gives the conclusion.

## 2  Related Research

The General Game Playing Competition [4] has been organised every year by AAAI since 2005. This competition focuses on turn-taking board games whose rules are not known. Monte Carlo tree search [3,13,14] has shown its powerful search ability in general game playing.

Bellemare et al. [1] created an Arcade Learning Environment which provides a platform to evaluate general, domain-independent AI technology. Naddaf [16] introduced two model free AI agents to play Atari 2600 console games in his master's thesis. One AI agent uses reinforcement learning while the other uses Monte Carlo tree search. Two handcrafted game state features are introduced. Hausknecht [6,7] presented a HyperNeat-based general video game player to play Atari 2600 video games. It uses a high-level game state representation. HyperNeat is said to be able to exploit geometric regularities.

In 2015, DeepMind [15] presented a Deep Q-Network which combines deep learning and reinforcement learning to play Atari 2600 video games, achieving human-level game play in many games. Deep Q-Network did so with minimal prior knowledge, receiving only visual images (raw pixels) and game scores as input. Guo et al. [5] trained both a neural network and a deep neural network with the action choice returned from Monte Carlo tree search as ground truth. Monte Carlo tree search is assumed to make correct decisions if it has enough searching depth. In 2016, AlphaGo [18] created by DeepMind has beaten Lee Sedol, a top-level Go player. AlphaGo relies on two different components: A tree search procedure and convolutional neural networks to guide the tree search. Two convolutional neural networks are trained: one is a policy neural network and the other is a value neural network, which are also trained using Reinforcement Learning.

GVG-AI [2] is another platform to test the general AI technology. Based on this platform, General Video Game AI Competition has been held every year since 2014. This competition explores the problem of creating controllers for general video game playing. Perez et al. [17] put forward a knowledge-based Fast Evolutionary Monte Carlo tree search method, where the reward function adapts to the knowledge and distance change. Jia et al. [8] presented a video game player based on genetic programming. Three handcrafted game state representations were used. Three trees were evolved based on Genetic Programming whose values were used to determine the game agent's movement in the horizontal/vertical direction and the shooting behaviour.

## 3   Materials

### 3.1   Games

The games used to evaluate the algorithm in this paper are run on the GVG-AI game engine. This game engine is able to run many games similar to old Atari 2600 games. We test our video game player on three different games: Space Invaders, Frogger and Missile Command. Screen grabs from the three games are shown in Fig. 1. Space Invaders is a classic arcade game. An alien invasion is coming in from above. The player has to control a small gun which is able to shoot vertically at the incoming space ships. Frogger is a classic game where a small frog needs to cross a road. The player essentially needs to move from the bottom of the screen to a goal position located at the top of the screen. The game agent has to watch out for cars while crossing the road. In Missile Command, the player needs to use smart bombs to destroy incoming ballistic missiles. The player needs to decide at what location the next smart bomb will explode. A smart bomb will destroy all incoming missiles within a certain radius.

### 3.2   Handcrafted Game State Features

In our previous work, we presented a method which combines handcrafted game state features and Genetic Programming in order to play different video games. We first identify the positions of all objects shown in the game screen. Five different object classes are distinguished. For each class, we inform the game avatar about the position of the nearest object as illustrated in Table 1. The game state is represented by four terminal symbols $X_i$, $Y_i$, $D_i$, and $A_i$ with $i \in \{1, \ldots 5\}$. The position of the nearest object is available as x- and y-coordinates, via symbols $X_i$ and $Y_i$. The distance to the nearest object is available through $D_i$ and the angle is given by $A_i$.

Three trees T1, T2 and T3 are evolved by taking the game state features as input to determine the behaviours of the avatar. Among them, the value of T1 and T2 determine the avatar's moving direction in the horizontal and vertical direction, as illustrated in Fig. 2. The value of T3 determines whether the avatar will release a shooting action as illustrated in Table 2.

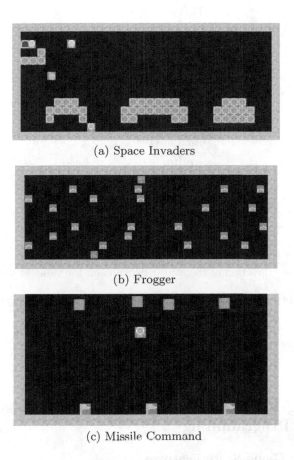

(a) Space Invaders

(b) Frogger

(c) Missile Command

**Fig. 1.** Three games created using the game engine GVG-AI.

**Table 1.** Terminal symbols [8]

| Terminal symbol | Description |
| --- | --- |
| $X_i$ | x coordinate of the nearest object for class $i$ relative to avatar |
| $Y_i$ | y coordinate of the nearest object for class $i$ relative to avatar |
| $D_i$ | Euclidean distance between avatar and the nearest object for class $i$ |
| $A_i$ | Angle between vector pointing from self to the nearest object and the horizontal axis |

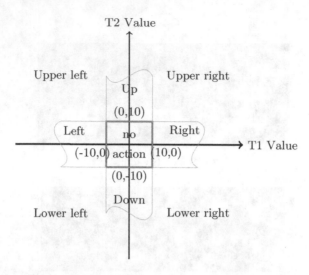

**Fig. 2.** The moving direction of the avatar depends on the value of tree 1 and tree 2.

**Table 2.** Depending on the value of tree 3, the button will be pressed or not.

| Value of tree 3 | Actions |
| --- | --- |
| $[0, \infty)$ | Press_button |
| $(-\infty, 0)$ | No_action |

# 4    Evolving Video Game State Visual Features Using Genetic Programming

## 4.1    Evolving Game State Features

It is a great challenge for a game player to find the proper game state features, especially for games with complex game play. In this paper, we use genetic programming [9,10] to evolve the game state visual features.

The computer programs evolved by genetic programming are traditionally represented as tree structures. Trees can be easily evaluated in a recursive manner. We used the ECJ package [11] to evolve the playing strategies in this paper. In our work, the game engine communicates with ECJ via the TCP/IP protocol. For each step, the game engine sends the game screen grab to ECJ. The evolved program then computes and returns back to game engine the actions which will be executed by the avatar. Once the game ends, irrespective of whether the game is won or lost, the game scores are passed on to ECJ, and are used to compute the fitness of the program.

The terminal set, that we have used for our experiments, is described in Table 3. All terminals return an object of type Image. Each terminal returns one channel of the down-scaled game screen grab: red channel, green channel, blue channel, yellow channel and grey channel. The red, green and blue channels are readily available from the screen grab. The other channels are computed from this data. During the learning process, the game screen grab is passed on to ECJ from the game engine. The size of the down-scaled screen grab is one sixteenth of its original size.

The elementary functions are shown in Table 4. From this table, we see that all the arguments and return values of these functions have the type Image. We have used arithmetic functions, such as addition, subtraction, multiplication and division. It should be noted that a protected division is used here.

They are used to combine features from different channels. We also applied a Gaussian filter and a non-local-maxima suppression function. If the latter two functions are applied in sequence, objects will be reduced to points, i.e. the local maxima. The attenuation function can be used to put an emphasis on objects close to the avatar. The attenuation function is a exponential function, whose value will attenuate with the object's distance to avatar. It can be used to put an emphasis on objects close to the avatar. The return value of the tree is an image which has the same size as the input image.

For each experiment, we perform 10 runs with different initialisation of the random seed. For each run, a population with 200 individuals is evolved for up to 100 generations. Crossover is applied with the probability 0.4. Mutation is applied with the probability 0.4. Reproduction is applied with the probability 0.2. Tournament selection is of size 3 is used as strategy. The ramped half-and-half tree building method (HalfBuilder) is used to initialise the first population of individuals.

**Table 3.** Terminal symbols I

| Terminal | Return type | Description |
|----------|-------------|-------------|
| imageGray | Img | Gray image |
| imageR | Img | Red channel |
| imageG | Img | Green channel |
| imageB | Img | Blue channel |
| imageY | Img | Yellow channel |

## 4.2  Voting for Actions

As described in the previous section, the return value of the genetic programming tree is an image. We search the resulting image in order to locate the position of the maximum value ($V_{max}$) and the minimum value ($V_{min}$). The point having the maximum value is regarded as the goal position of the game, i.e. a location on the screen that is of positive interest. The point having the minimum value is regarded as the position of a potential threat to the game avatar. The behaviour of the game avatar will be determined by these two points. It should always move towards the goal. However, it should also keep an eye on the potential threat. Once the location of the potential threat enters a certain area surrounding the avatar, then the avatar will move away from the threat. In this paper, we combine

**Table 4.** Function set

| Function | Output type | Description |
|---|---|---|
| add(Img a, Img b) | Img | $o(x,y) = a(x,y) + b(x,y)$ |
| subtract(Img a, Imgb) | Img | $o(x,y) = a(x,y) - b(x,y)$ |
| multiply(Img a, Img b) | Img | $o(x,y) = a(x,y) \cdot b(x,y)$ |
| divide(Img a, Img b) | Img | If $b(x,y) \neq 0$ then $o(x,y) = a(x,y)/b(x,y)$, otherwise $o(x,y) = 0$ |
| gaussian(Img a) | Img | Gaussian smoothing with standard deviation 1.1 |
| attenuation(Img a) | Img | Attenuates the image data $i(x,y)$ depending on its distance to the avatar. $o(x,y) = i(x,y) \cdot \exp^{-distance/50}$ |
| nlms(Img a) | Img | Non local maximum suppression with neighbourhood of $5 \times 5$ |
| gate(Img a, Img b, Img c) | Img | If $a(x,y) > 0$ then $o(x,y) = b(x,y)$, otherwise $o(x,y) = c(x,y)$ |
| max(Img a, Img b) | Img | If $a(x,y) > b(x,y)$, then $o(x,y) = a(x,y)$, otherwise $o(x,y) = b(x,y)$ |
| min(Img a, Img b) | Img | If $a(x,y) < b(x,y)$, then $o(x,y) = a(x,y)$, otherwise $o(x,y) = b(x,y)$ |
| upper(Img a) | Img | Keeps only values in the uppermost 25%. Let the pixel range be $[V_{\min}, V_{\max}]$. If $i(x,y) \geq V_{\max} - 0.25(V_{\max} - V_{\min})$ then $o(x,y) = i(x,y)$, otherwise $o(x,y) = 0$ |
| lower (Img a) | Img | Keeps only values in the lowest 25%. If $i(x,y) \leq V_{\min} + 0.25(V_{\max} - V_{\min})$ then $o(x,y) = i(x,y)$, otherwise $o(x,y) = 0$ |

the two behaviours by voting. If one action helps to move towards the goal, it will get a reward +1. Whereas, if the action will cause a threat to avatar, it will get a punishment score −5, if the action helps move away from the threat, it will get a reward +2 (Table 5). If neither happens, then the value will remain the same. The action with the largest value will be the one used to determine the direction of motion. The maximum value also determines the action of button, which will be pressed if and only if $V_{max} > V_{threshold}$. There are 18 actions which are same with the actions listed in Fig. 2 and Table 2.

**Table 5.** Avatar behaviour.

| Avatar behaviour | Action scores |
|---|---|
| Avatar moves towards goal | Score $+= 1$ |
| Avatar moves away from threat | Score $+= 2$ |
| Avatar is threatened | Score $-= 5$ |

# 5    Results

Figure 3 illustrates a sample tree which is used to extract game state features for Frogger. The input images are all extracted from down-scaled screen grabs. The return value of the tree is also an image, which is shown in Fig. 4. We search for the maximum and minimum points in the returned images. The two points are overlaid on the original screen grab. The maximum is marked with a green rectangle and the minimum point is marked with a red rectangle. In the Figure, we can see that the point having the maximum value is the desired home location of the frog. The point having the minimum value is the most dangerous car to the avatar.

Figure 5 shows the fitness of the best individual for these three games. We conduct 10 runs for each experiment. The average fitness of the best individual for each game is shown with a bold line. This algorithm is compared with our previous results using hand crafted features. Figure 6 shows the average best fitness for the two algorithms. The comparison is shown in Table 6. Mann Whitney U Test is used to compare the average best fitness in generation 9, 19 and 99. As shown in Table 7 the final results in generation 99 are not significantly different.

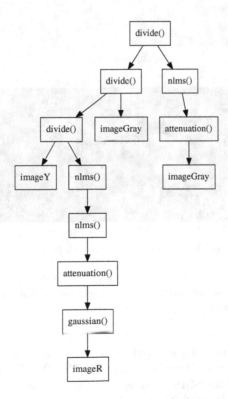

**Fig. 3.** A sample GP tree for extracting game state features of Frogger.

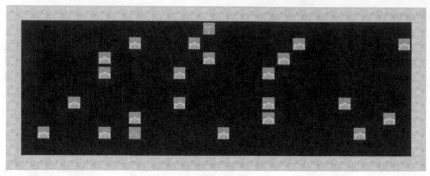

(a) Original screen grab

(b) Output image

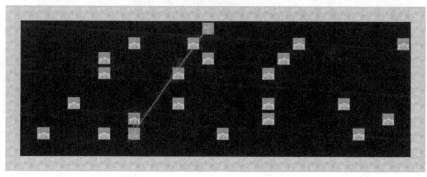

(c) Maximum and Minimum points

**Fig. 4.** Extracted game state features for Frogger. (a) the original screen grab. (b) output image from the sample GP tree. (c) maximum and minimum points overlaid on the original screen grab.

However, the algorithm using evolved game state visual features performs significantly better than the algorithm with hand crafter features for the games Frogger and Space Invaders. In other words, this algorithm is able to find good solutions faster than the original algorithm.

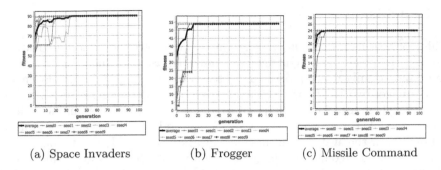

(a) Space Invaders          (b) Frogger          (c) Missile Command

**Fig. 5.** Best fitness for the three games. For each experiment, we perform 10 runs with different random seeds.

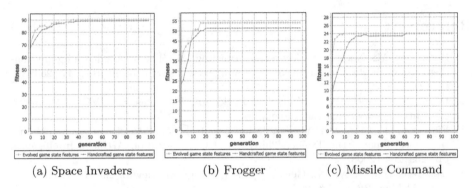

(a) Space Invaders          (b) Frogger          (c) Missile Command

**Fig. 6.** Comparison of the average best fitness between the two algorithms.

**Table 6.** Average scores over 10 runs obtained from the three games: Space invaders, Frogger, and Missile command. The maximum possible scores are: Space Invaders:90, Frogger:54 and Missile Command:24.

| Game | Average score | |
| --- | --- | --- |
| | GP+Handcrafted features | Evolved features+Voting |
| Space invaders | $89 \pm 2.84$ | $90 \pm 0$ |
| Frogger | $51.3 \pm 8.54$ | $54 \pm 0$ |
| Missile command | $24 \pm 0$ | $24 \pm 0$ |

**Table 7.** Comparison of the average best fitness using Mann Whitney U Test in generation 9, 19 and 99. $f_E$ is the average best fitness for the algorithm using evolved game state features. $f_H$ is the average best fitness for the algorithm using handcrafted game state features.

| Games | Hypothesis | p value | | |
| --- | --- | --- | --- | --- |
| | | Gen = 9 | Gen = 19 | Gen = 99 |
| Space Invaders | $H_o : f_E = f_H, H_1 : f_E > f_H$ | 0.0024 | 0.36 | 0.35 |
| Frogger | $H_o : f_E = f_H, H_1 : f_E > f_H$ | 0.02 | 0.22 | 0.36 |
| Missile Command | $H_o : f_E = f_H, H_1 : f_E > f_H$ | 0.42 | 0.38 | 0.5 |

## 6   Conclusion

In this work, we have used genetic programming to evolve the game state features from the raw pixels. The input image is provided as individual color channels. These color channels are processed to generate one output image. From the output image, two locations (maximum response and minimum response) are extracted. These locations correspond to a desired goal position and a position where a possible threat is located. A voting method is used to generate a movement action for the avatar from these two locations. Three different video games are used to evaluate our algorithm. The results show that this algorithm is able to find a good strategy faster when compared to an algorithm using hand crafted features.

## References

1. Bellemare, M., Naddaf, Y., Veness, J., Bowling, M.: The arcade learning environment: an evaluation platform for general agents. J. Artif. Intell. Res. **47**, 253–279 (2012)
2. Perez, D., Samothrakis, S., Togelius, J., Schaul, T., Lucas, S.: GVG-AI Competition. http://www.gvgai.net/index.php
3. Finnsson, H., Bjornsson, Y.: Simulation-based approach to general game playing. In: Proceedings of the Twenty-Third AAAI Conference on Artificial Intelligence, pp. 259–264 (2008)
4. Geneserech, M., Love, N.: General game playing: overview of the AAAI competition. AI Mag. **26**, 62–72 (2005)
5. Guo, X., Singh, S., Lee, H., Lewis, R., Wang, X.: Deep learning for real-time atari game play using offline monte-carlo tree search planning. Adv. Neural Inf. Process. Syst. **27**, 3338–3346 (2014)
6. Hausknecht, M., Khandelwal, P., Miikkulainen, R., Stone, P.: HyperNEAT-GGP: a HyperNEAT-based atari general game player. In: Genetic and Evolutionary Computation Conference(GECCO) (2012)
7. Hausknecht, M., Lehman, J., Miikkulainen, R., Stone, P.: A neuroevolution approach to general atari game playing. IEEE Trans. Comput. Intell. AI Games **6**, 355–366 (2013)
8. Jia, B., Ebner, M., Schack, C.: A GP-based video game player. In: Genetic and Evolutionary Computation Conference(GECCO) (2015)
9. Koza, J.R.: Genetic Programming: On the Programming of Computers by Means of Natural Selection. The MIT Press, Cambridge (1992)
10. Koza, J.R.: Genetic Programming II: Automatic Discovery of Reusable Programs. The MIT Press, Cambridge (1994)
11. Luke, S.: The ECJ Owner's Manual, 22nd edn. (2014)
12. Campbell, M., Hoane, A.J., Hsu, F.H.: Deep blue. Artif. Intell. **134**, 57–83 (2002)
13. Mehat, J., Cazenave, T.: Monte-Carlo Tree Search for General Game Playing. Technical report, LIASD, Dept. Informatique, Université Paris 8 (2008)
14. Mehat, J., Cazenave, T.: Combining UCT and nested monte-carlo search for single-player general game playing. IEEE Trans. Comput. Intell. AI Games **2**(4), 225–228 (2010)

15. Mnih, V., Kavukcuoglu, K., Silver, D., Rusu, A., Veness, J., Bellemare, M., Graves, A., Riedmiller, M., Fidieland, A., Ostrovski, G., Petersen, S., Beattie, C., Sadik, A., Antonoglou, I., King, H., Kumaran, D.: Human-level control through deep reinforcement learning. Nature **518**, 529–533 (2015)

16. Naddaf, Y.: Game-independent AI agents for playing atari 2600 console games. Master's thesis, University of Alberta (2010)

17. Perez, D., Samothrakis, S., Lucas, S.: Knowledge-based fast evolutionary MCTS for general video game playing. In: Proceedings of IEEE Conference on Computational Intelligence and Games, pp. 68–75 (2014)

18. Silver, D., Huang, A., Maddison, C., Guez, A., Sifre, L., van den Driessche, G., Schrittwieser, J., Antonoglou, I., Panneershelvam, V., Lanctot, M., Dieleman, S., Grewe, D., Nham, J., Kalchbrenner, N., Sutskever, I., Lillicrap, T., Leach, M., Kavukcuoglu, K., Graepel, T., Hassabis, D.: Mastering the game of go with deep neural networks and tree search. Nature **529**, 484–489 (2016)

# Emergent Tangled Graph Representations
# for Atari Game Playing Agents

Stephen Kelly[✉] and Malcolm I. Heywood

Dalhousie University, Halifax, NS, Canada
{skelly,mheywood}@cs.dal.ca

**Abstract.** Organizing code into coherent programs and relating different programs to each other represents an underlying requirement for scaling genetic programming to more difficult task domains. Assuming a model in which policies are defined by teams of programs, in which team and program are represented using independent populations and coevolved, has previously been shown to support the development of variable sized teams. In this work, we generalize the approach to provide a complete framework for organizing multiple teams into arbitrarily deep/wide structures through a process of continuous evolution; hereafter the Tangled Program Graph (TPG). Benchmarking is conducted using a subset of 20 games from the Arcade Learning Environment (ALE), an Atari 2600 video game emulator. The games considered here correspond to those in which deep learning was unable to reach a threshold of play consistent with that of a human. Information provided to the learning agent is limited to that which a human would experience. That is, screen capture sensory input, Atari joystick actions, and game score. The performance of the proposed approach exceeds that of deep learning in 15 of the 20 games, with 7 of the 15 also exceeding that associated with a human level of competence. Moreover, in contrast to solutions from deep learning, solutions discovered by TPG are also very 'sparse'. Rather than assuming that *all* of the state space contributes to every decision, each action in TPG is resolved following execution of a subset of an individual's graph. This results in significantly lower computational requirements for model building than presently the case for deep learning.

**Keywords:** Reinforcement learning · Task decomposition · Emergent modularity · Arcade learning environment

## 1 Introduction

Machine learning agents applied to a reinforcement learning (RL) task attempt to maximize the reward accrued over a training episode, during which a variable number of interactions with the task environment occur. In each interaction, the agent observes the state of the environment, takes an action, and receives feedback in the form of a reward signal. It is typically only the final reward, received when an end state (or max. interactions) is encountered, that quantifies the

© Springer International Publishing AG 2017
J. McDermott et al. (Eds.): EuroGP 2017, LNCS 10196, pp. 64–79, 2017.
DOI: 10.1007/978-3-319-55696-3_5

agent's performance relative to the task objective. The agent is said to represent a *policy* for maximizing this long-term reward.

Scaling RL to real-world tasks requires a representation that is: (1) Able to cope with high-dimensional sensor data; and (2) General enough to be applied to a wide variety of tasks without extensive parameter tuning. Video games provide an interesting test domain for scalable RL. In particular, they cover a diverse range of dynamic task environments that are designed to be challenging for humans, all through a common high-dimensional visual interface, or the game screen, e.g. [1].

In this work we propose a Genetic Programming (GP) framework to address the scaling problem through emergent modularity [2]. Specifically, we adopt a graph representation for decision-making policies, in which teams of programs are capable of growing and self-organizing into complex structures through interaction with the task environment. The framework is capable of:

- Adaptively dividing the task up into distinct sets of cooperating programs (or *teams*), an emergent process as there is no knowledge regarding the correct team size/complement. Teams represent the smallest stand-alone decision-making entity. As such, a team is synonymous with a *module* in the scope of this paper.
- Establishing how to select between/recombine teams into increasingly complex decision making structures, or a *policy graph*. Throughout this work, we refer to this process as emergent modularity because multiple independent teams are recombined with no prior knowledge regarding how many to include, which teams might work well together, or how to combine them.

Hence, the process starts off with very simple programs/teams and then adapts to develop graphs of connectivity between teams, creating policy graphs which are themselves subject to further development.

The concept is partly motivated by the intuition that (automatic) problem decomposition is an important learning skill for artificial agents, just as it is for humans. The algorithm proposed in this work, Tangled Program Graphs (TPG) is an extension of Symbiotic Bid-Based GP (SBB), a framework for automatic problem decomposition through coevolving teams of simple programs. Specifically, TPG facilitates a more open-ended approach, in which teams are incrementally organized into graphs of arbitrary topology and discovered using a single continuous cycle of evolution. Thus, more complex topologies can naturally emerge as soon as they perform better than simpler solutions.

The scope of this work is to introduce TPG and make a case for how the representation supports the development of strong yet simple policies. Empirical experiments are conducted in the Arcade Learning Environment (ALE) [1]. ALE provides a framework in which RL agents have access to hundreds of classic video games through a common sensory interface: the game screen as a high-dimensional *pixel matrix*. Moreover, actions are limited to those of the original Atari console joystick, and the ultimate feedback takes the form of game score. In short, RL is limited to the same set of experiences as a human (albeit without sound). We make a direct comparison with both Neuroevolution [3] and a recent Deep Reinforcement Learning architecture [4], and show that the TPG produces competitive agents at a fraction of the model complexity.

The remainder of this paper is organized as follows: We begin by summarizing related GP research regarding modularity in Sect. 2. The properties of the ALE Atari 2600 task that warrant its use as a challenging RL benchmark are then established in Sect. 3, as well as the specific representation assumed for the state space. Section 4 presents the framework for evolving Tangled Program Graphs, and the empirical study is performed in Sect. 5. Conclusions and future work are outlined in Sect. 6.

## 2    Background

Modular architectures are a recurring theme in GP, with early approaches such as Automatically Defined Functions [5] and Adaptive Representations through Learning [6], as well as Tag-Based Modules [7] all being motivated by the challenge of scaling GP to more complex tasks. From the perspective of modular task decomposition through teaming, previous studies have established guidelines for combining the contribution from individual team members [8] or enforcing island models [9]. Attempts have also been made to define fitness at the program as well as the team level [10,11]. In a final development, simple bidding mechanisms have been used to guarantee that task decomposition takes place between members of a team [10,12,13]. This latter approach also implies that each program establishes *context* for one discrete action (i.e. context and action are independent), and there is no need to a priori define the optimal team size or relevant distribution of actions, hence the composition of a team is now an entirely evolved property.

Recent work has established the utility of using team building through bidding (or SBB in particular) to build *hierarchical* decision making agents over two independent phases of evolution [14–18]. The first phase produces a library of diverse, specialist teams. The second phase attempts to build more general policies by reusing the library. While effective in many tasks, this approach makes several assumptions that potentially impact its generality:

1. Individuals at the second phase of evolution can only define actions in terms of the teams evolved at the first (in order to place plausible limits on the number of actions).
2. The number of phases of evolution/levels in the hierarchy is guessed a priori and to date has not passed two (a law of diminishing returns per phase of evolution is typically observed).
3. The computational budget (generation or evaluation limit) used for each phase of evolution needs defining a priori.

The work herein builds on these previous studies to propose a new team-based GP representation, or a *tangled* program graph, in which teams denote vertices, and programs identify which edges are traversed. The topology of a solution, i.e. the number of teams per graph and the connectivity between teams, is now entirely the result of open ended evolution.

In the case of the task domain used for benchmarking, we assume the Arcade Learning Environment (ALE) for the Atari 2600 console [1]. High scores in the ALE are currently dominated by neural network architectures, specifically Deep Reinforcement Learning (DQN) [3] and Neuroevolution (HyperNEAT) [4]. Both methods assume an exhaustive accounting for the entire input space regardless of the agent's experience in the environment, i.e. the convolution network in Deep Learning or the Compositional Pattern Producing Network in Hyper-NEAT. Conversely, TPG attempts to discover a suitable input representation through interacting with the task, while simultaneously discovering an appropriate decision making policy. In this initial study, we explicitly target the subset of 20 games for which DQN did not reach the threshold of human-level play [4].

## 3   The Arcade Learning Environment

Released in 1977, the Atari 2600 has been a popular home video game console that was capable of running a large variety of games, each stored on interchangeable ROM cartridges. Hundreds of games were compatible with the console, bringing the diversity of an Arcade experience into the home through a single device. As each game is designed to be unique and challenging for human players, the Atari 2600 provides an interesting test domain for general artificial decision making agents.

The Arcade Learning Environment (ALE) provides an Atari 2600 emulator with a common reinforcement learning interface [1]. In particular, ALE allows learning agents to interact with games over discrete time steps by extracting the current game state and score from ROM, and sending joystick commands to the emulator. The Atari 2600 joystick comprises a directional paddle and single push button, which ALE translates into 18 discrete actions, or all possible combinations of direction and push button state, including 'no action'. The task is particularly challenging because any learning agent is required to operate under the same conditions as a human player, i.e. sensory input (screen), action output (joystick), and game score.

### 3.1   Screen State Space Representation

Agents in this work observe the game via the most general state representation, or the raw Atari 2600 screen, which is a $210 \times 160$ pixel matrix, with 128 possible colour values for each pixel, updated at 60 Hz. In common with previous research, our agents interact with games at reduced frame rate, stochastically skipping $\approx 25\%$ of sequential frames, which is roughly the fastest that a human player can react. The most recent action is simply repeated in skipped frames[1]. Skipping frames in this manner implies that the environment is stochastic.

---

[1] ALE includes a parameter *repeat_action_probability*, for which we assumed the suggested value of 0.25.

The task is considered partially observable because agents only perceive one frame at a time, while game entities typically appear intermittently (flicker) over sequential frames[2]. Hence, it is often impossible to capture the complete game state from a single frame. While various methods for hand-crafting feature sets from the raw Atari screen frames are possible, including game-specific background and object detection [1,3], the focus of this work is learning from high-dimensional, task-independent sensory representation. However, we can reduce the dimensionality significantly by preprocessing frames based on the observation that most game entities are larger than a single pixel, and thus less resolution is required to capture important games events. Such an observation does not imply the use of image processing operators or adopting representations specifically appropriate for spatial representation, e.g. wavelets [19].

The screen quantization procedure assumes the following steps:

1. Each frame is subdivided into a $42 \times 32$ grid (Fig. 1(b)), in which only 50% of the pixels in each tile are considered (implies that most state information is redundant) and each pixel assumes an 8-colour SECAM[3] encoding.
2. Each tile is described by a single byte, in which each bit encodes the presence of one of eight SECAM colours within that tile.
3. The decimal value for each of the tile bytes is returned, so defining a sensory state space $x(t)$ of $42 \times 32 = 1344$ decimal features in the range of 0–255, visualized in Fig. 1(b) for a specific game at time step (frame) $t$.

This state representation is inspired by the Basic method defined in [1]. Note, however, that this method does not use a priori background detection or pairwise combinations of features.

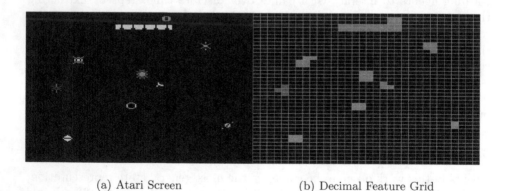

(a) Atari Screen                              (b) Decimal Feature Grid

**Fig. 1.** Screen quantization steps, reducing the raw Atari pixel matrix (a) to 1344 decimal sensor inputs (b).

---

[2] Partial observability can be mitigated by averaging pixel colours across each pair of sequential frames, a preprocessing step *not* used in this work.

[3] ALE provides SECAM as an alternative encoding to the default NSTC format.

# 4    Evolving Tangled Program Graphs

The proposed framework, Tangled Program Graphs (TPG), can be summarized by two key concepts: (1) Coevolving teams of programs, which represent single nodes of the graph, Fig. 2(b); and, (2) Emergent modularity, or the process by which the graph is incrementally constructed, Fig. 2(a).

## 4.1    Coevolving Teams of Programs

Evolution begins with a population of single independent teams. Each team is initialized with a stochastically chosen complement of programs over the interval $[2, \dots \omega]$. For example, Fig. 2(b) represents one such candidate team consisting of 5 programs.

A linear, or register machine, representation will be assumed for programs, where linear GP provides an efficient process for skipping intron code *during* execution [20]. Each program returns a single real number, the result of executing a sequence of instructions that operate on sensor inputs or internal registers, as illustrated in Algorithm 1. Each program is also associated with *one* task-specific 'atomic' action, selected from the set of discrete actions defined by the task domain (corresponding to 18 console directions with/without the 'fire' action and a 'no action', Sect. 3).

In RL tasks, all programs in a team will execute relative to the current state variables (screen) at each time step, $x(t)$. The team then deploys the action of the program with the highest output, or the **winning bid** [10,12,13]. Note that the winning bid merely defines the action to deploy at time step $t$. This potentially changes the state of the task, which may or may not represent the end of the training epoch (for which a measurable reward is received). In short, teams represent the minimal decision-making entity, in which the role of each program is to define a unique context for deploying its action given the current state of the environment. Finally, each GP program has to explicitly identify which *subset* of state variables to operate on. Each Program will typically index a different subset of the state variables, leading to teams emerging that make decisions based on specific sub-regions of the state space.

Unlike the previous hierarchical version of SBB [14,15,18], TPG maintains only one population of programs and one population of teams. Team development is driven by a breeder model of evolution such that a fixed fraction of the least desirable teams (*PopGap*) are deleted at each generation and replaced by the off-spring of surviving teams.[4] Team offspring are created by cloning the team along with all its programs, and then applying mutation based variation operators to the cloned team and programs, as parameterized in Table 1. Thus, evolution is driven by 'group-level' selection in which the team is judged as a whole rather

---

[4] Individuals in the team population merely index a subset of programs from the program population under a variable length representation. A valid team conforms to the constraint that it must index a minimum of 2 programs and have at least two different actions.

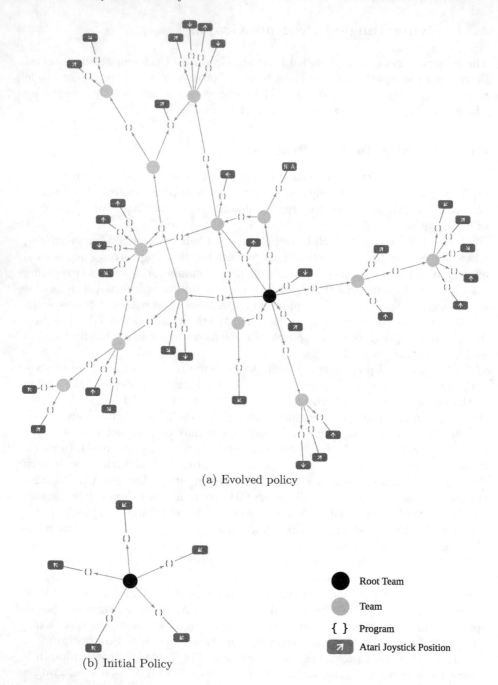

(a) Evolved policy

(b) Initial Policy

Root Team

Team

{ } Program

Atari Joystick Position

**Fig. 2.** TPG Policies. Decision making in each time step (frame) begins at the root team and follows the edge with the winning program bid (output) until an atomic action (Atari Joystick Position) is reached. The initial population contains only single-team policies (b). Multi-team graphs emerge as evolution progresses (a). The policies pictured are from a game that does not use the fire button. Thus, only directional joystick positions and 'No Action' appear.

**Algorithm 1.** Example program in which execution is sequential. Programs may include two-argument instructions of the form $R[i] \leftarrow R[x] \circ R[y]$ in which $\circ \in +, -, x, \div$; single-argument instructions of the form $R[i] \leftarrow \circ(R[y])$ in which $\circ \in cos, ln, exp$; and a conditional statement of the the form IF $(R[i] < R[y])$ THEN $R[i] \leftarrow -R[i]$. $R[i]$ is a reference to an internal register, while $R[x]$ and $R[y]$ may reference internal registers or state variables (sensor inputs). Determining *which* of the available sensor inputs are actually used in the program, as well as the number of instructions and their operations, are both emergent properties of the evolutionary process.

1. $R[0] \leftarrow R[0] - R[3]$
2. $R[1] \leftarrow R[0] \div R[7]$
3. $R[1] \leftarrow Log(R[1])$
4. IF $(R[0] < R[1])$ THEN $R[0] \leftarrow -R[0]$
5. RETURN $R[0]$

than by the performance of individual components. As such, programs have no individual fitness. However, at each generation, orphaned programs – those that are no longer a member of any team – are deleted, i.e. they were only associated with the worst performing teams.

## 4.2   Emergent Modularity

All programs are initialized with exclusively atomic actions, thus only single-team policies exist in the initial population, Fig. 2(b). As such, all initial teams represent 'graph' root nodes. In order to support code reuse and emergent modularity, search operators will occasionally mutate a program's action. The modified action has an equal probability of referencing either an atomic action or another team. Thus, search operators have the ability to *incrementally* construct multi-team graphs, Fig. 2(a). Naturally, decision making in multi-team graphs begins at the root node (team) and follows *one* path through the network until an atomic action is selected. Cycles may exist in the graph, but they are never followed during execution. That is, a team is never visited twice per decision. The single visit constraint is enforced by testing whether the action of the program with highest output (winning bid established during team evaluation, Sect. 4.1) corresponds to a previously visited team. If so, the next highest bid is selected, and the validation step repeats. Each team maintains at least one program with an atomic action, hence guaranteeing cycles never appear.

The number of unique policies in the population at any given generation is equal to the number of root teams, i.e. teams that are not referenced as any program's action. This number fluctuates, as root teams are sometimes 'subsumed' by another graph. For example, variation operators may mutate a program's action to point to the root team of an existing policy graph, in which case there would be one less policy in the population. Conversely, a graph may be separated through the reverse process, resulting in a new root team/policy.

Only root teams are subject to modification by the variation operators. Internal nodes (teams and individual programs) are essentially cached blocks of code, which may appear in more than one policy graph and at more than one location in the same graph. However, a team is never simultaneously a root *and* an internal node in another policy graph. In short, graphs of multiple teams emerge through a continuous process of development. Most importantly, as programs composing a team typically index different subsets of the state space (i.e., the screen), the resulting policy graph will incrementally index more of the state space *and* prioritize the types of decisions made in different regions.

**Table 1.** Parameterization of Team and Program populations. For the team population, $p_{mx}$ denotes a mutation operator in which: $x \in \{d, a\}$ are the prob. of deleting or adding a program respectively; $x \in \{m, n\}$ are the prob. of creating a new program or changing the program action respectively. $\omega$ is the max. initial team size. For the program population, $p_x$ denotes a mutation operator in which $x \in \{delete, add, mutate, swap\}$ are the probabilities for deleting, adding, mutating, or reordering instructions within a program.

| Team population | | | |
|---|---|---|---|
| Parameter | Value | Parameter | Value |
| *PopSize* | 360 | *PopGap* | 50% of *Root* Teams |
| $p_{md}, p_{ma}$ | 0.7 | $\omega$ | 5 |
| $p_{mm}$ | 0.2 | $p_{mn}$ | 0.1 |
| Program population | | | |
| *numRegisters* | 8 | *maxProgSize* | 96 |
| $p_{delete}, p_{add}$ | 0.5 | $p_{mutate}, p_{swap}$ | 1.0 |

### 4.3   Diversity Maintenance

Searching for *good* behaviours as well as *different kinds* of behaviours has two key benefits: (1) diversity helps prevent premature (team) convergence; and (2) when developing a library of reusable code, a diverse population represents a versatile toolbox for subsequent reuse [18]. In this work, diversity is maintained by ensuring that each program's bidding behaviour is unique w.r.t the rest of the program population. To achieve this, a global archive of the most recent 50 state observations is maintained at all times, where each observation is simply a vector of integers representing a single quantized game frame (See Sect. 3.1), as experienced by some member of the team population. When a new program is created or an existing program is modified, its *profile*[5] of bids over the archive is required to be unique relative to the rest of the program population [13]. Such a definition is task-independent.

---

[5] A vector of 50 double-precision values, or the program's output when executed relative to each unique state stored in the archive.

# 5    Empirical Experiments

The goal of this research is to establish the baseline capability of TPG over a diverse selection of Atari 2600 video games [1]. These games are particularly interesting because they are known to be challenging for both humans and learning algorithms. For this initial study, we concentrate on the 20 games in which Deep Reinforcement Learning, the algorithm with the most high scores to date, failed to reach human-level play [4].

## 5.1    Experimental Setup

We conducted 5 independent runs of TPG in each game, for as many generations as possible given our resource time constraint[6]. The same parameterization for TPG was used for all games, Table 1. The only information provided to the agents is the number of atomic actions available for each game, the raw pixel screen matrix at each frame (time step) during play, and the final game score. Each episode continues until the simulator returns a 'game over' signal or a maximum of 18,000 frames is reached. Policy graphs are evaluated in 5 episodes per generation, up to a maximum of 10 episodes per lifetime. This allows weak policies to be identified and replaced after only 5 evaluations in one generation, while promising policies are verified with an additional 5 evaluations. Team fitness is simply the mean game score over all evaluations for that team. The single champion from each run was identified as the individual with the highest training reward, or mean score over at least 5 evaluations. As per established test conditions [4], champions are evaluated in 30 test games for a maximum of 5 min of play (max. 18,000 frames) in each game.

## 5.2    Results

Test results for TPG, along with game scores for a human professional video game tester (from [4]) and two comparison algorithms, are reported in Table 2. To the best of our knowledge, Deep Reinforcement Learning (DQN) [4] and HyperNeat (NEAT) [3] are the only two algorithms, working from the 'pixel' state representation, that previously held the highest (machine-learning) score in at least one of the games considered. It is apparent that TPG is competitive with both methods, achieving a new highest score in 14 of the 20 games. In 7 of these, TPG also outperforms the human professional video game tester.

## 5.3    Solution Analysis

**Model Complexity.** An evolved TPG policy is essentially a directed graph in which vertices are teams and edges are programs (see Fig. 2(a)). The run-time

---

[6] All experiments were conducted on a shared cluster with a maximum run-time of 2 weeks. The nature of some games allowed for >1000 generations, while others limited evolution to the order of a few hundred.

efficiency of TPG policies is a factor of how many instructions are executed in order to make a decision in any single time step. Recall from Sect. 4 that each decision requires following a *single* path from the root team to atomic action. The column 'Ins' in Table 2 reports the average number of instructions executed along this path in each time step over the 30 test games.

Two observations regarding the run-time complexity of evolved policy graphs appear: (1) There is significant diversity across different game titles, thus policy complexity scales based on the requirements of each environment. (2) The overall complexity level, requiring between 116 (Double Dunk) and 1036 (Ms. Pac-Man) instructions on average, is significantly less than both comparison algorithms. For example, both DQN and Hyper-NEAT employ neural network architectures consisting of >800,000 weights, all of which are computed for *every* decision (See Appendix A in [3] and 'Model Architecture' under Methods in [4]). The simplicity of TPG's solutions is made possible by the modular nature of the architecture, which allows for adaptive, environment-driven complexification.

**Table 2.** Results for TPG along with the top learning algorithms in the ALE literature which use a raw pixel state representation. Also reported for TPG is the number of teams in each champion policy (Tms), the average number of instructions required for the policy to make a decision in each time step (Ins), and the proportion of the input space covered by the policy (%IP). Scores in bold indicate the highest score for a learning algorithm for that game, while a gray cell indicates the score was also better than the human professional video game tester (Hum).

| Game | DQN | HNEAT | Hum | TPG | Tms | Ins | %IP |
|------|------|-------|-----|-----|-----|-----|-----|
| Alien | 3069(±1093) | 1586 | 6875 | **3382.7**(±1364) | 46 | 455 | 56 |
| Amidar | **739.5**(±3024) | 184.4 | 1676 | 398.4(±91) | 63 | 812 | 69 |
| Asterix | **6012**(±1744) | 2340 | 8503 | 2400(±505) | 42 | 414 | 51 |
| Asteroids | 1629(±542) | 1694 | 13157 | **3050.7**(±947) | 13 | 346 | 23 |
| BankHeist | 429.7(±650) | 214 | 734.4 | **1051**(±56) | 58 | 572 | 65 |
| BattleZone | 26300(±7725) | 36200 | 37800 | **47233.4**(±11924) | 4 | 123 | 11 |
| Bowling | 42.4(±88) | 135.8 | 154.8 | **223.7**(±1) | 56 | 585 | 57 |
| Centipede | 8309(±5237) | 25275.2 | 11963 | **34731.7**(±12333) | 28 | 516 | 39 |
| C.Command | 6687(±2916) | 3960 | 9882 | **7010**(±2861) | 51 | 280 | 58 |
| DoubleDunk | -18.1(±2.6) | 2 | -15.5 | 2(±0) | 4 | 116 | 6 |
| Frostbite | 328.3(±250.5) | 2260 | 4335 | **8144.4**(±1213) | 21 | 382 | 28 |
| Gravitar | 306.7(±223.9) | 370 | 2672 | **786.7**(±503) | 13 | 496 | 36 |
| M'sRevenge | 0 | 0 | 4367 | 0(±0) | 18 | 55 | 28 |
| Ms.Pac-Man | 2311(±525) | 3408 | 15693 | **5156**(±1089) | 111 | 1036 | 83 |
| PrivateEye | 1788(±5473) | 10747.4 | 69571 | **15028.3**(±24) | 59 | 938 | 60 |
| RiverRaid | **8316**(±1049) | 2616 | 13513 | 3884.7(±566) | 67 | 660 | 64 |
| Seaquest | **5286**(±1310) | 716 | 20182 | 1368(±443) | 22 | 392 | 37 |
| Venture | 380(±238.6) | NA | 1188 | **576.7**(±192) | 3 | 165 | 7 |
| WizardOfWor | 3393(±2019) | 3360 | 4757 | **5196.7**(±2550) | 17 | 247 | 31 |
| Zaxxon | 4977(±1235) | 3000 | 9173 | **6233.4**(±1018) | 20 | 424 | 33 |

**Emergent Modularity.** All policies in TPG are initialized as a single team of programs (See Fig. 2(b)), which represents the simplest possible decision-making entity in TPG. The rate of growth into more complex multi-team graph structures is driven by interaction with the task environment, i.e. the development of complex policies is only possible if simpler policies are out performed.

Figure 3 depicts the development of modularity for TPG polices throughout evolution in 4 games where TPG ultimately achieved the best score of any learning algorithm. Clearly, different game environments result in different levels of complexity in the champion policies. Ms. Pac-Man is known to be a challenging game [21,22] and, perhaps not surprisingly, benefits form relatively complex structures. On the other hand, TPG managed to reach a high level of play in Asteroids with simple policies containing 7 teams. The trajectory denoted by 'Random' in Fig. 3 refers to a run in which policies were assigned random fitness values. The lack of development confirms that complex policies emerge by selective pressure rather than drift or other potential biases.

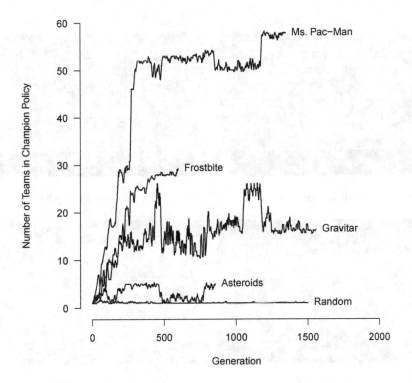

**Fig. 3.** Number of teams per champion TPG policy throughout evolution in 4 games where TPG ultimately achieved the best score of any learning algorithm. 'Random' is the control experiment, referring to a run in which policies were assigned random fitness values. Each line depicts the median over 5 runs. The diverse nature of each game implies that the cost of evaluating policies varies. Thus, given the same time constraint, a different number of generations are possible in each game environment.

**Adapted Visual Field.** Each Atari game presents a unique graphical environment in which important game events occur in different areas of the screen and at different resolutions. Part of the challenge with high-dimensional visual input data is determining what information is relevant to the task. TPG begins with single teams, thus minimal screen coverage, and incrementally explores more of the screen through complexification. By doing so, the utility of policies with complex screen coverage is continually checked against simpler alternatives, only persisting when they prove useful. Indeed, in 8 of the 14 games for which TPG achieved the highest score for a learning algorithm, it did so while indexing less that 50% of the screen, minimizing the number of instructions required per decision. In contrast, neural network representations as applied to the ALE task define sub-fields that are fully connected to the input space (e.g. through a convolution applied to all locations of the state space). In short, the architecture is predefined as opposed to developed through interaction with the environment.

For example, Fig. 4 shows the Adapted Visual Field (AVF) of champion TPG agents in Ms. Pac-Man and Battle Zone. In the case of Ms. Pac-Man, the game

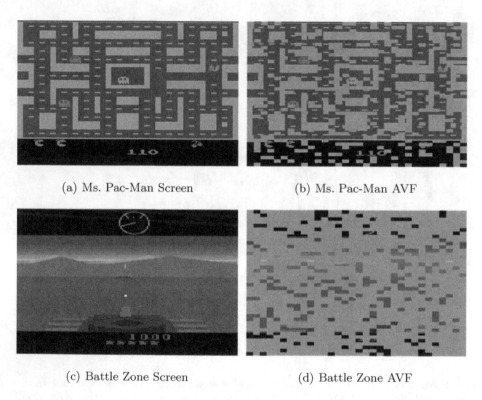

(a) Ms. Pac-Man Screen          (b) Ms. Pac-Man AVF

(c) Battle Zone Screen          (d) Battle Zone AVF

**Fig. 4.** Adapted Visual Field (AVF) of champion TPG policies in Ms. Pac-Man and Battle Zone, two games in which TPG achieved the highest score of any learning algorithm. The champion Battle Zone policy also scored higher than the human professional video game tester. Grey regions indicate areas of the screen not indexed by the policy.

defines a 2-dimensional maze environment that the player navigates in order to collect pills, where the pills are evenly distributed throughout the maze. Relatively high resolution is required in order to distinguish objects such as the agent's avatar and pills from the maze walls, and near-complete screen coverage is required to locate all the active pills and guide the avatar to/from any maze location. On the other hand, Battle Zone is a first person shooter game in which the agent can swivel left or right to position targets at centre-screen before shooting. While even screen coverage is helpful in locating targets and determining the direction to swivel, targets are large and thus low-resolution coverage is sufficient. Interestingly, Battle Zone includes a high-resolution global radar view in the top centre of the screen, which the champion policy's low-resolution AVF did not make efficient use of. Nonetheless, the bare-bones policy was able to out-perform the human video game tester without this advantage.

# 6 Conclusion and Future Work

A Tangled Program Graph (TPG) representation is proposed for discovering deep combinations of programs that collectively define policies in high-dimensional reinforcement learning tasks. Benchmarking is conducted under the subset of 20 games from the Atari 2600 ALE challenge for which the Deep Reinforcement Learning (DQN) framework did not reach the threshold for human-level play. In terms of score ranking for learning algorithms, TPG returns the best score in 14 games, DQN in 4 games, with HyperNeat and TPG tied in 1 game. Moreover, in 7 of the 15 games for which TPG provides a better strategy than DQN, TPG is also better than the threshold for human-level play.

Key to TPG is support for emergent modularity. That is to say, the ability to identify decisions local to different sub-regions of the state–action space and then organize such decisions hierarchically. This makes for a very efficient decision making process that completely decouples the overall complexity of a candidate solution (total number of programs) from the number of programs actually executed to make a decision (or size of the graph versus the proportion of the graph traversed to make a decision).

Such an approach is much more efficient than the representations assumed to date for deep learning, in which specialized (GPU) hardware support is necessary. Specifically, DQN assumes a fixed neural topology from the outset (i.e., hidden layer contains >800,000 weights and this cost is independent of game title) and a specific association with the state space (a computationally costly deep learning correlation step). Moreover, DQN assumes a correlation filter to discover encoded representations of game state. This introduces millions of calculations per frame whereas TPG merely subsamples the original frame information. The capacity of TPG to make decisions efficiently while not compromising on the quality of the resulting policies might also open up additional application areas to GP (e.g. real-time interpretation of video for obstacle avoidance in autonomous cars).

Having established the baseline capability of TPG, future work will investigate the utility of explicit diversity maintenance through multi-objective fitness regularization. In particular, switching between *multiple* diversity measures (in combination with fitness-based selection) has been shown to impact the development of modularity [23,24]. Future work is also likely to expand the set of test games, further validating the generality of the approach.

**Acknowledgments.** S. Kelly gratefully acknowledges support from the Nova Scotia Graduate Scholarship program. M. Heywood gratefully acknowledges support from the NSERC Discovery program. All runs were completed on cloud computing infrastructure provided by ACENET, the regional computing consortium for universities in Atlantic Canada. The TPG code base is not in any way parallel, but in adopting ACENET the five independent runs for each of the 20 games were conducted in parallel.

# References

1. Bellemare, M.G., Naddaf, Y., Veness, J., Bowling, M.: The arcade learning environment: an evaluation platform for general agents. J. Artif. Intell. Res. **47**, 253–279 (2013)
2. Nolfi, S.: Using emergent modularity to develop control systems for mobile robots. Adapt. Behav. **5**(3–4), 343–363 (1997)
3. Hausknecht, M., Lehman, J., Miikkulainen, R., Stone, P.: A neuroevolution approach to general Atari game playing. IEEE Trans. Comput. Intell. AI in Games **6**(4), 355–366 (2014)
4. Mnih, V., Kavukcuoglu, K., Silver, D., Rusu, A.A., Veness, J., Bellemare, M.G., Graves, A., Riedmiller, M., Fidjeland, A.K., Ostrovski, G., Petersen, S., Beattie, C., Sadik, A., Antonoglou, I., King, H., Kumaran, D., Wierstra, D., Legg, S., Hassabis, D.: Human-level control through deep reinforcement learning. Nature **518**(7540), 529–533 (2015)
5. Koza, J.R.: Genetic Programming: On the Programming of Computers by Means of Natural Selection. MIT Press, Cambridge (1992)
6. Rosca, J.: Towards automatic discovery of building blocks in genetic programming. In: Working Notes for the AAAI Symposium on Genetic Programming, AAAI, pp. 78–85, 10–12 1995
7. Spector, L., Martin, B., Harrington, K., Helmuth, T.: Tag-based modules in genetic programming. In: Proceedings of the 13th Annual Conference on Genetic and Evolutionary Computation, pp. 1419–1426. ACM (2011)
8. Brameier, M., Banzhaf, W.: Evolving teams of predictors with linear genetic programming. Genet. Program. Evolvable Mach. **2**(4), 381–407 (2001)
9. Imamura, K., Soule, T., Heckendorn, R.B., Foster, J.A.: Behavioural diversity and probabilistically optimal GP ensemble. Genet. Program. Evolvable Mach. **4**(3), 235–254 (2003)
10. Wu, S.X., Banzhaf, W.: Rethinking multilevel selection in genetic programming. In: Proceedings of the ACM Genetic and Evolutionary Computation Conference, pp. 1403–1410 (2011)
11. Thomason, R., Soule, T.: Novel ways of improving cooperation and performance in ensemble classifiers. In: Proceedings of the ACM Genetic and Evolutionary Computation Conference, pp. 1708–1715 (2007)
12. Lichodzijewski, P., Heywood, M.I.: Managing team-based problem solving with symbiotic bid-based genetic programming. In: Proceedings of the ACM Genetic and Evolutionary Computation Conference, pp. 863–870 (2008)

13. Lichodzijewski, P., Heywood, M.I.: Symbiosis, complexification and simplicity under GP. In: Proceedings of the ACM Genetic and Evolutionary Computation Conference, pp. 853–860 (2010)
14. Kelly, S., Heywood, M.I.: On diversity, teaming, and hierarchical policies: observations from the keepaway soccer task. In: Nicolau, M., Krawiec, K., Heywood, M.I., Castelli, M., García-Sánchez, P., Merelo, J.J., Rivas Santos, V.M., Sim, K. (eds.) EuroGP 2014. LNCS, vol. 8599, pp. 75–86. Springer, Heidelberg (2014). doi:10.1007/978-3-662-44303-3_7
15. Kelly, S., Heywood, M.I.: Genotypic versus behavioural diversity for teams of programs under the 4-v-3 keepaway soccer task. In: Proceedings of the AAAI Conference on Artificial Intelligence, pp. 3110–3111 (2014)
16. Lichodzijewski, P., Heywood, M.I.: The Rubik cube and GP temporal sequence learning: an initial study. In: Riolo, R., McConaghy, T., Vladislavleva, E. (eds.) Genetic Programming Theory and Practice VIII, 35–54. GEC. Springer, Heidelberg (2011)
17. Doucette, J.A., Lichodzijewski, P., Heywood, M.I.: Hierarchical task decomposition through symbiosis in reinforcement learning. In: Proceedings of the ACM Genetic and Evolutionary Computation Conference, pp. 97–104 (2012)
18. Kelly, S., Lichodzijewski, P., Heywood, M.I.: On run time libraries and hierarchical symbiosis. In: IEEE Congress on Evolutionary Computation, pp. 3245–3252 (2012)
19. Steenkiste, S., Koutník, J., Driessens, K., Schmidhuber, J.: A wavelet-based encoding for neuroevolution. In: Proceedings of the ACM Genetic and Evolutionary Computation Conference, pp. 517–524 (2016)
20. Brameier, M., Banzhaf, W.: Linear Genetic Programming, 1st edn. Springer, Heidelberg (2007)
21. Pepels, T., Winands, M.H.M.: Enhancements for monte-carlo tree search in Ms Pac-Man. In: IEEE Symposium on Computational Intelligence in Games, pp. 265–272 (2012)
22. Schrum, J., Miikkulainen, R.: Discovering multimodal behavior in Ms. Pac-Man through evolution of modular neural networks. IEEE Trans. Comput. Intell. AI in Games 8(1), 67–81 (2016)
23. Kashtan, N., Noor, E., Alon, U.: Varying environments can speed up evolution. Proc. Nat. Acad. Sci. 104(34), 13711–13716 (2007)
24. Parter, M., Kashtan, N., Alon, U.: Facilitated variation: how evolution learns from past environments to generalize to new environments. PLoS Comput. Biol. 4(11), e1000206 (2008)

# A General Feature Engineering Wrapper for Machine Learning Using ε-Lexicase Survival

William La Cava$^{(\boxtimes)}$ and Jason Moore

Institute for Biomedical Informatics, University of Pennsylvania,
Philadelphia, PA 19104, USA
{lacava,jhmoore}@upenn.edu

**Abstract.** We propose a general wrapper for feature learning that interfaces with other machine learning methods to compose effective data representations. The proposed feature engineering wrapper (FEW) uses genetic programming to represent and evolve individual features tailored to the machine learning method with which it is paired. In order to maintain feature diversity, ε-lexicase survival is introduced, a method based on ε-lexicase selection. This survival method preserves semantically unique individuals in the population based on their ability to solve difficult subsets of training cases, thereby yielding a population of uncorrelated features. We demonstrate FEW with five different off-the-shelf machine learning methods and test it on a set of real-world and synthetic regression problems with dimensions varying across three orders of magnitude. The results show that FEW is able to improve model test predictions across problems for several ML methods. We discuss and test the scalability of FEW in comparison to other feature composition strategies, most notably polynomial feature expansion.

**Keywords:** Genetic programming · Feature selection · Representation learning · Regression

## 1 Introduction

The success of machine learning (ML) algorithms in generating predictive models depends completely on the representation of the data used to train them. For this reason, and given the accessibility of many standard ML implementations to today's researchers, feature selection and synthesis are quickly becoming the most pressing issues in machine learning. The goal of feature engineering, i.e. feature learning or representation learning [2], is to learn a transformation of the attributes that improves the predictions made by a learning algorithm. Formally, given $N$ paired examples of $d$ attributes from the training set $\mathcal{T} = \{(\mathbf{x}_i, y_i), i = 1 \ldots N\}$, we wish to find a $P$-dimensional feature mapping $\Phi(\mathbf{x}) : \mathbb{R}^d \to \mathbb{R}^P$ for a regression model $\hat{y}(\Phi(\mathbf{x})) : \mathbb{R}^P \to \mathbb{R}$ that performs better than the model $\hat{y}(\mathbf{x}) : \mathbb{R}^d \to \mathbb{R}$ formed directly from $\mathbf{x}$. Doing so is a significant challenge, since it is not straightforward to determine useful nonlinear transformations of the raw feature set that may prove consequential in predicting

© Springer International Publishing AG 2017
J. McDermott et al. (Eds.): EuroGP 2017, LNCS 10196, pp. 80–95, 2017.
DOI: 10.1007/978-3-319-55696-3_6

the phenomenon in question. It is known that optimal feature selection (that is, selecting the optimal subset of features from a dataset) is itself hard [4], not to mention the process of determining an optimal feature representation. Although approaches to feature expansion are well known, for example polynomial basis expansion, these methods must be paired with a feature selection or parameter regularization scheme to control model complexity [5]. As we argue and show in this paper, feature engineering using genetic programming (GP) can be competitive in this task of optimizing the data representation for several ML methods, leading to more predictive regression models with a controlled number of features.

Typically GP is applied to regression by constructing and optimizing a population of symbolic models. Symbolic regression is a useful method for generating meaningful model structures for real-world applications [13] due to its flexibility in model representation. This comes at the expense of high computational cost, due to the expansive search space required to search model structures and parameters. In comparison, many ML approaches only attempt optimize the parameters of a single, fixed model structure with respect to a loss function, such as the mean squared error (linear regression) or the $\epsilon$-insensitive loss (support vector regression). Recent studies in GP [1,3] suggest that computation time can be saved by narrowing the scope of search to the space of model building-blocks rather than models themselves, and then generating a model via linear regression over the population outputs. Such an approach presents new challenges, since the population should no longer be pressured to converge on a single model, but rather to reach a set of transformations that are more or less orthogonal to each other. In this paper, we extend previous work by introducing a general framework for feature engineering that is agnostic with respect to the ML method used, and that modifies a recent parent selection technique [14] to maintain an uncorrelated population of features. The proposed feature engineering wrapper (FEW) is a GP-based method that interfaces with Scikit-learn [21] to provide generic feature engineering for its entire suite of learners. It is available open-source[1] as a Python package via the Python Package Index (PyPi)[2].

In this paper we demonstrate FEW in conjunction with Lasso [25], linear and non-linear support vector regression (SVR), K-Nearest Neighbors (KNN) regression, and decision tree (DT) regression. We show that in many cases, FEW is able to significantly improve the learners with which it is paired. We also demonstrate that for problems with large numbers of attributes, FEW is able to effectively search the feature space to generate models with a smaller set of optimized features, a task that is infeasible for some brute-force feature expansion methods. Central to the ability of FEW to learn a set of features rather than a single redundant feature is its use of a new survival method, known as $\epsilon$-lexicase survival, which is introduced in Sect. 2.1. $\epsilon$-lexicase suvival is a variant of $\epsilon$-lexicase selection [14] that maintains a population of individuals with unique outputs and

---

[1] https://github.com/lacava/few.
[2] https://pypi.python.org/pypi/FEW.

that survive based on their performance in the context of both the population semantics and the difficulty of the training cases.

We describe FEW in detail in Sect. 2, including a description of the survival algorithm as well as a discussion the scalability of this approach compared to other feature composition strategies. We review related work in Sect. 3, both in GP and in representation learning. The experiments are described in Sect. 4 and include a hyper-parameter optimization study of FEW and an analysis of FEW on a number of real-world problems and ML algorithms in comparison to polynomial basis expansion. The results are presented in Sect. 5 in terms of model accuracy on test sets and the rankings of methods across problems. We discuss the results and future work in Sect. 6 and conclude in Sect. 7.

## 2    Feature Engineering Wrapper

FEW uses GP optimize a population of feature transformations that are used to generate a predictive model each generation using a user-defined ML method. It continues to optimize the feature transformations as a GP population while building a single predictive model at the beginning of each subsequent generation. The steps of FEW are shown in Fig. 1.

FEW begins by fitting a model using a given ML method and the original attributes. This model is stored, along with its score on an internal validation set, and updated whenever a model is found with a better cross-validation (CV) score. This guarantees that the model generated by FEW performs at least as well on the internal validation set as the underlying ML method with which it is paired. FEW then constructs an initially random population of engineered features represented as $\Phi^k(\mathbf{x})$ in Fig. 1. This initial population is seeded with any original attributes with non-zero coefficients (for ML methods that use coefficients); for example, $\Phi^k$ may equal $[x_1, x_2, \phi_3, \ldots, \phi_P]$, where $x_1$ and $x_2$ are seeded from the initial ML model, and $\phi_3 \ldots \phi_p$ are initialized randomly. FEW then loops through the processes of selection, variation, fitness, survival, and ML model construction.

The selection phase is treated as an opportunity for feedback from the ML model on the current population of features, if any, to bias the parents chosen for variation. For example, some learners (e.g. Lasso, SVR) implement $\ell_1$ regularization to minimize feature coefficients as a form of feature selection. This information is incorporated during selection by removing individuals with coefficients of zero. Otherwise, selection for variation is random; note that selection is designed to be generally weak in order to prevent population convergence. The selected population is represented by $\Phi^{k'}$ in Fig. 1.

Following selection, the individuals in the population are varied using subtree crossover and point mutation to produce individuals $\Phi^{\text{offspring}}$. The fitness of each feature is then assessed. It is tempting to tailor the fitness function to each individual ML algorithm, but in the interest of time and scope, fairly conventional fitness metrics are used here. For the ML methods designed to minimize mean squared error (MSE), the MSE metric is used in fitness assessment. For

SVR, where the $\epsilon$-insensitive loss function is minimized, the mean absolute error (MAE) is used.

After fitness assessment, the set consisting of the parents and offspring ($\Phi^{k'}$ and $\Phi^{\text{offspring}}$) compete for survival, resulting in a reduction of the population back to its original size. This updated population of features, $\Phi^{k+1}$ in Fig. 1, is then used to fit a new ML model and the process repeats.

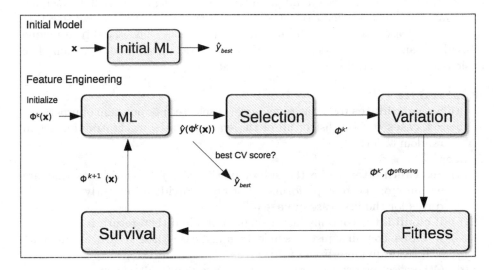

**Fig. 1.** Steps in the FEW process. After fitting an initial ML model, a set of feature transformations are initialized in a GP population. These engineered features are used to fit a new ML model each generation. The engineered features are then updated via selection, variation, fitness assessment and survival. The final model returned is that with the highest internal validation score among models generated.

Defining a GP system in such a way presents unique challenges. The individuals in the population not only have to compete to be effective features to solve the problem, but must work well in combination with the rest of the population. As a result, it is undesirable to have redundant i.e. correlated features, or to have the population converge to a single behavior. The goal instead is to have a population of features that essentially minimize the residuals of the outputs of other individuals in the population. As such, we are motivated to use $\epsilon$-lexicase selection [14] as the survival scheme for individuals, since this method rewards individuals for performing well on unique subsets of the training samples. This method shifts selection pressure continually to training cases that have are most difficult for the population to solve, as defined by population performance. It also has been shown to maintain very high semantic diversity in the population [8, 14], which should result in un-correlated engineered features.

## 2.1   ε-lexicase Survival

Ideally, the population would consist of uncorrelated features that complement each other in the construction of a single model. Central to this goal is maintaining diversity among individual outputs in the population while continuing to improve model prediction. With this in mind, we propose a new method for the survival routine called ε-lexicase survival that chooses individuals for survival based on their performance on unique subsets of training cases. It implements a slight alternation of ε-lexicase selection [14], a recent method for parent selection based on lexicase selection [8,23], in order to (1) make ε-lexicase return unique individuals and (2) make ε-lexicase work for survival rather than selection. The basic ε-lexicase survival strategy is as follows:

1. Initialize
   (a) Set **candidates** to be the remaining (unselected) population of programs.
   (b) Set **cases** to be a list of all of the training cases in the training set in random order.
2. Loop
   (a) Set **candidates** to be the subset of the current **candidates** that are within ε of the best performance of any individual currently in **candidates** for the first case in **cases**.
   (b) If **candidates** contains just a single individual then return it.
   (c) If **cases** contains just a single training case then return a randomly selected individual from **candidates**.
   (d) Otherwise remove the first case from **cases** and go to Loop.

This routine is repeated until the surviving population matches the user-defined population size. Several methods of determining ε were presented in the original paper [14]; we use the most successful one, which defines ε based on the median absolute deviation of fitnesses on the training case in question. In other words, for each training case $i$, $\epsilon_i$ is defined as

$$\epsilon_i = \text{median}\left(|\mathbf{e}_i - \text{median}(\mathbf{e}_i)|\right) \tag{1}$$

where $\mathbf{e}_i \in \mathbb{R}^{|2P|}$ is the fitness on training case $i$ of every individual in the set consisting of parents and their offspring. This definition of ε is nice because it is parameter free, adjusts to accommodate differences in hardness of training cases and population performance, and has been demonstrated to perform better than user-defined versions [13]. Two other differences between ε-lexicase *survival* and *selection* should be noted. First, unlike ε-lexicase selection, once an individual has been chosen for survival, it is removed from the survival pool in step 1.(a). This is to prevent the same individual from being selected multiple times, which would lead to redundant features in the population. In addition, 2.(a) in ε-lexicase survival is more similar to lexicase selection [23] in that the best error is defined relative to the current candidates rather than the whole population. In ε-lexicase selection, this was not the case.

## 2.2    Scaling

It is worth considering the complexity of the proposed method in the context of other feature engineering methods, namely basis function expansions and kernel transformations. Basis function expansions may scale exponentially; for example a $n$-degree polynomial expansion of $d$ features grows as $O(d^n)$. Kernel representations, meanwhile, scale with the number of samples; the complexity of computing a kernel is $O(N^2)$ and computing the solution to a kernelized regression problem with parameter regularization is $O(N^3)$ [5]. In comparison, FEW scales linearly with $N$, and otherwise scales according to the population size $P$ and the number of generations $g$ set by the user. This scaling can be advantageous compared to basis function expansion when the number of features is large, since the user controls the number of features to be included in the ML model ($P$), as well as the training time via $g$. For large sample sizes, FEW may be advantageous with respect to kernel expansion as well. The survival method used determines how FEW scales with respect to $P$. $\epsilon$-lexicase survival has a worst-case complexity of $O(P^2 N)$ in selecting a suvivor, compared to $O(PN)$ for tournament selection. In practice, however, it should be noted that no significant difference in run-time has been observed between tournament and $\epsilon$-lexicase selection [14], due to the fact that $\epsilon$-lexicase selection normally only uses a small set of training examples for each selection event. Although FEW can be expected to scale independently of the number of attributes, it should be noted that larger population sizes or more generations may be necessary to achieve satisfactory results for larger data sets.

## 3    Related Work

Hybrid learning methods that incorporate constant optimization in GP are quite common [9,12,26]. Others have proposed two-step approaches designed to first generate a set of features using GP and then construct a deterministic ML model. For example, evolutionary constructive induction methods [17–19] have been proposed to learn features for decision trees. Another method, FFX [16], applies a non-evolutionary, two-step technique that generates randomized basis functions for linear regression. FFX was coupled with a third symbolic regression stage in later research [10]. FEW's wrapper-based approach is inspired by evolutionary feature synthesis (EFS) [1], an algorithm that iterates upon a population of synthesized features while using them to fit a Lasso model. EFS is perhaps better classified as a population-based stochastic hill climbing method since it does not incorporate typical crossover or selection stages. Nevertheless it demonstrated the ability of feature space to be searched with EC-inspired search operators. FEW differs from EFS in its reliance on $\epsilon$-lexicase survival to maintain population diversity, and its use of crossover as a search operator. It also opts for a single population size parameter to minimize the number of hyper-parameters. Unlike the methods mentioned above, FEW can be paired with any ML estimator in Scikit-learn [21], whereas previous approaches have been tailored to use a fixed ML method (linear regression, Lasso or DT).

In the context of non-evolutionary methods, other feature engineering/ selection techniques are common, such as basis function expansions, feature subset selection via forward and backward selection [6], and regularization. Typically the feature space is augmented via the basis function expansion, and then either a feature selection method is used to select a subset of features prior to learning, or regularization pressure is applied to perform implicit feature selection during learning by minimizing the coefficients of some features to zero [5]. This approach is of course quite different from what is proposed in FEW: first, basis expansion relies on enumeration of the expanded set features (unless the kernel trick is used, in which case interpretability is lost); second, it either relies on a greedy approach to solving the feature selection problem [4] or relies on the regularization penalty to provide an acceptable level of sparseness in the resultant coefficients, which is not guaranteed.

In the context of meta-learning, FEW can be viewed as a wrapper concerned specifically with feature construction for a given estimator rather than a hyper-optimization strategy such as TPOT [20], which optimizes machine learning pipelines, or HyperOpt [11], which optimizes parameters of ML pipelines. It would be straightforward to include FEW as an estimator in either of these systems.

## 4    Experimental Analysis

A set of 7 real-world and 1 synthetic problems were used to analyze FEW, ranging in sample size (506–21048) and number of features (5–529). As a first step, we conducted a hyper-parameter optimization experiment, varying the population size and survival method for FEW. This hyper-parameter tuning step is conducted on three of the studied problems with Lasso set as the ML method. To make the population size somewhat normalized to the tested problem, we specified population size as a fraction of the problem dimensions; e.g. 0.5× indicates a population size equal to half the number of raw features and 4x indicates a population size equal to four times the number of raw features. In the second step, each ML method was tested with and without FEW, and compared in terms of the accuracy of the generated models on the test set as measured by the coefficient of determination:

$$R^2(y, \hat{y}) = 1 - \frac{\sum_{i \in \mathcal{T}} (y_i - \hat{y}_i)^2}{\sum_{i \in \mathcal{T}} (y_i - \bar{y})^2}$$

where $\bar{y}$ is the mean of the target output $y$. Note that $R^2$ in this case can be negative, since the variance of the residual of the model prediction can exceed the variance in the data. The FEW settings and problem descriptions are summarized in Table 1.

Each paired ML method uses the default settings from Scikit-learn. For Lasso, the LassoLarsCV() method is used, which is a least-angle regression implementation of Lasso that uses internal cross-validation to tune the regularization parameter. Support vector regression defaults to using a radial basis function kernel. KNN defaults to $k = 5$. The following section describes the set of problems used.

**Table 1.** FEW problem settings. The bold values indicate those chosen for the main analysis based on hyper-parameter validation.

| Setting | Value | | |
|---|---|---|---|
| Population size | **0.5x**, 1x, 2x, **3x**, 4x | | |
| Survival method | tournament, $\epsilon$-**lexicase** | | |
| Program depth limit | 2 | | |
| Generations | 100 | | |
| Crossover/mutation | 50/50% | | |
| Elitism | keep best | | |
| Trials | 30 shuffled 70/30% splits | | |
| Machine learner | Lasso, Linear SVR, KNN, SVR, DT | | |
| Terminal set | $\{\mathbf{x}, +, -, *, /, \sin(\cdot), \cos(\cdot), \exp(\cdot), \log(|\cdot|), (\cdot)^2, (\cdot)^3\}$ | | |
| Problem | Dimension | Training samples | Test samples |
| UBall5D | 5 | 1024 | 5000 |
| ENH | 8 | 538 | 230 |
| ENC | 8 | 538 | 230 |
| Housing | 14 | 354 | 152 |
| Tower | 25 | 2195 | 940 |
| Crime | 128 | 1395 | 599 |
| UJI_long | 529 | 14734 | 6314 |
| UJI_lat | 529 | 14734 | 6314 |

## 4.1 Problems

The UBall5D problem, also known as Vladislavleva-4, is the one synthetic problem studied whose solution is of the form $y = \frac{10}{5 + \sum_{i=1}^{5}(x_i-3)^2}$, which is known to be difficult to solve exactly [30]. The second and third problem tasks are to estimate the energy efficiency of heating (ENH) and cooling (ENC) requirements for various simulated buildings [28]. The housing data set [7] seeks a model to estimate Boston housing prices. The Tower problem consists of 15-minute averaged time series data taken from a chemical distillation tower, with the goal of predicting propelyne concentration. The Tower problem and UBall5D were chosen from the benchmark suite suggested by White et al. [31].

FEW is tested on three problems with large numbers of features. For these problems a population size of 0.5× is used in order to achieve dimensionality reduction in the model. The Crime problem consists of 128 community features used to estimate the per capita violent crime rates across the United States [22].

The UJI_long and UJI_lat problems are benchmark data sets for predicting the latitude and longitude, respectively, of indoor users from WLAN signals [27]. The dimensions of all data sets are shown in Table 1. The data sets were divided 70/30 into training and testing sets for 30 trials.

## 5   Results

The results of the hyper-parameter optimization are presented first, for which the FEW settings for the main set of problem comparisons are chosen. Afterwords we present a comparison of model fitness using the raw features, using polynomial feature expansion, and using FEW for each of the ML methods.

### 5.1   Hyper-Parameter Optimization

The test $R^2$ values of the models generated using each combination of hyper-parameters are shown in boxplot form in Fig. 2. As the population size is increased, and therefore the number of features presented to the ML model, the models produced tend to perform better, as expected. This performance improvement appears to level off for the Housing and Tower problem with a population size of 3×. For this reason we choose 3× as the population size in our subsequent experiments. Also note that $\epsilon$-lexicase survival outperforms tournament survival for larger population sizes, including 3×. This observation complements other experiments with lexicase selection that show its performance to be dependent on the number of training cases [15]. $\epsilon$-lexicase's good performance may be explained in part by the increase in output diversity among the engineered features, shown in Fig. 3. Here diversity is measured as one minus the average feature correlations, such that a higher value indicates less correlated features.

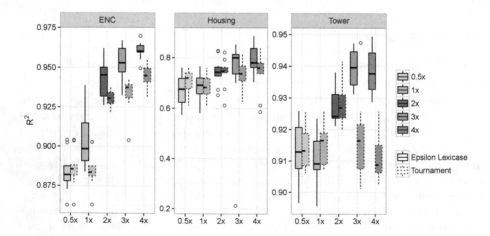

**Fig. 2.** Comparison of test set accuracy for various population sizes and selection methods.

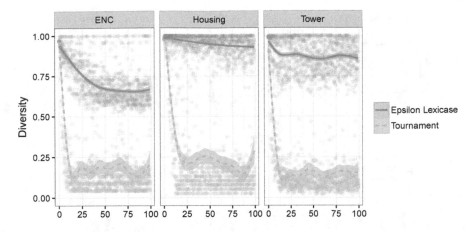

**Fig. 3.** Diversity of engineered features using $\epsilon$-lexicase survival or tournament survival with $3\times$ population size.

## 5.2   Problem Performance

The test scores on each problem are shown in Fig. 4 with respect to the ML algorithm pairing (Lasso, LinearSVR, KNN, SVR, DT) and the feature set used (raw, 2-degree polynomial, FEW). The final subplot in Fig. 4 shows the ranking statistics for each method across all problems. It is important to note that the polynomial feature expansion ran out of memory for the UJIIndoor problems, and hence no results for that method are included for those two problems. The memory error for this problem is reasonable given that a 2-degree expansion of 529 features would result in more than 279,000 features.

With the exception of its use with Lasso, FEW outperforms polynomial feature expansion for each ML method in terms of rankings across problems. The polynomial expansion worsens the ML rankings for KNN, SVR and DT compared to using the raw features, suggesting that these methods overfit with polynomial features. Conversely, FEW improves the model rankings across problems for all ML methods except for DT, on which it performs roughly the same as the raw features. Over all problems and methods, SVR paired with FEW ranks the best, followed by KNN paired with FEW.

## 5.3   Statistical Analysis

We conduct pairwise Wilcoxon rank-sum tests to compare the feature representation methods for each ML method and problem, correcting for multiple comparisons. The detailed results are presented in Table 2. In comparison to using the raw features, FEW works best in conjunction with linear and non-linear SVR, in which cases it significantly improves model accuracy for 6 out of 8 problems. FEW improves Lasso on 5 problems, KNN on 3 problems, and

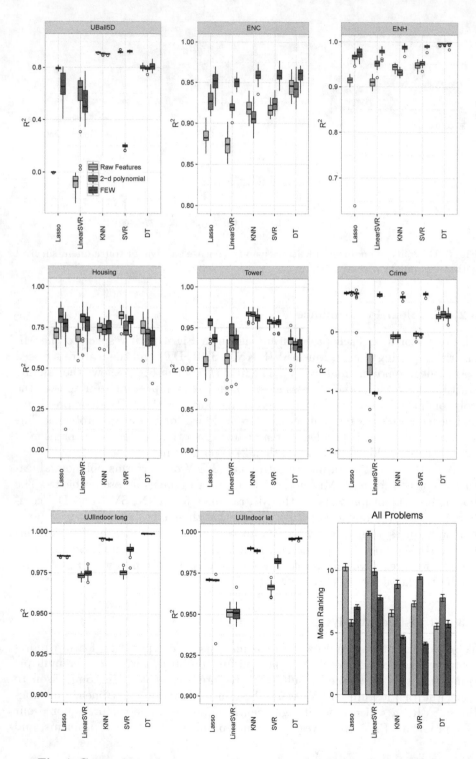

**Fig. 4.** Comparison of results for ML methods and FEW-enabled methods.

**Table 2.** $p$-values based on a Wilcoxon rank-sum test with Holm correction for multiple comparisons. Pairwise comparisons of the feature methods (FEW, Raw, and 2d poly) are grouped by the ML method and problem. Yellow highlighting indicates that the first feature method is significantly better than the second, whereas gray highlighting indicates that the first feature method is significantly worse than the second ($p < 0.05$).

| | UBall5D | ENC | ENH | Housing | Tower | Crime | UJIIndoor Long | UJIIndoor Lat |
|---|---|---|---|---|---|---|---|---|
| | | | | Lasso | | | | |
| FEW - Raw | 0.0000 | 0.0000 | 0.0000 | 0.0004 | 0.0000 | 0.0287 | 0.0714 | 0.0208 |
| 2d poly - Raw | 0.0000 | 0.0000 | 0.0000 | 0.0000 | 0.0000 | 0.0105 | N/A | N/A |
| FEW- 2d poly | 0.0000 | 0.0000 | 0.0024 | 0.0020 | 0.0000 | 0.0001 | N/A | N/A |
| | | | | Linear SVR | | | | |
| FEW - Raw | 0.0000 | 0.0000 | 0.0000 | 0.0000 | 0.0000 | 0.0000 | 0.0564 | 0.0346 |
| 2d poly - Raw | 0.0000 | 0.0000 | 0.0000 | 0.0000 | 0.0000 | 0.0000 | N/A | N/A |
| FEW- 2d poly | 0.1646 | 0.0000 | 0.0000 | 0.0428 | 0.0584 | 0.0000 | N/A | N/A |
| | | | | KNN | | | | |
| FEW - Raw | 0.0000 | 0.0000 | 0.0000 | 0.4471 | 0.0004 | 0.0000 | 0.0000 | 0.0000 |
| 2d poly - Raw | 0.0000 | 0.0005 | 0.0000 | 0.3219 | 0.6152 | 0.9528 | N/A | N/A |
| FEW- 2d poly | 0.0002 | 0.0000 | 0.0000 | 0.1473 | 0.0016 | 0.0000 | N/A | N/A |
| | | | | SVR | | | | |
| FEW - Raw | 0.0149 | 0.0000 | 0.0000 | 0.0025 | 0.2973 | 0.0000 | 0.0000 | 0.0000 |
| 2d poly - Raw | 0.0000 | 0.0071 | 0.0321 | 0.0000 | 0.2369 | 0.7901 | N/A | N/A |
| FEW- 2d poly | 0.0000 | 0.0000 | 0.0000 | 0.0006 | 0.1563 | 0.0000 | N/A | N/A |
| | | | | DT | | | | |
| FEW - Raw | 0.5444 | 0.0000 | 0.1646 | 0.0005 | 0.0114 | 0.2625 | 0.1331 | 0.0073 |
| 2d poly - Raw | 0.0230 | 0.1359 | 0.0011 | 0.0048 | 0.0002 | 0.0399 | N/A | N/A |
| FEW- 2d poly | 0.0148 | 0.0000 | 0.0005 | 0.4688 | 0.9882 | 0.0644 | N/A | N/A |

DT on 2 problems. The effects of FEW on KNN and DT are mixed, since FEW performs worse than the raw features for 4 and 2 problems, respectively. In comparison to 2d polynomial feature expansion, FEW is less effective when paired with Lasso, performing better in 2 problems and worse for 4 problems out of 6, all of which are statistically significant. However, when paired with linear and nonlinear SVR, FEW performs better than polynomial features on 3 and 5 problems, respectively. FEW also performs significantly better than polynomial features when paired with KNN, and DT, in terms of numbers of problems.

We also conduct an analysis of variance (ANOVA) test by sampling the mean rankings of each ML and feature method for each trial over all problems (i.e. for the data corresponding to the bottom right plot in Fig 4). In this case the UJIIndoor results are excluded since the polynomial method failed to produce models. The ANOVA suggests significant differences in the ML methods ($p = 0.02117$) and feature representations ($p = 0.0015$). A post-hoc test (Tukey's multiple comparisons of means) is conducted to determine the source of differences in feature representations; the results are shown in Table 3. The test shows

**Table 3.** Tukey multiple comparisons of means, 95% family-wise confidence level. Test of significance for feature engineering rankings across problems and methods.

| Comparison | difference | lower | upper | adjusted $p$-value |
|---|---|---|---|---|
| FEW-Raw Features | -3.428 | -5.652 | -1.205 | 0.00119 |
| 2-d polynomial-Raw Features | -1.121 | -3.345 | 1.102 | 0.45463 |
| FEW-2-d polynomial | -2.308 | -4.531 | -0.084 | 0.04011 |

significant differences between FEW and raw features ($p = 0.0019$), FEW and 2-d polynomial ($p = 0.0401$), and no significant difference between 2-d polynomial and raw features ($p = 0.4546$).

# 6 Discussion

The results suggest that FEW can provide improved model predictions by learning a data representation for several problems and ML methods. Research questions remain, for example: how can FEW be improved when paired with specific learners? and: how can the number of features be more precisely controlled? Concerning the first question, it is highly likely that FEW could be more tailored to each learner with which it is paired, in the following ways: (i), by changes to the fitness function for each engineered feature; (ii), by adjusting the GP parameters of FEW for each ML method; (iii), by biasing the GP operators used to construct features such that they produce individuals more likely to benefit a given learner. For example to address (i), it may be advantageous to use an entropy criterion for fitness when FEW is paired with DT in a classification problem. Doing so is not straightforward in combination with $\epsilon$-lexicase survival, due to the de-aggregated fitness values that lexicase methods expect when selecting individuals. One solution is to sample the entropy values over subsets of the training cases.

Concerning the second question, it would also be advantageous to de-couple the number of engineered features from the population size without having to introduce more parameters as was necessary in [1]. The motivation for de-coupling the two is that they form a trade-off: fewer features create a more readable model, yet a smaller population size restricts the sampling of the search space. With regularization available in many ML methods, it is tempting to use a large population size, although this results in longer run-times. All this must be considered in the context of the goals of the problem at hand. Ideally the population size would adapt automatically to fit the needs of the problem. There is some precedent for dynamic population size management in GP that may be applicable in this regard [24, 29].

As a final comment, FEW deserves a treatment on problems other than regression. For classification problems, a fitness function for continuous-valued features must be analyzed, since their outputs do not explicitly map to categorical labels. In addition, the options for engineered features may include boolean transformations for certain problems. FEW should be tested with popular ML methods in that domain, notably logistic regression with $\ell_1$ regularization.

# 7    Conclusions

In this paper, we introduced FEW, a GP-based feature learning method that interfaces with Scikit-learn suite of standard ML methods to improve data representation, a pressing issue in ML today. FEW represents engineered features as individuals in a population which are used in conjuction by the chosen ML algorithm to create a model. In order to evolve complimentary features, a new survival technique, $\epsilon$-lexicase survival, was presented that maintains a population of un-correlated features by pressuring them to perform well on unique subsets of training cases. In comparison to polynomial feature expansion, FEW is better able to improve model predictions for the problems and ML methods tested.

**Acknowledgments.** This work was supported by the Warren Center for Network and Data Science at the University of Pennsylvania, as well as NIH grants P30-ES013508, AI116794 and LM009012.

# References

1. Arnaldo, I., O'Reilly, U.M., Veeramachaneni, K.: Building predictive models via feature synthesis, pp. 983–990. ACM Press (2015)
2. Bengio, Y., Courville, A., Vincent, P.: Representation learning: a review and new perspectives. IEEE Trans. Pattern Anal. Mach. Intell. **35**(8), 1798–1828 (2013)
3. De Melo, V.V.: Kaizen programming, pp. 895–902. ACM Press (2014)
4. Foster, D., Karloff, H., Thaler, J.: Variable selection is hard. In: Proceedings of The 28th Conference on Learning Theory, pp. 696–709 (2015)
5. Friedman, J., Hastie, T., Tibshirani, R.: The elements of statistical learning. Springer series in statistics, vol. 1. Springer, Berlin (2001)
6. Guyon, I., Elisseeff, A.: An introduction to variable and feature selection. J. Mach. Learn. Res. **3**, 1157–1182 (2003)
7. Harrison, D., Rubinfeld, D.L.: Hedonic housing prices and the demand for clean air. J. Environ. Econ. Manage. **5**(1), 81–102 (1978)
8. Helmuth, T., Spector, L., Matheson, J.: Solving uncompromising problems with lexicase selection. IEEE Trans. Evol. Comput. **PP**(99), 1–1 (2014)
9. Iba, H., Sato, T.: Genetic Programming with Local Hill-Climbing. Tech. Rep. ETL-TR-94-4, Electrotechnical Laboratory, 1-1-4 Umezono, Tsukuba-city, Ibaraki, 305, Japan (1994). http://www.cs.ucl.ac.uk/staff/W.Langdon/ftp/papers/Iba_1994_GPlHC.pdf
10. Icke, I., Bongard, J.C.: Improving genetic programming based symbolic regression using deterministic machine learning. In: IEEE Congress on Evolutionary Computation (CEC), 2013, pp. 1763–1770. IEEE (2013)
11. Kamath, U., Lin, J., De Jong, K.: SAX-EFG: an evolutionary feature generation framework for time series classification, pp. 533–540. ACM Press (2014)
12. Kommenda, M., Kronberger, G., Winkler, S., Affenzeller, M., Wagner, S.: Effects of constant optimization by nonlinear least squares minimization in symbolic regression. In: Blum, C., Alba, E., Bartz-Beielstein, T., Loiacono, D., Luna, F., Mehnen, J., Ochoa, G., Preuss, M., Tantar, E., Vanneschi, L. (eds.) GECCO 2013 Companion, pp. 1121–1128. ACM, Amsterdam (2013)

13. La Cava, W., Danai, K., Spector, L., Fleming, P., Wright, A., Lackner, M.: Automatic identification of wind turbine models using evolutionary multiobjective optimization. Renew. Energy Part 2 **87**, 892–902 (2016)
14. La Cava, W., Spector, L., Danai, K.: Epsilon-Lexicase Selection for Regression, pp. 741–748. ACM Press (2016)
15. Liskowski, P., Krawiec, K., Helmuth, T., Spector, L.: Comparison of semantic-aware selection methods in genetic programming. In: Proceedings of the Companion Publication of the 2015 Annual Conference on Genetic and Evolutionary Computation, GECCO Companion 2015, pp. 1301–1307. ACM, New York (2015)
16. McConaghy, T.: FFX: fast, scalable, deterministic symbolic regression technology. In: Riolo, R., Vladislavleva, E., Moore, J.H. (eds.) Genetic Programming Theory and Practice IX. Genetic and Evolutionary Computation, pp. 235–260. Springer, New York (2011)
17. Muharram, M., Smith, G.D.: Evolutionary constructive induction. IEEE Trans. Knowl. Data Eng. **17**(11), 1518–1528 (2005)
18. Muharram, M.A., Smith, G.D.: The effect of evolved attributes on classification algorithms. In: Gedeon, T.T.D., Fung, L.C.C. (eds.) AI 2003. LNCS (LNAI), vol. 2903, pp. 933–941. Springer, Heidelberg (2003). doi:10.1007/978-3-540-24581-0_80
19. Muharram, M.A., Smith, G.D.: Evolutionary feature construction using information gain and gini index. In: Keijzer, M., O'Reilly, U.-M., Lucas, S., Costa, E., Soule, T. (eds.) EuroGP 2004. LNCS, vol. 3003, pp. 379–388. Springer, Heidelberg (2004). doi:10.1007/978-3-540-24650-3_36
20. Olson, R.S., Bartley, N., Urbanowicz, R.J., Moore, J.H.: Evaluation of a tree-based pipeline optimization tool for automating data science. arXiv preprint (2016). http://arxiv.org/abs/1603.06212
21. Pedregosa, F., Varoquaux, G., Gramfort, A., Michel, V., Thirion, B., Grisel, O., Blondel, M., Prettenhofer, P., Weiss, R., Dubourg, V., et al.: Scikit-learn: machine learning in Python. J. Mach. Learn. Res. **12**, 2825–2830 (2011)
22. Redmond, M., Baveja, A.: A data-driven software tool for enabling cooperative information sharing among police departments. Eur. J. Oper. Res. **141**(3), 660–678 (2002)
23. Spector, L.: Assessment of problem modality by differential performance of lexicase selection in genetic programming: a preliminary report. In: Proceedings of the Fourteenth International Conference on Genetic and Evolutionary Computation Conference Companion, pp. 401–408 (2012)
24. Tan, K.C., Lee, T.H., Khor, E.F.: Evolutionary algorithms with dynamic population size and local exploration for multiobjective optimization. IEEE Trans. Evol. Comput. **5**(6), 565–588 (2001)
25. Tibshirani, R.: Regression shrinkage and selection via the lasso. J. R. Stat. Soc. Ser. B (Methodological) **58**, 267–288 (1996)
26. Topchy, A., Punch, W.F.: Faster genetic programming based on local gradient search of numeric leaf values. In: Proceedings of the Genetic and Evolutionary Computation Conference (GECCO-2001), pp. 155–162 (2001)
27. Torres-Sospedra, J., Montoliu, R., Martnez-Us, A., Avariento, J.P., Arnau, T.J., Benedito-Bordonau, M., Huerta, J.: UJIIndoorLoc: A new multi-building and multi-floor database for WLAN fingerprint-based indoor localization problems. In: 2014 International Conference on Indoor Positioning and Indoor Navigation (IPIN), pp. 261–270. IEEE (2014)
28. Tsanas, A., Xifara, A.: Accurate quantitative estimation of energy performance of residential buildings using statistical machine learning tools. Energy Build. **49**, 560–567 (2012)

29. Vanneschi, L., Cuccu, G.: A study of genetic programming variable population size for dynamic optimization problems. In: International Conference on Evolutionary Computation (ICEC 2009), pp. 119–126. Madeira, Portugal (2009)
30. Vladislavleva, E., Smits, G., Hertog, D.: Order of nonlinearity as a complexity measure for models generated by symbolic regression via pareto genetic programming. IEEE Trans. Evol. Comput. **13**(2), 333–349 (2009)
31. White, D.R., McDermott, J., Castelli, M., Manzoni, L., Goldman, B.W., Kronberger, G., Jakowski, W., O'Reilly, U.M., Luke, S.: Better GP benchmarks: community survey results and proposals. Genet. Program. Evolvable Mach. **14**(1), 3–29 (2012)

# Visualising the Search Landscape
# of the Triangle Program

William B. Langdon[1]([✉]), Nadarajen Veerapen[2], and Gabriela Ochoa[2]

[1] CREST, Computer Science, UCL, London WC1E 6BT, UK
W.Langdon@cs.ucl.ac.uk
[2] Computing Science and Mathematics, University of Stirling, Stirling FK9 4LA, UK

**Abstract.** High order mutation analysis of a software engineering benchmark, including schema and local optima networks, suggests program improvements may not be as hard to find as is often assumed. (1) Bit-wise genetic building blocks are not deceptive and can lead to all global optima. (2) There are many neutral networks, plateaux and local optima, nevertheless in most cases near the human written C source code there are hill climbing routes including neutral moves to solutions.

**Keywords:** Genetic improvement · Genetic algorithms · Genetic programming · Software engineering · Heuristic methods · Test equivalent higher order mutants · Fitness landscape · Local search

## 1 Genetic Improvement

Genetic Improvement [1–3] can be thought of as the use of Search Based Software Engineering (SBSE) [4] techniques, principally genetic programming [5–7], to the optimisation of existing (human written) software. GI is often applied to non-functional properties of software but perhaps it is most famous for improving program's functionality, e.g. by removing bugs [8–16] or adding to its abilities [17–22]. Non-functional improvements that have been considered or results reported include: faster code [23,24], code which uses less energy [25–34] or less memory [35], and automatic parallelisation [36–38] and automatic porting [39] and embedded systems [25,40–45] as well as refactorisation [46], reverse engineering [47,48] and software product lines [49,50]. There is very much a GI flavour in the air with a three-fold increase in GI publications (as measured by GI papers in the genetic programming bibliography) since the first GI workshop [51] was first mooted (October, 7 2014)[1]. Nonetheless there remains a deal of scepticism in software engineering circles.

One criticism of genetic improvement is that the results are empirical and there is little theoretical underpinning [52]. One of the prejudices holding back software engineering is the assumption that software is fragile. It has been shown in a small number of cases that this fear has been overplayed. Whilst many

---

[1] http://geneticimprovementofsoftware.com/?page_id=13 (accessed Oct, 9 2016).

© Springer International Publishing AG 2017
J. McDermott et al. (Eds.): EuroGP 2017, LNCS 10196, pp. 96–113, 2017.
DOI: 10.1007/978-3-319-55696-3_7

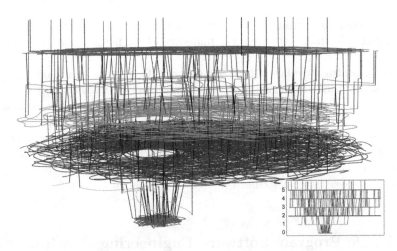

**Fig. 1.** Local optima network of the Triangle Program using 100 random starts (see Sect. 4.4). Edges are coloured if they start and end at the same fitness. Insert shows fitness levels edge on. Best (bottom) red 0 (pass all tests), pink 1 (fail only one test), green 2, purple 3, orange 4, brown 5. (Color figure online)

mutations are highly deleterious, in the few cases reported, many others are not [53]. If presented differently, this idea is not news to software engineers: the failure of software engineering to widely adopt mutation testing [54] is in part due to the presence of many equivalent mutants. But an equivalent mutant is simply a mutation which has no effect, which is exactly what GI has reported! Until recently [55] mutation testing rarely considered more than one change to the source code at a time, whereas GI typically allows high order mutations (which make multiple changes simultaneously). We consider up to $17^{\text{th}}$ order mutations.

Mostly genetic improvement considers mutations to source code (typically C, C++ or Java [56]) but similar empirical results have been reported at byte code [57], assembly [58] and indeed machine code [59,60] levels.

Due to the dearth of theoretical genetic improvement analysis, we shall study the GI search landscape [61,62]. Although small (39 lines of C code) we chose the Triangle Program as it is a well known software engineering benchmark and we have used it with mutation testing [55]. Briefly, in mutation testing [54] errors (mutations) like those a human programmer might make are deliberately injected into the program to see if the program's test suite can detect them. Our mutations were to replace numeric comparison operators, e.g., replace == with <=. In the next section we describe a cut down version where there are only two choices for each of the seventeen comparison sites in the Triangle Program. Section 3 presents a schema [63] analysis of the simplified landscape which shows *none* of the high order schema are deceptive [64]. This suggests that it might be easy for a genetic algorithm (GA) to find solutions. (Section 4.2 adds to this by showing none are strongly deceptive when larger moves are allowed.) We also see (Sect. 3.3) that there are large plateaux where neighbours have identical fitness.

Section 4 returns to allowing all six possible C numeric comparison operations and shows that although solutions are extremely rare, the vastly increased search space is still easy for genetic improvement in the sense that a local search hill climbing algorithm can reach a global optima from almost any low order mutation provided neutral moves are allowed. In contrast search from a random point seldom finds a program that can pass all the test cases. This supports the idea that genetic improvement, where search may start near the good part of the search space, is easier than GP, where search may start from a random point.

## 2   Triangle Program Software Engineering Benchmark

The Triangle Program is well studied software engineering benchmark. It can be thought of as a model of unit testing. It classifies triangles as scalene, isosceles, equilateral or not a triangle. We have previously used it to study high order mutation, concentrating particularly on injecting faults which change the numeric comparison operators (<, <=, ==, !=, => and >) [55]. We now consider mutation of all 17 of the comparison operators in the Triangle Program as a genetic algorithm fitness landscape. Taking as our fitness the number of tests [55, Table 2] which the modified code fails. A test equivalent mutant is one that passes all the tests and so has the best fitness value, which is zero. We consider all possible simultaneous changes. For the Triangle Program there are $6^{17} - 1 = 16\,926\,659\,444\,735$ mutations, of which 9215 are test equivalent, i.e. pass all the test cases.

## 3   Binary Representation: Replacing Comparisons with One Alternative

At first, instead of allowing all possible combinations, we study allowing each numerical comparison in the Triangle Program to be replaced by only one other. Table 1 is taken from [55]. It shows hard to detect mutations of the Triangle Program. The source code of the unmutated Triangle Program contains only <=, == and > comparisons. The last three lines of Table 1 gives their replacements in the commonest lower order hard to detect mutations. In mutation testing these mutations are known as the hardest to "kill". Therefore the substitutions given in last three lines of Table 1 are the ones we use. Since there are six comparison operators and 17 potential mutation sites, this reduces the search space from $6^{17}$ to $2^{17}$. (We shall return to the original problem in Sect. 4.) We evaluate all possible mutants.

**Table 1.** Hardest to detect mutations of the Triangle Program [55, Fig. 3]. The first column contains the number of times the individual changes shown appear in $1^{st}$, $2^{nd}$, $3^{rd}$ and $4^{th}$ order test equivalent mutations.

| | |
|---|---|
| 354 | == replaced by >= |
| 576 | <= replaced by < |
| 708 | == replaced by <= |
| 1062 | > replaced by != |
| 1992 | <= replaced by == |

**Table 2.** Mean and standard deviation of number of tests failed for highest order binary schema of the Triangle Program (excluding 22 with average means). Last column is estimated population size needed for a random sample to distinguish between competing pairs of schema.

| | | |
|---|---|---|
| -4 | 3.719 ±1.328 | 1.9 |
| 4 | 4.969 ±1.075 | |
| -5 | 4.062 ±1.478 | 4.7 |
| 5 | 4.625 ±1.166 | |
| -6 | 3.812 ±1.509 | 2.4 |
| 6 | 4.875 ±0.927 | |
| -11 | 3.438 ±1.273 | 1.1 |
| 11 | 5.250 ±0.661 | |
| -14 | 4.312 ±1.424 | 43.5 |
| 14 | 4.375 ±1.293 | |
| -16 | 4.188 ±1.550 | 8.6 |
| 16 | 4.500 ±1.118 | |

### 3.1 High Order Binary Schema Are Not Deceptive

There are 2048 global optima (shown in white in Fig. 2). On average each mutant fails only $4.344 \pm 1.360$ tests. The worst mutant only fails six of the 14 tests (Fig. 3).

Of the 34 high order schema[2], 22 have exactly average fitness and contain exactly half the global optima. The other 12 schema either contain no solutions or all of them. In the six schema which contain solutions, on average individuals are better than the average of the whole space. In the other six, the schema average is worse than the average of the whole space. That is, 22 schema have no signal and the remaining 12 are not deceptive. In the best schema (i.e. $-11$) mutants pass on average $1.813 \pm 1.015$ more tests than its opposite (11) (see also Table 2).

### 3.2 Binary Schema Predict *All* Solutions of the Triangle Program

As the previous section showed, there are twenty two 16-order schema that have exactly average fitness. These correspond to $22/2 = 11$ variable gene locations. I.e. locations of *s, where either alternative can be used. Together they can be represented as a $17 - 11 = 6^{th}$ order schema by taking their union. This sixth

---

[2] A $16^{th}$ order schema has 16 defined positions [64, page 29], and one variable * position (length = 17). Whereas a $1^{st}$ order mutation is identical to the original except for one change.

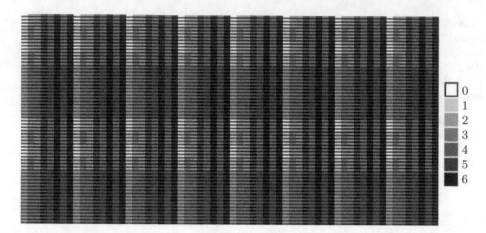

**Fig. 2.** Fitness landscape of binary comparison improvement of Triangle Program. First 9 bits (512) horizontal, last 8 bits (256) vertical. 2048 test equivalent mutants (fitness=0) in white. The regular pattern of individuals with the same fitness indicates short building blocks. E.g. the vertical strips 8 pixels wide indicates the first three bits do not impact fitness. In contrast the last but one bit divides the figure into four horizontal stripes, two contain 50 176 mutants which fail 4 or more tests (dark pixels) whilst the others hold all the solutions (white). Fitness distance correlation is 0.45

order schema is shown in Table 3. These 11 * (don't cares) give 2048 combinations ($2^{11} = 2048$) each of which is one of the solutions!

**Table 3.** 6th order binary schema giving 2048 test equivalent mutations of the Triangle Program. * indicates don't care but a 0 (or 1) means that only the one numeric comparison shown in the next row can be used. The bottom two rows gives the $2^{11}$ alternatives (for the $17 - 6 = 11$ *s) which pass all the tests.

| Position | 1 | 2 | 3 | 4 | 5 | 6 | 7 | 8 | 9 | 10 | 11 | 12 | 13 | 14 | 15 | 16 | 17 |
|---|---|---|---|---|---|---|---|---|---|---|---|---|---|---|---|---|---|
| Schema | * | * | * | 0 | 0 | 0 | * | * | * | * | 0 | * | * | 0 | * | 0 | * |
| Code | <= <= <=<br>== == == | | | == == == | | | == <= <= <=<br><= == == == | | | | == ><br>> | | > | <br>> == | == | > | | | | |

An alternative way of looking at this is, once we fix the six mutation sites in the C source code corresponding to the better than average schema in Table 2 we are free to mutate all the others (using our restricted mutation operator, last three rows of Table 1) and the new program will return the correct answer for all of the tests (Table 3).

## 3.3 Local Search Landscape of the Binary Space

After full enumeration, we modelled the corresponding landscape as a network, with nodes being mutants and edges linking mutants with only one difference between them. Plateaus, i.e. a connected collection of solutions with the same fitness, were identified using code adapted from [65], giving Fig. 4. In Fig. 4 each

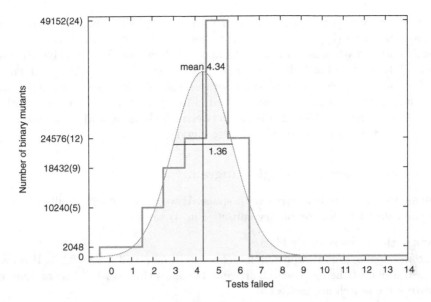

**Fig. 3.** Fitness distribution in the binary comparison version of the Triangle Program. Gaussian fit to mean and standard deviation shown in background.

rectangular box is a plateau of mutants, whose width is proportional to the number of mutants. Lines between boxes indicate pairs of mutants which differ only by a single bit flip. An edge's width is proportional to the number of such mutants. In the context of this simplified binary version, all mutants can reach at least one other mutant that fails fewer test cases in a single step. There is therefore no point in traversing any plateau to try to escape it.

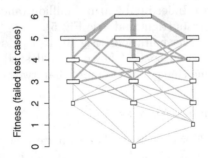

**Fig. 4.** Plateaux and their connections in the landscape of the binary comparison version of the Triangle Program.

## 4   Original All Comparisons

Sections 2 and 3 described how in [52] we simplified the Triangle Program Software Engineering benchmark to ease analysis of its schema and fitness landscape.

From here on we return to the original formulation [55]. Next we evaluate all possible mutations (Figs. 5, 6, 7 and 8). In Sect. 4.2 we repeat the binary schema analysis and find when larger moves are allowed there may be deceptive schema. In Sect. 4.3 we run a local hill climber both, from every part of the search space near the original program, and from samples of higher order mutations, and show that improvements are easy to find. Finally in Sect. 4.4 we use another local iterated search, which alternates between hillclimbing and random moves, to map the local optima network, see also Fig. 1 (page 1).

## 4.1   Fitness Space of Triangle Program

We evaluated the whole of the search space. If we compare the results in Fig. 5 with the same data for the binary subset (Fig. 3) we see:

- Firstly the space is vastly bigger.
- The number of global solutions increases by a factor of 4.5 to 9215. However as a fraction of the total search space it becomes tiny ($5\,10^{-10}$) so random or brute force search are ineffective.
- However the overall shape of the distribution of fitness values of high order mutations remains similar. (The mean increases slightly from 4.34 to 5.42 and the standard deviations are similar, 1.36 v. 1.05 in the full search space.) Note, for example, in both cases there is a large peak of mutated programs that fail exactly five tests. Indeed most mutants still pass most tests.

    As we shall see in Sect. 4.3, the large fraction of the search space with fitness of exactly five contributes to a huge neutral network. That is, there are many neighbouring programs (i.e. they differ in exactly one comparison) which fail exactly the same number of tests.
- The fraction of mutants which fail more than half the tests remains low (albeit finite rather than zero). Indeed only 40 in a million randomly sampled mutations fail all fourteen tests (right end side Fig. 5).
- Considering only first order mutations (Fig. 6), 16% pass all the tests. Indeed, as with the smaller search space (shown with dashed line in Fig. 6) most programs with only one change fail no more than two tests.

## 4.2   High Order Schema Analysis

There are $17 \times 6 = 102$ 16-order schema. We estimated their fitness by randomly sampling each one a million times. The results are given in Table 4 and Fig. 7. Whereas when considering only two options (Sect. 3.1, Table 2) in eleven of the 17 mutation sites both alternative schema had the same average fitness, in only three locations (4, 5 and 6) might all six schema be said to be within sampling noise of the mean fitness of the whole space. Now that we are considering moves that take us further from the original code, we do find weakly deceptive schema. For locations 11, 12, 14, 15 and 17, although none of the $5 \times 6 = 30$ schema differ strongly from the average, all of the solutions occur in *below* average schema. (See cluster in centre of Fig. 7.) Also in three more locations (3, 8 and 9), although

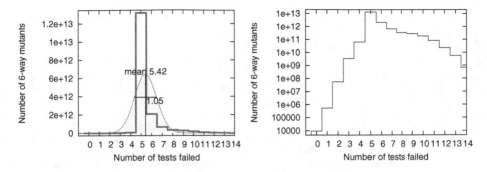

**Fig. 5.** Fitness distribution of all 16 926 659 444 735 possible comparison mutations of the Triangle Program. Note most (78%) mutants fail five tests which corresponds with the peak in the two options only subspace, Fig. 3. Right: same data plotted on log scale. Considering all six comparison operations increases the number of changes which still pass the whole test suite from 2048 to 9215.

one or more schema containing solutions are above average, the strongest schema does not contain any solutions. Only in the first two locations is there a strong signal (i.e. >0.1) leading to any of the solutions. In the remaining four locations (7, 10, 13 and 16) there are solutions in above average schema, but there isn't a strong signal leading to any of them.

### 4.3   Local Search for the Triangle Program

Since all first order mutations are by definition one move away from a solution, in all cases it is possible to hillclimb from any $1^{st}$ order mutation to a solution (Fig. 9). In the case of $2^{nd}$ order mutations, a hill climber can find a program which passes all the test cases in all but two cases (both these local optima fail two tests). For third order there are 133 (0.15% of $3^{rd}$ order mutations) and for $4^{th}$ 3623 (0.24%) points in the search space from which a solution cannot be reached by hill climbing. For $5^{th}$ and $6^{th}$ its about 4%, after which the fraction of higher order mutations from which a solution can be reached progressively falls towards zero. Even so in most cases a program which fails only one or two test cases can be found by hill climbing. I.e., in almost all cases hill climbing can improve a mutant from failing five test cases to failing two (Figs. 8 and 9).

### 4.4   Local Optima Networks

Local optima networks are a compact representation of fitness landscapes that can be used for analysis and visualisation [66]. A solution is a *local optimum* if none of its neighbours have a better fitness value. A full enumeration of the local optima for the Triangle Program is clearly unmanageable. The networks are therefore based on a sample of high-quality local optima in the search space. This is much more of a practical search algorithm than in the previous section, where we treat finding any hill climbing route to a solution as a problem of

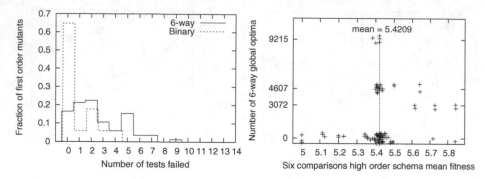

**Fig. 6.** Fitness distribution of all 85 possible first order comparison mutations of the Triangle Program. (I.e. rescaled leftmost line of Fig. 8) Corresponding data for the binary version of the search space shown plotted with dashed line for comparison.

**Fig. 7.** Fitness of all 102 high order schema and the number of solutions they contain for the Triangle Program with six comparison mutations. Schema contain either none, 1/3, 1/2 or all the solutions. Vertical noise added to separate data. All 16-schema containing a solution have fitness > mean − 0.0368.

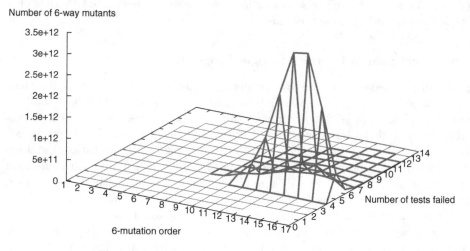

**Fig. 8.** Fitness distribution of all 16 926 659 444 735 possible comparison mutations of the Triangle Program. (Summary data plotted in Fig. 5.) Values above $10^9$ plotted in bold. 39% of mutations are either $14^{th}$ or $15^{th}$ order which fail five tests, whilst the fitness distance correlation is near zero at −0.070837.

searching a directed graph and so we include backing up and trying again if it appears no forward progress is possible.

The sampling algorithm is an Iterated Local Search (ILS) which starts from a locally optimal solution and then alternates between a random mutation and a best-improvement hill-climber. The termination criterion is a fixed number

**Table 4.** Average fitness of high order schema and number of solutions they contain for all six comparisons in the Triangle Program. See also Fig. 7

| Schema | | mean | sd | above average | | Sol | Schema | | mean | sd | above average | | Sol |
|---|---|---|---|---|---|---|---|---|---|---|---|---|---|
| 1 | < | 5.843567 | 1.423400 | 0.4227 | ±0.0014 | 3072 | 10 | < | 5.481349 | 1.104830 | 0.0604 | ±0.0011 | 0 |
| 1 | <= | 5.727581 | 1.349277 | 0.3067 | ±0.0013 | 3071 | 10 | <= | 5.422098 | 1.055740 | 0.0012 | ±0.0011 | 4607 |
| 1 | == | 5.726547 | 1.349000 | 0.3056 | ±0.0013 | 3072 | 10 | == | 5.421187 | 1.056004 | 0.0003 | ±0.0011 | 4608 |
| 1 | != | 5.114173 | 0.318021 | -0.3067 | ±0.0003 | 0 | 10 | != | 5.419925 | 1.032951 | -0.0010 | ±0.0010 | 0 |
| 1 | >= | 5.000000 | 0.000000 | -0.4209 | ±0.0000 | 0 | 10 | >= | 5.357749 | 0.994258 | -0.0632 | ±0.0010 | 0 |
| 1 | > | 5.113825 | 0.317599 | -0.3071 | ±0.0003 | 0 | 10 | > | 5.420015 | 1.033797 | -0.0009 | ±0.0010 | 0 |
| 2 | < | 5.842667 | 1.471239 | 0.4218 | ±0.0015 | 3072 | 11 | < | 5.385125 | 0.943758 | -0.0358 | ±0.0009 | 0 |
| 2 | <= | 5.614215 | 1.319723 | 0.1933 | ±0.0013 | 3071 | 11 | <= | 5.455657 | 1.102885 | 0.0348 | ±0.0011 | 0 |
| 2 | == | 5.616813 | 1.323260 | 0.1959 | ±0.0013 | 3072 | 11 | == | 5.349893 | 0.897228 | -0.0710 | ±0.0009 | 0 |
| 2 | != | 5.228102 | 0.473518 | -0.1928 | ±0.0005 | 0 | 11 | != | 5.493383 | 1.185560 | 0.0725 | ±0.0012 | 0 |
| 2 | >= | 5.000000 | 0.000000 | -0.4209 | ±0.0000 | 0 | 11 | >= | 5.457862 | 1.142620 | 0.0370 | ±0.0011 | 0 |
| 2 | > | 5.228345 | 0.473535 | -0.1926 | ±0.0005 | 0 | 11 | > | 5.384137 | 0.979864 | -0.0368 | ±0.0010 | 9215 |
| 3 | < | 5.840933 | 1.576470 | 0.4200 | ±0.0016 | 0 | 12 | < | 5.401587 | 1.013621 | -0.0193 | ±0.0010 | 0 |
| 3 | <= | 5.499540 | 1.233664 | 0.0786 | ±0.0012 | 4607 | 12 | <= | 5.407264 | 1.023709 | -0.0136 | ±0.0010 | 4608 |
| 3 | == | 5.500957 | 1.236198 | 0.0801 | ±0.0012 | 4608 | 12 | == | 5.399789 | 1.014724 | -0.0211 | ±0.0010 | 4607 |
| 3 | != | 5.343792 | 0.582205 | -0.0771 | ±0.0006 | 0 | 12 | != | 5.440903 | 1.078283 | 0.0200 | ±0.0011 | 0 |
| 3 | >= | 5.000000 | 0.000000 | -0.4209 | ±0.0000 | 0 | 12 | >= | 5.439726 | 1.081894 | 0.0188 | ±0.0011 | 0 |
| 3 | > | 5.342161 | 0.582014 | -0.0787 | ±0.0006 | 0 | 12 | > | 5.435198 | 1.072511 | 0.0143 | ±0.0011 | 0 |
| 4 | < | 5.421830 | 1.059801 | 0.0009 | ±0.0011 | 0 | 13 | < | 5.393600 | 1.005139 | -0.0273 | ±0.0010 | 0 |
| 4 | <= | 5.421768 | 1.041495 | 0.0009 | ±0.0010 | 0 | 13 | <= | 5.420620 | 1.052961 | -0.0003 | ±0.0011 | 0 |
| 4 | == | 5.419896 | 1.025397 | -0.0010 | ±0.0010 | 9215 | 13 | == | 5.419422 | 1.051208 | -0.0015 | ±0.0011 | 0 |
| 4 | != | 5.422214 | 1.078773 | 0.0013 | ±0.0011 | 0 | 13 | != | 5.421373 | 1.044389 | 0.0005 | ±0.0010 | 4608 |
| 4 | >= | 5.421401 | 1.050374 | 0.0005 | ±0.0011 | 0 | 13 | >= | 5.446829 | 1.087792 | 0.0259 | ±0.0011 | 0 |
| 4 | > | 5.420884 | 1.038998 | -0.0000 | ±0.0010 | 0 | 13 | > | 5.421251 | 1.042247 | 0.0004 | ±0.0010 | 4607 |
| 5 | < | 5.421382 | 1.060194 | 0.0005 | ±0.0011 | 0 | 14 | < | 5.407715 | 1.022905 | -0.0132 | ±0.0010 | 0 |
| 5 | <= | 5.421160 | 1.033338 | 0.0003 | ±0.0010 | 0 | 14 | <= | 5.413683 | 1.033827 | -0.0072 | ±0.0010 | 0 |
| 5 | == | 5.420832 | 1.012703 | -0.0001 | ±0.0010 | 9215 | 14 | == | 5.400401 | 1.016918 | -0.0205 | ±0.0010 | 9215 |
| 5 | != | 5.420481 | 1.085840 | -0.0004 | ±0.0010 | 0 | 14 | != | 5.441626 | 1.079657 | 0.0207 | ±0.0011 | 0 |
| 5 | >= | 5.420653 | 1.051449 | -0.0002 | ±0.0011 | 0 | 14 | >= | 5.433470 | 1.069622 | 0.0126 | ±0.0011 | 0 |
| 5 | > | 5.419979 | 1.044165 | -0.0009 | ±0.0010 | 0 | 14 | > | 5.426515 | 1.057726 | 0.0056 | ±0.0011 | 0 |
| 6 | < | 5.419947 | 1.040180 | -0.0010 | ±0.0010 | 0 | 15 | < | 5.394988 | 1.007207 | -0.0259 | ±0.0010 | 0 |
| 6 | <= | 5.418500 | 1.017661 | -0.0024 | ±0.0010 | 0 | 15 | <= | 5.429938 | 1.064344 | 0.0090 | ±0.0011 | 0 |
| 6 | == | 5.420138 | 1.021691 | -0.0008 | ±0.0010 | 9215 | 15 | == | 5.430083 | 1.064206 | 0.0092 | ±0.0011 | 0 |
| 6 | != | 5.421374 | 1.094645 | 0.0005 | ±0.0011 | 0 | 15 | != | 5.410615 | 1.028593 | -0.0103 | ±0.0010 | 4608 |
| 6 | >= | 5.420922 | 1.058717 | 0.0000 | ±0.0011 | 0 | 15 | >= | 5.448421 | 1.090379 | 0.0275 | ±0.0011 | 0 |
| 6 | > | 5.420671 | 1.048630 | -0.0002 | ±0.0010 | 0 | 15 | > | 5.410818 | 1.030977 | -0.0101 | ±0.0010 | 4607 |
| 7 | < | 5.716858 | 1.387539 | 0.2960 | ±0.0014 | 0 | 16 | < | 5.414233 | 1.035495 | -0.0067 | ±0.0010 | 0 |
| 7 | <= | 5.641402 | 1.279383 | 0.2205 | ±0.0013 | 4608 | 16 | <= | 5.426977 | 1.059495 | 0.0061 | ±0.0011 | 0 |
| 7 | == | 5.643428 | 1.283482 | 0.2225 | ±0.0013 | 4607 | 16 | == | 5.407243 | 1.030172 | -0.0137 | ±0.0010 | 4607 |
| 7 | != | 5.200235 | 0.626614 | -0.2207 | ±0.0006 | 0 | 16 | != | 5.434739 | 1.066421 | 0.0138 | ±0.0011 | 0 |
| 7 | >= | 5.125798 | 0.485354 | -0.2951 | ±0.0005 | 0 | 16 | >= | 5.427646 | 1.060224 | 0.0067 | ±0.0011 | 4608 |
| 7 | > | 5.199680 | 0.624128 | -0.2212 | ±0.0006 | 0 | 16 | > | 5.414871 | 1.035411 | -0.0060 | ±0.0010 | 0 |
| 8 | < | 5.482451 | 1.101441 | 0.0616 | ±0.0011 | 0 | 17 | < | 5.404515 | 1.012176 | -0.0164 | ±0.0010 | 0 |
| 8 | <= | 5.437518 | 1.066325 | 0.0166 | ±0.0011 | 4607 | 17 | <= | 5.438065 | 1.070895 | 0.0172 | ±0.0011 | 0 |
| 8 | == | 5.436342 | 1.063603 | 0.0154 | ±0.0011 | 4608 | 17 | == | 5.430419 | 1.063981 | 0.0095 | ±0.0011 | 0 |
| 8 | != | 5.406059 | 1.027887 | -0.0148 | ±0.0010 | 0 | 17 | != | 5.410146 | 1.028519 | -0.0108 | ±0.0010 | 4608 |
| 8 | >= | 5.358139 | 0.993148 | -0.0628 | ±0.0010 | 0 | 17 | >= | 5.439023 | 1.080056 | 0.0181 | ±0.0011 | 0 |
| 8 | > | 5.404727 | 1.026829 | -0.0162 | ±0.0010 | 0 | 17 | > | 5.403619 | 1.026794 | -0.0173 | ±0.0010 | 4607 |
| 9 | < | 5.467923 | 1.101697 | 0.0470 | ±0.0011 | 0 | | | | | | | |
| 9 | <= | 5.406438 | 1.051033 | -0.0145 | ±0.0011 | 4607 | | | | | | | |
| 9 | == | 5.419790 | 1.052137 | -0.0011 | ±0.0011 | 4608 | | | | | | | |
| 9 | != | 5.420790 | 1.035139 | -0.0001 | ±0.0011 | 0 | | | | | | | |
| 9 | >= | 5.373288 | 0.997365 | -0.0476 | ±0.0010 | 0 | | | | | | | |
| 9 | > | 5.437700 | 1.045860 | 0.0168 | ±0.0010 | 0 | | | | | | | |

of iterations. At each hill climbing step, only non-worsening local moves are accepted. Both the hill-climber and mutation consider the immediate neighbourhood, i.e., a program which differs only in a single comparison.

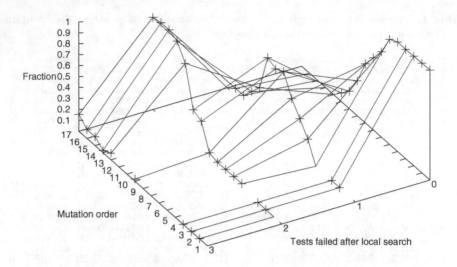

**Fig. 9.** Best fitness reached by local search hillclimber starting from high order mutations. + non-zero value. Up to $9^{\text{th}}$ order mutations, in most cases a program which passes all the tests can be found. (Data by enumerating $1$–$4^{\text{th}}$ order mutations and sampling others, more than $1.5\ 10^{12}$ fitness evaluations.)

Edges are directed and based on the mutation operation. There is an *escape edge* from local optimum $i$ to local optimum $j$ if $j$ is obtained from the mutation of $i$ followed by hill-climbing. The *local optima network* is the graph where the nodes are the local optima and the edges are the escape edges.

To sample the local optima, we ran an ILS, from 1000 random starting points, with a budget of 10 000 iterations. This generated 2 372 805 unique local minima. Figure 10 presents the proportion of those runs that find a solution that passes all test cases. Figure 10 shows that in many cases the landscape can be easily traversed to reach solutions that pass all test cases.

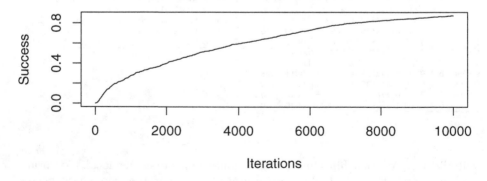

**Fig. 10.** Proportion of 1000 Iterated Local Search (ILS) runs that find a solution that passes all test cases versus the number of iterations.

In Fig. 11, the fitness and Hamming distance of each local minima to the unmutated Triangle Program is presented as a sunflower plot. We can observe that solutions need to maintain some similarity to the original program to pass all test cases. In addition, almost all local optima pass >50% of the test cases.

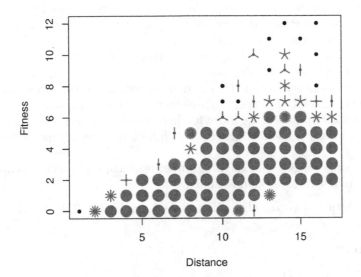

**Fig. 11.** Fitness vs Hamming distance of 2 372 805 unique local minima presented as a sunflower plot. The number of petals is proportional to the number of points at each coordinate. Note positive correlation between local optima fitness and distance to the original program. Also many mutants that differ from the original program in up to 12 out of 17 mutation points pass all test cases.

Figure 1 (page 1) shows the local optima network obtained from observing a subset of the sampling described before. We only consider 100 ILS runs and their first 1000 iterations. Since this process generated more than 60 000 nodes and edges, only the edges are plotted for the sake of clarity. Edges are coloured if they start and end at the same fitness. Other edges are painted black. The points are positioned in the $x$-$y$ plane using a force-directed layout algorithm and the fitness is used for the $z$ axis [67]. The fact that many nodes are at the same level does not necessarily mean that they are part of the same plateau. Nevertheless, we can observe that the fitness levels for 2, 3 and 5 failed test cases are densely populated. In addition, the numerous clear paths between fitness levels provide visual evidence of the relative ease with which the network can be traversed.

## 5   Conclusions

Although the Triangle Program is small, the number of possible Triangle Programs is huge. We have fully explored a regular subset of it. We reduced the size of its search space by considering only potential improvements to the existing code made by replacing its comparisons. Firstly we restricted the comparator mutations. This enabled us to analyse a systematic subset of the whole improvement fitness landscape. Solutions in the subset are still solutions in the full problem. There are many solutions all of which are readily found by high order schema analysis. Since we use only the number of tests passed there are few fitness levels and as expected there are large plateaux of neutral moves.

Secondly we returned to allowing all possible comparisons. This greatly increases the size of the search space. By allowing more moves we are further from the human written starting point and now schema analysis suggests global search (e.g. via a genetic algorithm) may be deceived. On the other hand, local search (including neutral moves) remains easy near the start point but its chance of finding a solution falls as we move further from the human code. Nonetheless even from a totally random starting point, hill climbing can improve test based fitness.

These results suggest that the program improvement fitness landscape is not as difficult to search as is often assumed.

**Datasets.** http://www.cs.ucl.ac.uk/staff/W.Langdon/egp2017/triangle/.

# References

1. Langdon, W.B.: Genetically improved software. In: Gandomi, A.H., et al. (eds.) Handbook of Genetic Programming Applications, pp. 181–220. Springer, New York (2015). http://www.cs.bham.ac.uk/~wbl/biblio/gp-html/langdon_2015_hbgpa.html
2. Langdon, W.B.: Genetic improvement of software for multiple objectives. In: Barros, M., Labiche, Y. (eds.) SSBSE 2015. LNCS, vol. 9275, pp. 12–28. Springer, Heidelberg (2015). doi:10.1007/978-3-319-22183-0_2. http://www.cs.bham.ac.uk/~wbl/biblio/gp-html/Langdon_2015_SSBSE.html
3. Petke, J.: Preface to the special issue on genetic improvement. Genet. Program. Evolvable Mach. (2017). Editorial Note, http://www.cs.bham.ac.uk/~wbl/biblio/gp-html/Petke_2016_GPEM.html
4. Harman, M., Jones, B.F.: Search based software engineering. Inf. Softw. Technol. **43**(14), 833–839 (2001). http://dx.doi.org/10.1016/S0950-5849(01)00189-6
5. Koza, J.R.: Genetic Programming: On the Programming of Computers by Natural Selection. MIT Press, Cambridge (1992). http://www.cs.bham.ac.uk/~wbl/biblio/gp-html/koza_book.html
6. Banzhaf, W., Nordin, P., Keller, R.E., Francone, F.D.: Genetic Programming: An Introduction On the Automatic Evolution of Computer Programs and its Applications. Morgan Kaufmann, San Francisco (1998). http://www.cs.bham.ac.uk/~wbl/biblio/gp-html/banzhaf_1997_book.html
7. Poli, R., Langdon, W.B., McPhee, N.F.: A field guide to genetic programming. Published via http://lulu.com and freely available at http://www.gp-field-guide.org.uk (2008). (With contributions by Koza, J.R.), http://www.cs.bham.ac.uk/~wbl/biblio/gp-html/poli08_fieldguide.html

8. Arcuri, A., Yao, X.: A novel co-evolutionary approach to automatic software bug fixing. In: Wang, J. (ed.) 2008 IEEE World Congress on Computational Intelligence, Hong Kong, pp. 162–168. IEEE (2008). http://www.cs.bham.ac.uk/~wbl/biblio/gp-html/Arcuri_2008_cec.html

9. Weimer, W., Nguyen, T., Le Goues, C., Forrest, S.: Automatically finding patches using genetic programming. In: Fickas, S. (ed.) 2009 International Conference on Software Engineering (ICSE), Vancouver, pp. 364–374 (2009). http://www.cs.bham.ac.uk/~wbl/biblio/gp-html/Weimer_2009_ICES.html

10. Forrest, S., Nguyen, T., Weimer, W., Le Goues, C.: A genetic programming approach to automated software repair. In: Raidl, G., et al. (eds.) GECCO, Montreal, pp. 947–954. ACM (2009). Best paper, http://www.cs.bham.ac.uk/~wbl/biblio/gp-html/DBLP_conf_gecco_ForrestNWG09.html

11. Weimer, W., Forrest, S., Le Goues, C., Nguyen, T.: Automatic program repair with evolutionary computation. Commun. ACM. **53**(5), 109–116 (2010). http://www.cs.bham.ac.uk/~wbl/biblio/gp-html/Weimer_2010_ACM.html

12. Le Goues, C., Dewey-Vogt, M., Forrest, S., Weimer, W.: A systematic study of automated program repair: fixing 55 out of 105 bugs for \$8 each. In: Glinz, M. (ed.) 34th International Conference on Software Engineering (ICSE 2012), Zurich, pp. 3–13 (2012). http://www.cs.bham.ac.uk/~wbl/biblio/gp-html/LeGoues_2012_ICSE.html

13. Le Goues, C., Nguyen, T., Forrest, S., Weimer, W.: GenProg: a generic method for automatic software repair. IEEE Trans. Softw. Eng. **38**(1), 54–72 (2012). http://www.cs.bham.ac.uk/~wbl/biblio/gp-html/DBLP_journals_tse_GouesNFW12.html

14. Le Goues, C., Forrest, S., Weimer, W.: Current challenges in automatic software repair. Softw. Qual. J. **21**, 421–443 (2013). http://www.cs.bham.ac.uk/~wbl/biblio/gp-html/legouesWFSQJO2013.html

15. Ke, Y., Stolee, K.T., Le Goues, C., Brun, Y.: Repairing programs with semantic code search. In: Grunske, L., Whalen, M. (eds.) 30th IEEE/ACM International Conference on Automated Software Engineering (ASE 2015), Lincoln, Nebraska, USA (2015). http://www.cs.bham.ac.uk/~wbl/biblio/gp-html/Ke_2015_ASE.html

16. Kocsis, Z.A., Drake, J.H., Carson, D., Swan, J.: Automatic improvement of Apache Spark queries using semantics-preserving program reduction. In: Petke, J., et al. (eds.) 2016 Workshop on Genetic Improvement, Denver, pp. 1141–1146. ACM (2016). http://www.cs.bham.ac.uk/~wbl/biblio/gp-html/Kocsis_2016_GI.html

17. Petke, J., Harman, M., Langdon, W.B., Weimer, W.: Using genetic improvement and code transplants to specialise a C++ program to a problem class. In: Nicolau, M., Krawiec, K., Heywood, M.I., Castelli, M., García-Sánchez, P., Merelo, J.J., Rivas Santos, V.M., Sim, K. (eds.) EuroGP 2014. LNCS, vol. 8599, pp. 137–149. Springer, Heidelberg (2014). doi:10.1007/978-3-662-44303-3_12. http://www.cs.bham.ac.uk/~wbl/biblio/gp-html/Petke_2014_EuroGP.html

18. Marginean, A., Barr, E.T., Harman, M., Jia, Y.: Automated transplantation of call graph and layout features into kate. In: Barros, M., Labiche, Y. (eds.) SSBSE 2015. LNCS, vol. 9275, pp. 262–268. Springer, Heidelberg (2015). doi:10.1007/978-3-319-22183-0_21. http://www.cs.bham.ac.uk/~wbl/biblio/gp-html/Marginean_2015_SSBSE.html

19. Barr, E.T., Harman, M., Jia, Y., Marginean, A., Petke, J.: Automated software transplantation. In: Xie, T., Young, M. (eds.) International Symposium on Software Testing and Analysis, ISSTA 2015, Baltimore, Maryland, USA, pp. 257–269. ACM (2015). ACM SIGSOFT Distinguished Paper Award, http://www.cs.bham.ac.uk/~wbl/biblio/gp-html/Barr_2015_ISSTA.html

20. Harman, M., Jia, Y., Langdon, W.B.: Babel Pidgin: SBSE can grow and graft entirely new functionality into a real world system. In: Goues, C., Yoo, S. (eds.) SSBSE 2014. LNCS, vol. 8636, pp. 247–252. Springer, Heidelberg (2014). doi:10.1007/978-3-319-09940-8_20. http://www.cs.bham.ac.uk/~wbl/biblio/gp-html/Harman_2014_Babel.html

21. Jia, Y., Harman, M., Langdon, W.B., Marginean, A.: Grow and serve: growing django citation services using SBSE. In: Barros, M., Labiche, Y. (eds.) SSBSE 2015. LNCS, vol. 9275, pp. 269–275. Springer, Heidelberg (2015). doi:10.1007/978-3-319-22183-0_22. http://www.cs.bham.ac.uk/~wbl/biblio/gp-html/jia_2015_gsgp.html

22. Langdon, W.B., White, D.R., Harman, M., Jia, Y., Petke, J.: API-constrained genetic improvement. In: Sarro, F., Deb, K. (eds.) SSBSE 2016. LNCS, vol. 9962, pp. 224–230. Springer, Heidelberg (2016). doi:10.1007/978-3-319-47106-8_16. http://www.cs.bham.ac.uk/~wbl/biblio/gp-html/Langdon_2016_SSBSE.html

23. Langdon, W.B., Harman, M.: Optimising existing software with genetic programming. IEEE Trans. Evol. Comput. 19(1), 118–135 (2015). http://www.cs.bham.ac.uk/~wbl/biblio/gp-html/Langdon_2013_ieeeTEC.html

24. Langdon, W.B., Lam, B.Y.H., Modat, M., Petke, J., Harman, M.: Genetic improvement of GPU software. Genet. Program. Evolvable Mach. (2017). Online first, http://www.cs.bham.ac.uk/~wbl/biblio/gp-html/Langdon_2016_GPEM.html

25. White, D.R., Arcuri, A., Clark, J.A.: Evolutionary improvement of programs. IEEE Trans. Evol. Comput. 15(4), 515–538 (2011). http://www.cs.bham.ac.uk/~wbl/biblio/gp-html/White_2011_ieeeTEC.html

26. Bruce, B.R., Petke, J., Harman, M.: Reducing energy consumption using genetic improvement. In: Silva, S., et al. (eds.) GECCO, Madrid, Spain, ACM, pp. 1327–1334 ACM (2015). http://www.cs.bham.ac.uk/~wbl/biblio/gp-html/bruce2015reducing.html

27. Bruce, B.R.: Energy optimisation via genetic improvement a SBSE technique for a new era in software development. In: Langdon, W.B., et al. (eds.) 2015 Workshop on Genetic Improvement, Madrid, pp. 819–820. ACM (2015). http://www.cs.bham.ac.uk/~wbl/biblio/gp-html/Bruce_2015_gi.html

28. Burles, N., Bowles, E., Brownlee, A.E.I., Kocsis, Z.A., Swan, J., Veerapen, N.: Object-oriented genetic improvement for improved energy consumption in Google Guava. In: Barros, M., Labiche, Y. (eds.) SSBSE 2015. LNCS, vol. 9275, pp. 255–261. Springer, Heidelberg (2015). doi:10.1007/978-3-319-22183-0_20. http://www.cs.bham.ac.uk/~wbl/biblio/gp-html/Burles_2015_SSBSE.html

29. Burles, N., Bowles, E., Bruce, B.R., Srivisut, K.: Specialising Guava's cache to reduce energy consumption. In: Barros, M., Labiche, Y. (eds.) SSBSE 2015. LNCS, vol. 9275, pp. 276–281. Springer, Heidelberg (2015). doi:10.1007/978-3-319-22183-0_23. http://www.cs.bham.ac.uk/~wbl/biblio/gp-html/Burles_2015_SSBSEa.html

30. Bokhari, M., Wagner, M.: Optimising energy consumption heuristically on android mobile phones. In: Petke, J., et al. (eds.) Genetic Improvement 2016 Workshop, Denver, pp. 1139–1140. ACM (2016). http://www.cs.bham.ac.uk/~wbl/biblio/gp-html/Bokhari_2016_GI.html

31. Haraldsson, S.O., Woodward, J.R.: Genetic improvement of energy usage is only as reliable as the measurements are accurate. In: Langdon, W.B., et al. (eds.) Genetic Improvement 2015 Workshop, Madrid, pp. 831–832. ACM (2015). http://www.cs.bham.ac.uk/~wbl/biblio/gp-html/Haraldsson_2015_gi.html

32. Langdon, W.B., Petke, J., Bruce, B.R.: Optimising quantisation noise in energy measurement. In: Handl, J., Hart, E., Lewis, P.R., López-Ibáñez, M., Ochoa, G., Paechter, B. (eds.) PPSN 2016. LNCS, vol. 9921, pp. 249–259. Springer, Heidelberg (2016). doi:10.1007/978-3-319-45823-6_23. http://www.cs.bham.ac.uk/~wbl/biblio/gp-html/Langdon_2016_PPSN.html

33. Schulte, E., Dorn, J., Harding, S., Forrest, S., Weimer, W.: Post-compiler software optimization for reducing energy. In: Proceedings of the 19th International Conference on Architectural Support for Programming Languages and Operating Systems, ASPLOS 2014, Salt Lake City, Utah, USA, pp. 639–652. ACM (2014). http://www.cs.bham.ac.uk/~wbl/biblio/gp-html/schulte2014optimization.html

34. Wagner, M.: Speeding up the proof strategy in formal software verification. In: Petke, J., et al. (eds.) Genetic Improvement 2016 Workshop, Denver, pp. 1137–1138 ACM (2016). http://www.cs.bham.ac.uk/~wbl/biblio/gp-html/Wagner_2016_GI.html

35. Wu, F., Weimer, W., Harman, M., Jia, Y., Krinke, J.: Deep parameter optimisation. In: Silva, S., et al. (eds.) GECCO, Madrid, pp. 1375–1382. ACM (2015). http://www.cs.bham.ac.uk/~wbl/biblio/gp-html/Wu_2015_GECCO.html

36. Walsh, P., Ryan, C.: Automatic conversion of programs from serial to parallel using genetic programming - the paragen system. In: D'Hollander, E.H., et al. (eds.) Proceedings of ParCo 1995, Volume 11 of Advances in Parallel Computing, Gent, Belgium, pp. 415–422. Elsevier (1995). http://www.cs.bham.ac.uk/~wbl/biblio/gp-html/ryan_1995_paragen.html

37. Williams, K.P.: Evolutionary algorithms for automatic parallelization. Ph.D. thesis, Department of Computer Science, University of Reading, Whiteknights Campus, Reading (1998). http://www.cs.bham.ac.uk/~wbl/biblio/gp-html/williams98.html

38. Williams, K.P., Williams, S.A.: Genetic compilers: a new technique for automatic parallelisation. In: 2nd European School of Parallel Programming Environments (ESPPE 1996), L'Alpe d'Hoez, France, pp. 27–30 (1996). http://citeseerx.ist.psu.edu/viewdoc/summary?doi=10.1.1.49.3499

39. Langdon, W.B., Harman, M.: Evolving a CUDA kernel from an nVidia template. In: Sobrevilla, P., (ed.) 2010 IEEE World Congress on Computational Intelligence, Barcelona, pp. 2376–2383. IEEE (2010). http://www.cs.bham.ac.uk/~wbl/biblio/gp-html/langdon_2010_cigpu.html

40. White, D.R., Clark, J., Jacob, J., Poulding, S.M.: Searching for resource-efficient programs: low-power pseudorandom number generators. In: Keijzer, M. et al., (eds.) GECCO, Atlanta, GA, USA, pp. 1775–1782. ACM (2008). http://www.cs.bham.ac.uk/~wbl/biblio/gp-html/White2_2008_gecco.html

41. White, D.R.: Genetic programming for low-resource systems. Ph.D. thesis, Department of Computer Science, University of York, UK (2009). http://www.cs.bham.ac.uk/~wbl/biblio/gp-html/White_thesis.html

42. Yeboah-Antwi, K., Baudry, B.: Embedding adaptivity in software systems using the ECSELR framework. In: Langdon, W.B., et al. (eds.): Genetic Improvement 2015 Workshop, Madrid, pp. 839–844. ACM (2015). http://www.cs.bham.ac.uk/~wbl/biblio/gp-html/Yeboah-Antwi_2015_gi.html

43. Mrazek, V., Vasicek, Z., Sekanina, L.: Evolutionary approximation of software for embedded systems: median function. In: Langdon, W.B., et al. (eds.) Genetic Improvement 2015 Workshop, Madrid, pp. 795–801. ACM (2015). http://www.cs.bham.ac.uk/~wbl/biblio/gp-html/Mrazek_2015_gi.html

44. Burles, N., Swan, J., Bowles, E., Brownlee, A.E.I., Kocsis, Z.A., Veerapen, N.: Embedded dynamic improvement. In: Langdon, W.B., et al. (eds.) Genetic Improvement 2015 Workshop, Madrid, pp. 831–832. ACM (2015). http://www.cs. bham.ac.uk/~wbl/biblio/gp-html/Swan_2015_gi.html

45. Vasicek, Z., Mrazek, V.: Trading between quality and non-functional properties of median filter in embedded systems. Genet. Program. Evolvable Mach. (2017). Online first, http://www.cs.bham.ac.uk/~wbl/biblio/gp-html/Vasicek_2016_GPEMa.html

46. Petke, J.: Genetic improvement for code obfuscation. In: Petke, J., et al. (eds.) Genetic Improvement 2016 Workshop, Denver, pp. 1135–1136. ACM (2016). http://www.cs.bham.ac.uk/~wbl/biblio/gp-html/Petke_2016_GI.html

47. Harman, M., Jia, Y., Langdon, W.B., Petke, J., Moghadam, I.H., Yoo, S., Wu, F.: Genetic improvement for adaptive software engineering. In: Engels, G. (ed.) 9th International Symposium on Software Engineering for Adaptive and Self-Managing Systems (SEAMS 2014), Hyderabad, India, pp. 1–4. ACM (2014). Keynote, http:// www.cs.bham.ac.uk/~wbl/biblio/gp-html/Harman_2014_seams.html

48. Landsborough, J., Harding, S., Fugate, S.: Removing the kitchen sink from software. In: Langdon, W.B., et al. (eds.) Genetic Improvement 2015 Workshop, Madrid, pp. 833–838. ACM (2015). http://www.cs.bham.ac.uk/~wbl/biblio/gp-html/Landsborough_2015_gi.html

49. Harman, M., Jia, Y., Krinke, J., Langdon, W.B., Petke, J., Zhang, Y.: Search based software engineering for software product line engineering: a survey and directions for future work. In: 18th International Software Product Line, SPLC 2014, Florence, Italy, pp. 5–18 (2014). Invited keynote, http://www.cs.bham.ac. uk/~wbl/biblio/gp-html/Harman_2014_SPLC.html

50. Lopez-Herrejon, R.E., Linsbauer, L., Assuncao, W.K.G., Fischer, S., Vergilio, S.R., Egyed, A.: Genetic improvement for software product lines: an overview and a roadmap. In: Langdon, W.B., et al. (eds.) Genetic Improvement 2015 Workshop, Madrid, pp. 823–830. ACM (2015). http://www.cs.bham.ac.uk/~wbl/biblio/gp-html/Lopez-Herrejon_2015_gi.html

51. Langdon, W.B., Petke, J., White, D.R.: Genetic improvement 2015 chairs' welcome. In: Langdon, W.B., et al. (eds.) Genetic Improvement 2015 Workshop, Madrid, pp. 791–792. ACM (2015). http://www.cs.bham.ac.uk/~wbl/biblio/gp-html/langdon_2015_gi.html

52. Langdon, W.B., Harman, M.: Fitness landscape of the triangle program. In: Veerapen, N., Ochoa, G. (eds.) PPSN-2016 Workshop on Landscape-Aware Heuristic Search, Edinburgh (2016). Also available as UCL RN/16/05, http://www.cs.bham. ac.uk/~wbl/biblio/gp-html/langdon_2016_PPSNlandscape.html

53. Langdon, W.B., Petke, J.: Software is not fragile. In: Parrend, P., et al. (eds.) CS-DC 2015, pp. 203–211. Springer, Cham (2015). Invited talk, http://www.cs.bham. ac.uk/~wbl/biblio/gp-html/langdon_2015_csdc.html

54. Jia, Y., Harman, M.: An analysis and survey of the development of mutation testing. IEEE Trans. Softw. Eng. 37(5), 649–678 (2011)

55. Langdon, W.B., Harman, M., Jia, Y.: Efficient multi-objective higher order mutation testing with genetic programming. J. Syst. Softw. 83(12), 2416–2430 (2010). http://www.cs.bham.ac.uk/~wbl/biblio/gp-html/langdon_2010_jss.html

56. Cody-Kenny, B., Lopez, E.G., Barrett, S.: locoGP: improving performance by genetic programming Java source code. In: Langdon, W.B., et al. (eds.) Genetic Improvement 2015 Workshop, Madrid, pp. 811–818. ACM (2015). http://www.cs. bham.ac.uk/~wbl/biblio/gp-html/Cody-Kenny_2015_gi.html

57. Orlov, M., Sipper, M.: Flight of the FINCH through the Java wilderness. IEEE Trans. Evol. Comput. **15**(2), 166–182 (2011). http://www.cs.bham.ac.uk/~wbl/biblio/gp-html/Orlov_2011_ieeeTEC.html
58. Schulte, E., Forrest, S., Weimer, W.: Automated program repair through the evolution of assembly code. In: Proceedings of the IEEE/ACM International Conference on Automated Software Engineering, Antwerp, pp. 13–316. ACM (2010). http://www.cs.bham.ac.uk/~wbl/biblio/gp-html/schulte10_autom_progr_repair_evolut_assem_code.html
59. Schulte, E., Fry, Z.P., Fast, E., Weimer, W., Forrest, S.: Software mutational robustness. Genet. Program. Evolvable Mach. **15**(3), 281–312 (2014). http://www.cs.bham.ac.uk/~wbl/biblio/gp-html/Schulte_2014_GPEM.html
60. Schulte, E., Weimer, W., Forrest, S.: Repairing COTS router firmware without access to source code or test suites: a case study in evolutionary software repair. In: Langdon, W.B., et al. (eds.) Genetic Improvement 2015 Workshop, Madrid, pp. 847–854. ACM (2015). Best Paper, http://www.cs.bham.ac.uk/~wbl/biblio/gp-html/Schulte_2015_gi.html
61. Wright, S.: The roles of mutation, inbreeding, crossbreeding and selection in evolution. In: Proceedings of the Sixth Annual Congress of Genetics, pp. 356–366 (1932). http://www.blackwellpublishing.com/ridley/classictexts/wright.pdf
62. Reidys, C.M., Stadler, P.F.: Combinatorial landscapes. SIAM Rev. **44**(1), 3–54 (2002). http://dx.doi.org/10.1137/S0036144501395952
63. Holland, J.H.: Genetic algorithms and the optimal allocation of trials. SIAM J. Comput. **2**, 88–105 (1973). http://dx.doi.org/10.1137/0202009
64. Goldberg, D.E.: Genetic Algorithms in Search Optimization and Machine Learning. Addison-Wesley, Boston (1989)
65. Daolio, F., Tomassini, M., Verel, S., Ochoa, G.: Communities of minima in local optima networks of combinatorial spaces. Phys. A: Stat. Mech. Appl. **390**(9), 1684–1694 (2011). http://dx.doi.org/10.1016/j.physa.2011.01.005
66. Ochoa, G., Verel, S., Daolio, F., Tomassini, M.: Local optima networks: a new model of combinatorial fitness landscapes. In: Richter, H., Engelbrecht, A. (eds.) Recent Advances in the Theory and Application of Fitness Landscapes, pp. 233–262. Springer, Berlin (2014). http://dx.doi.org/10.1007/978-3-642-41888-4_9
67. Ochoa, G., Veerapen, N.: Additional dimensions to the study of funnels in combinatorial landscapes. In: GECCO, pp. 373–380. ACM (2016) http://dx.doi.org/10.1145/2908812.2908820

# RANSAC-GP: Dealing with Outliers in Symbolic Regression with Genetic Programming

Uriel López[1], Leonardo Trujillo[1(✉)], Yuliana Martinez[1], Pierrick Legrand[2,3,4],
Enrique Naredo[5], and Sara Silva[6,7]

[1] Posgrado en Ciencias de la Ingenieria, Instituto Tecnológico de Tijuana,
Unidad Otay, Blvd. Industrial, Ave. ITR. Tijuana S/N, Mesa de Otay,
C.P. 22500 Tijuana, B.C., Mexico
{uriel.lopez,leonardo.trujillo}@tectijuana.edu.mx, ysaraimr@gmail.com
[2] University of Bordeaux, 3ter Place de la Victoire, 33076 Bordeaux, France
pierrick.legrand@u-bordeaux.fr
[3] IMB, UMR CNRS 5251, 351 Cours de la Libération, 33405 Talence, France
[4] INRIA Bordeaux Sud-Ouest, 200 Rue Vieille Tour, 33405 Talence, France
[5] Laboratorio Nacional de Geointeligencia (GeoINT),
Centro de Investigación en Geografía y Geomática (CentroGeo),
Aguascalientes, Mexico
enaredo@centrogeo.edu.mx
[6] BioISI - Biosystems & Integrative Sciences Institute,
Departamento de Informática, Faculdade de Ciências,
Universidade de Lisboa, 1749-016 Lisboa, Portugal
sara@fc.ul.pt
[7] CISUC, Department of Informatics Engineering, University of Coimbra,
Coimbra, Portugal

**Abstract.** Genetic programming (GP) has been shown to be a powerful tool for automatic modeling and program induction. It is often used to solve difficult symbolic regression tasks, with many examples in real-world domains. However, the robustness of GP-based approaches has not been substantially studied. In particular, the present work deals with the issue of outliers, data in the training set that represent severe errors in the measuring process. In general, a datum is considered an outlier when it sharply deviates from the true behavior of the system of interest. GP practitioners know that such data points usually bias the search and produce inaccurate models. Therefore, this work presents a hybrid methodology based on the RAndom SAmpling Consensus (RANSAC) algorithm and GP, which we call RANSAC-GP. RANSAC is an approach to deal with outliers in parameter estimation problems, widely used in computer vision and related fields. On the other hand, this work presents the first application of RANSAC to symbolic regression with GP, with impressive results. The proposed algorithm is able to deal with extreme amounts of contamination in the training set, evolving highly accurate models even when the amount of outliers reaches 90%.

**Keywords:** Genetic programming · RANSAC · Robust regression · Outliers

© Springer International Publishing AG 2017
J. McDermott et al. (Eds.): EuroGP 2017, LNCS 10196, pp. 114–130, 2017.
DOI: 10.1007/978-3-319-55696-3_8

# 1   Introduction

One of the most common application domains of genetic programming (GP) is to solve regression problems (or real-valued learning problems), with an approach referred to as symbolic regression. Unlike other regression approaches, the search/learning process is not focused on determining the best fit parameters for a pre-specified model. The problem is stated more generally, such that GP searches for both the structure (symbolic expression) and the optimal parametrization of the model that best describes a set of learning or training data. Indeed, GP has produced a variety of successful results in this domain.

However, one problem that has not received an adequate amount of attention is the impact that outliers (gross errors) in the training set can have on the quality of the solution found. Furthermore, almost no research work has been devoted to developing GP-based symbolic regression that is robust to the presence of outliers in the training data. To the best of the authors knowledge, [8] is the only work that deals with the problem of outlier detection in a GP-based system, but does not propose a general approach for robust symbolic regression in such scenarios.

While robust regression, as is reviewed next, has been the focus of large amounts of work in standard regression literature, symbolic regression has not followed suit so far. In general, most symbolic regression research works under the assumption (even if not explicitly stated) that the input data is "clean" (without outliers), giving complete confidence on the error estimates of the evolved solutions with respect to the training data. This assumption is realistic, even in real-world scenarios. If a small amount of outliers are present in the data, then preprocessing or filtering approaches might be able to remove outliers from the training set before running the symbolic regression system. However, this assumption will fail when the contamination is severe, above what is commonly referred to as the breakdown point for a robust regression method, when 50% or more of the data is in fact outliers. Other issues that might limit the usefulness of pre-processing methods is when the data sampling is sparse and non-uniform, making it difficult to apply filters that require accurate estimates of local signal statistics.

Therefore, this work presents an initial study to fill this research gap, with the following main contributions. First, we explore the effect that outlier contamination has on the performance of symbolic regression models evolved with GP. In particular, we evaluate standard GP using a typical error measure, as well as robust error estimates that are widely used in linear regression tasks when outliers are present in the data. Moreover, we evaluate a recent set of fitness case sampling methods, that do not use all of the training data instances at each fitness evaluation. In all cases, we conclusively show that these approaches fail when the amount of outliers is large, particularly above the breakdown point. Second, we propose a hybrid approach for robust symbolic regression modeling with GP, based on the RANdom SAmpling Consensus (RANSAC) algorithm [3,18]. While RANSAC has been shown to be a very robust approach for parameter estimation [19], particularly popular in the computer vision community, used to determine the epi-polar geometry for stereo reconstruction [10,20]. However, despite its success the application of RANSAC for symbolic regression with GP has not been studied before. Third, the results presented in this paper are extremely encouraging, with our proposed hybrid RANSAC-GP algorithm identifying highly accurate models (with a testing error $\epsilon = 0.01$) even when data contamination greatly exceeds the breakdown point, with as much as 90% contamination of the training data with outliers.

The remainder of this paper proceeds as follows. Section 2 presents a quick overview of related background. An introduction to robust linear regression is presented in Sect. 3, while also discussing fitness case sampling methods that are hypothesized to be useful in dealing with outlier contamination in the training set. Then, the RANSAC algorithm and the proposed RANSAC-GP hybrid are presented in Sect. 4. Section 5 presents our experimental work and summarizes our main results. Finally, concluding remarks are given in Sect. 6.

## 2    Background

Let us begging by framing the basic regression task, where given a training dataset $\mathbb{T} = \{(\mathbf{x}_i, y_i); i = 1, \ldots, n\}$, the goal is to derive a model that predicts $y_i$ based on $\mathbf{x}_i$, where $\mathbf{x}_i \in \Re^p$ and $y_i \in \Re$. In GP literature we can refer to each input/output pair $(\mathbf{x}_i, y_i)$ as a fitness case. For linear regression, the model is expressed as

$$y_i = \beta_1 x_{i1} + \cdots + \beta_p x_{ip} + \varepsilon_i \quad i = 1, \ldots, n \tag{1}$$

where the model parameters $\beta = (\beta_1, \ldots, \beta_p) \in \Re^p$, are estimated by $\widehat{\beta}_1, \ldots, \widehat{\beta}_p$ using the least squares method [16], which can be expressed as

$$(\widehat{\beta}_1, \ldots, \widehat{\beta}_p) \leftarrow \arg\min_{\beta \in \Re^p} \sum_{i=1}^{n} r_i^2, \tag{2}$$

to find the best fit parameters of the linear model, where $r_i$ denotes the residuals $r_i(\widehat{\beta}_1, \ldots, \widehat{\beta}_p) = y_i - (\widehat{\beta}_1 x_{i1} + \cdots + \widehat{\beta}_p x_{ip})$ and the errors $\varepsilon_i$ have an expected value of zero [1]; if the summation in Eq. 2 is divided by $n$, the error measure that must be minimized is the mean squared error (MSE).

Conversely, the symbolic regression problem solved with standard GP[1] can be expressed as

$$K^o \leftarrow \arg\min_{K \in \mathbb{G}} f(K(x_i), y_i) \, with \, i = 1, \ldots, n, \tag{3}$$

where the goal is to find the best model $K^o$ that minimizes the error computed by the fitness function $f$ between the expected output $y_i$ and the estimate given by each model (program) $K(\mathbf{x}_i)$, with $\mathbb{G}$ representing the space of all possible models. In practice, $f$ is usually expressed by an error measure such as the one of Eq. 2 (MSE), or other error measures such as the mean absolute error (MAE) or the root mean squared error (RMSE).

---

[1] All results presented in this work are based on a standard GP implementation, using a tree representation, subtree crossover and subtree mutation.

## 2.1    Outliers

As stated above, GP has been shown to be very competitive in symbolic regression tasks, with many real-world examples and even commercial GP-based software tools. However, the effect that outliers have on GP performance has not been studied in depth. First, lets define outliers as follows:

**Definition 1.** *An outlier is a measurement of a system that is anomalous with respect to the behavior of the system.*

Unlike other definitions given in the literature [15], we do not focus on the dataset that results after a measuring session of a system of interest, and instead focus on the behavior of the actual system under observation. We do so because it is entirely possible that a given dataset might be severely corrupted with more than 50% of the data representing outliers. One may ask, if the outliers are a majority in a dataset, then are they truly outliers? That is why it is important to distinguish between a given measurement (observation) and the true value of a variable of interest. It is therefore reasonable for a dataset to be contaminated by a majority of outliers, which can be produced by several factors including measurement errors, equipment malfunction, human errors or missing data points. All these are common problems found in real-world settings, that are often left unaccounted for in most GP-based symbolic regression research.

In particular, for simplicity we will focus on unidimensional problems, such that $\mathbf{x}_i \in \Re$. In such cases, we can say that a fitness case $(x_i, y_i)$ is an outlier if

$$| y_i - y_o | > t\zeta \tag{4}$$

where $y_i$ is the value to be characterized, $y_o$ is a reference value, $\zeta$ is a measure of data variation, and $t$ is a user defined threshold that controls to what extent the deviation of a particular fitness case can be regarded as "anomalous" and therefore be considered to be an outlier [15]. For instance, Eq. 4 can be used to identify and remove outliers from a dataset using the Hampel identifier [15], where a moving window $W$ centered on $x_i$ is used to compute $y_o$ and $\zeta$. In particular, $y_o$ is set to the median of all $y_j$ in window $W$ and $\zeta$ is given by 1.4826 × MAD (Mean Absolute Deviation) within $W$. In the case of outlier removal, when $y_i$ is determined to be an outlier it is replaced by $y_o$ such that the new fitness case is now $(x_i, y_o)$. Filters such as the Hampel identifier can be used to pre-process a dataset before applying linear or symbolic regression, however there are several drawbacks. First, if the percentage of outliers is above 50% then the newly inserted value $y_o$ should not be used. Moreover, it is difficult to set the size of the moving window $W$ when the training set does not provide a uniform sampling of the independent variable. Finally, a bigger issue lies in the fact that such methods are not easily extended to multidimensional spaces.

Let us provide a quick example of the effect that even a small number of outliers can have on a simple symbolic regression problem, as depicted in Fig. 1. In this example, the training set is contaminated with 2 outliers, representing only 10% of the entire training set. Even with this small amount of outliers, we can see how the GP search is biased by the outliers and is unable to find the real underlying model of the data.

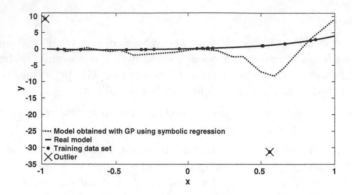

**Fig. 1.** Comparison of the model found by GP using symbolic regression (shown in dashed line) using a training set $\mathbb{T}$ (shown in dots) with two outliers (crosses), compared against the real model (shown in a solid line).

## 3   Robust Regression

In robust linear regression [5, 14–16] the most common approach to deal with outliers in the training data is to substitute the objective function of the least squares problem. In particular, measures such as the MSE or RMSE are very sensitive to outliers in the data. The sensitivity of a least squares method is measured by the breakdown point [15], which is reached when a certain percentage of outliers is present in the training set which produces an arbitrarily large bias in the final model. For instance, for standard linear least squares regression the breakdown point is 0%, a single outlier can induce arbitrarily large bias, which is also evidenced in GP symbolic regression as shown in Fig. 1. In fact, the highest breakdown point for a robust least squares method is 50% [1,5,14,15]. Therefore, in this work we will experimentally test two popular robust estimators (with a 50% breakdown point) on symbolic regression with GP; these measures are described next, based on [14].

*Least Median Squares.* The first robust measure is the Least Median Squares (LMS), given by

$$(\widehat{\beta}_1, \ldots, \widehat{\beta}_p) \leftarrow \underset{\beta_1, \ldots, \beta_p}{arg\ min}\ med\ \{r_1^2, \ldots, r_n^2\} \tag{5}$$

where *med* represents the median, such that the summation average of the MSE method is substituted by the median of the residuals. One attractive feature of this method is that it is relatively simple and efficient to implement, only requiring a sorting of the residuals.

*Least Trimmed Squares.* The second robust approach is the Least Trimmed Squares (LTS) minimization problem, given by

$$(\widehat{\beta_1}, \ldots, \widehat{\beta_p}) \leftarrow \underset{\beta_1, \ldots, \beta_p}{arg\ min} \sum_{i=1}^{hp} \{r_1^2, \ldots, r_n^2\}_{i:n}. \tag{6}$$

The method is called "trimmed" because it will search for the best combination (subsample), from among $\binom{n}{hp}$, of the complete training set with the smallest summation of least squares errors [5], where $hp$ is the trimmed proportion of the training set. This method achieves a breakdown point of 50% when $hp = \frac{n}{2} + 1$.

*Fitness Case Sampling Methods.* These methods can be roughly defined as those that use only a portion of the total fitness cases when computing the fitness of an individual, which can be done in different ways, including using a subset of fitness cases at every other generation [6], or by using a single fitness case when selecting individuals for reproduction [9,17]. While these methods have been derived for different purposes, they have been extensively evaluated recently in [11,12], showing that they can improve performance relative to a standard GP. Moreover, based on the LTS measure, in this work we hypothesize that they might be useful in dealing with outliers, since they rely on a similar general approach, to bias the search based on a subsample of the training data. In particular, the fitness case sampling methods tested in this work are[2]:

1. Interleaved Sampling (IS): Use all the fitness cases in every odd numbered generations, and use a randomly chosen fitness case otherwise [6].
2. Randon Interleaved Sampling (RIS): Uses all or one random fitness case based on a random decision at each generation [6].
3. Keep Worst-Interleaved Sampling (KW-IS): Use all the fitness cases in odd numbered generations and a subset of the most difficult ones on the rest [11].
4. Keep Best-Interleaved Sampling (KB-IS): Simliar to KW-IS, but instead focusing on the easiest fitness cases every other generation.
5. Lexicase Selection (Lex): Randomly order the fitness cases for each parent selection event, and then sequentially discard individuals based on each fitness case until a single individual is left [17].
6. $\epsilon$-Lexicase ($\epsilon$-Lex): Similar to Lex, but uses a threshold to compare individuals for real-valued symbolic regression [9].

# 4   Proposed RANSAC-GP

*Random Sample Consensus.* RANSAC is a random sampling algorithm that is used to solve problems where data contamination is expected to exceed the 50% breakdown point of standard robust regression methods. Originally proposed in [3] by Fischler and Bolles, it has become a standard technique in modern computer vision systems [20], widely used to solve complex regression problems. However, RANSAC was originally intended, and is currently used, for parameter estimation. It is of note that, to the authors knowledge, RANSAC is not widely

---

[2] An extensive discussion of these methods is beyond the scope of this work.

**Algorithm 1.** RANSAC pseudo-code.

1 Take a random $MSS_j$ of size $m$ from the training set $\mathbb{T}$
2 Build a model $K_j$ using the data in $MSS_j$.
3 Compute the residuals $r_j$ for all the data points in $\mathbb{T}$.
4 Build the consensus set $CS_j$ with all the data points in $\mathbb{T}$ for which $r_j < t$
5 If $|CS_j| \geq v$ then return $K_j$ as the final model.
6 Repeat $|1$ trough $4|$ until a maximum number of iterations, otherwise return $K_j$ with maximum $|CS_j|$.

used in areas outside computer vision, and in particular it has not been applied on symbolic regression tasks.

While achieving strong results in this difficult problem formulation (regression fitting with more than 50% of outliers), it is in fact a very simple and intuitive algorithm, with four user defined parameters. RANSAC assumes that while the training set can be heavily contaminated by outliers, it nonetheless contains sufficient inlier's so as to reconstruct the underlying model of the "true" data. To do so, it iteratively and randomly samples the training set, and uses each sample to build a model. The size of this sample set $m$ is the first RANSAC parameter, and the set is called the Minimal Sample Set ($MMS \subset \mathbb{T}$). It then evaluates the model with all remaining data in the training set ($\mathbb{T} \setminus MMS$), and computes what is referred to as the consensus set (CS); i.e. the set of all data points in $\mathbb{T} \setminus MMS$ that agree or are consistent with the particular model generated with the $MMS$. This is done by considering the residuals $r_j$, and marking a data point as an inlier if the corresponding residual falls below a threshold $t$, the second RANSAC parameter. This is done until the size of set $CS$ reaches the estimated total of inliers, the third user parameter $v$, or when a maximum number of iterations $l$ is reached, the fourth parameter. The entire pseudo-code of RANSAC [2,20] is given in Algorithm 1. We must stress that in this work we are considering the original RANSAC formulation. More recent and in some sense improved versions will be studied in future work, such as the M-estimator Sample Consensus (MSAC) or the Maximum Likelihood Estimation SAmple and Consensus (MLESAC) [19], or more recent methods such as the Optimal RANSAC algorithm [7].

### 4.1 Proposal

The proposal in this work is very straightforward, to build non-linear symbolic regression models with GP within RANSAC. Therefore, we only need to modify step 2 in Algorithm 1, where the model $K$ is derived using a standard GP search. Indeed, one of the most attractive aspects of RANSAC is the ease with which it can be adapted to other modeling approaches.

## 5 Experiments and Results

The goal of the experimental work is twofold. First, we want to experimentally evaluate standard GP using the MSE as a fitness function, as well as the reviewed robust estimators (LMS and LTS) and all the fitness case sampling methods (IS,

RIS, KW-IS, KB-IS, Lex and $\epsilon$-Lex), on benchmark problems contaminated with different amounts of outliers. We consider a wide range of contaminations, from 10% to 90% (in 10% increments), using relatively simple benchmark problems, to fully illustrate how even a small amount of outliers in the training set can bias the resulting model[3]. The common parameters of the GP system for all these experiments are summarized in Table 1.

**Table 1.** GP parameters used for the benchmark symbolic regression problems.

| Parameter | Description |
|---|---|
| Population size | 100 Individuals |
| Generations | 200 Generations |
| Initialization | *Ramped Half-and-Half*, with maximum depth level 6 |
| Operator probabilities | Crossover $p_c = 0.9$, Mutation $p_\mu = 0.1$ |
| Function set | $(+, -, \times, \div, sin, cos)$ |
| Terminal set | $x, randint(-1, 1)$ |
| Maximum tree depth | 17 levels |
| Selection | Tournament size 3 (except in Lex and $\epsilon$-Lex) |
| Elitism | Best individual always survives |

The second goal of our work is to show how the proposed RANSAC-GP method can easily handle large amounts of outliers in the training set. In particular, we will present a detailed analysis of the models found for the most difficult scenarios, when the contamination is at 90%. All of our experiments and algorithms were coded using the Distributed Evolutionary Algorithms in Python library (DEAP) [4], a Python library for evolutionary computation. However, we begin this section by explaining how the training data is constructed and artificially contaminated by outliers.

*Benchmark Problems.* For the experimental work of this paper, five benchmark problems are used from [13], described in Table 2. These problems are chosen for the following reasons. First, preliminary runs of our GP algorithm were able to consistently find optimal solutions with nearly perfect training and testing errors on all of them. This requirement was considered to be important for this work, to properly evaluate the effect that outlier contamination has on the performance of the symbolic regression task. If the modeling fails (large testing or generalization error), we want to be certain that this is due to the presence of outliers in the training set and not due to the underlying difficulty of the problem for the chosen GP algorithm. Second, since these are unidimensional problems, it is straightforward to properly contaminate the training set with outliers, as will be described next.

---

[3] For all practitioners, it is intuitively evident that 10% of outliers can have drastic effects in the modeling process.

**Table 2.** Benchmark problems used in this work, where U[a,b,c] denotes $c$ uniform random samples drawn from $a$ to $b$, that specifies how the training and testing sets are constructed, consisting solely of inliers.

| Objective function | Test set | Function set |
|---|---|---|
| $x^4 + x^3 + x^2 + x$ | U[-1, 1, 20] | U[-1, 1, 20] |
| $x^5 - 2x^3 + x$ | U[-1, 1, 20] | U[-1, 1, 20] |
| $x^3 + x^2 + x$ | U[-1, 1, 20] | U[-1, 1, 20] |
| $x^5 + x^4 + x^3 + x^2 + x$ | U[-1, 1, 20] | U[-1, 1, 20] |
| $x^6 + x^5 + x^4 + x^3 + x^2 + x$ | U[-1, 1, 20] | U[-1, 1, 20] |

*Training Set Contamination with Outliers.* The proposed approach to contaminate the data is to use the inverse of the Hample identifier discussed in Sect. 2.1 and defined in Eq. 4. In particular, to turn a particular fitness case $(x_i, y_i)$ into an outlier, we must first solve Eq. 4 for $y_i$, such that

$$y_i > y^o + t\zeta$$
$$\text{or} \quad y_i < y^o - t\zeta. \tag{7}$$

In particular, we randomly choose a percentage of fitness cases (for a different contamination percentage) and then randomly add or substract $\zeta$ from the ground truth $y_i$. The value of $t$ is set randomly within the range [10, 100] to guarantee a large amount of deviance from the original data, and $\zeta$ was computed by the median of all $y_i$ within the domain of each symbolic regression benchmark.

## 5.1   Results

The results are presented in two parts. First, we consider the testing performance given by the median MSE over 30 runs of the standard GP using the fitness functions MSE, robust fitness functions (LMS and LTS), and all of the sampling methods. The results are presented as plots of the median testing error relative to the amount of outlier contamination on each benchmark. Second, the results of the proposed RANSAC-GP are presented, focusing on the most extreme cases of outlier contamination.

*Robust Fitnesss Measures and Sampling Methods.* The results are summarized in Fig. 2, where MSE corresponds to standard GP with the MSE fitness function. The median performance (vertical axis) is plotted in a logarithmic scale, since the testing error of most methods reaches quite large values. In fact, most methods perform very poorly even for the smallest amount of outliers considered here (10%). It is clear that standard MSE and all the fitness case sampling methods are strongly biased by the presence of outliers. There are some good news though, when the contamination is equal or under 50% the robust fitness measures perform quite well. In particular, LTS shows the best performance of

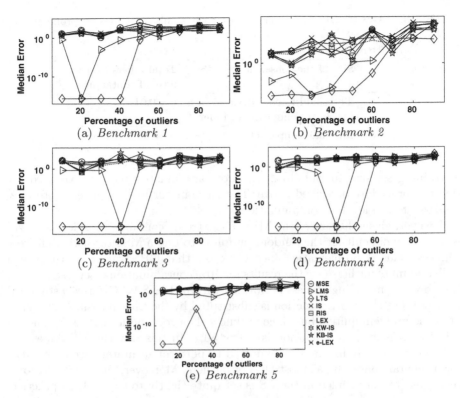

**Fig. 2.** Comparison of the median error of 30 runs for the five benchmarks at different levels of contamination for all methods.

all the methods, with LMS also achieving strong results. This behavior, however, does not hold above the breakdown point of 50%, above which all methods perform poorly.

Finally, it is important to consider that the LTS measure is computationally much more complex than the one used by LMS, since it needs to cycle across all possible combinations of the training set. Therefore, as a first conclusion, it is reasonable to state that when faced with a real-world symbolic regression problem that might be contaminated by outliers, it is preferable to use LTS only when the size of the training set is relatively small, otherwise LMS should be preferred. This can improve the chances of achieving an accurate and general model, but only when the contamination is expected to be below 50%.

*RANSAC-GP.* Here we present the results for the proposed RANSAC-GP algorithm, with some important considerations discussed first. Regarding the size $m$ of the MSS, it is assumed that the MSS should contain at least as much data as required by the modeling process to find an accurate model. For instance, for a linear model in a one dimensional problem two points would be sufficient. However, this is not defined for the studied benchmarks. Moreover, we increase

**Table 3.** RANSAC-GP parameters used for the symbolic regression problems.

| Parameter | Description | Value |
|---|---|---|
| $v$ | Size of the consensus set $CS$ | Total inliers for each level of contamination |
| $t$ | Threshold to consider a $r_j$ of a fitness training case an inlier | $t = 0.01$ |
| $m$ | Minimal sample set | $m = 50\%$ of inliers |

the training set size 10-fold (to 200), and contaminate this new extended set using the procedure described above. Therefore in our work, $m$ is equal to 50% of inliers at each level of contamination.

However, given the size of the MSS, for extreme contamination levels it will be extremely difficult for the random sampling to find a MSS consisting entirely of inliers. This is problematic, since we know that GP struggles with even a small amount of outliers (see the results for 10% contamination discussed above). Therefore, the modeling process performed within RANSAC-GP uses a standard GP search but the fitness function is substituted by the LMS measure. This simplifies the problem quite a bit, since we know that good performance is achieved by LMS when the contamination is below 50%. In this way the MSS does not need to contain only inliers, it is expected to perform accurately even when the MSS is contaminated by at most 50% of outliers. Moreover, the threshold for a fitness case to be included in the CS is set quite tightly to $t = 0.01$. In practice, such levels of accuracy might not be necessary but our intention is to test the algorithm in a difficult scenario. The configuration and parameters of RANSAC-GP are summarized in Table 3, and the rest of the parameters for the embedded GP search are the same as in Table 1 with the exception of the number of generations which is now set to 300 and the function set only includes arithmetic operators. Finally, for these experiments only a single execution of RANSAC-GP is reported for each level of data contamination, for the following reasons. First, we already know that GP is quite accurate and stable across multiple runs on widely used benchmarks, including when LMS is used as the fitness function (as shown above). Second, for higher levels of contamination the number of required iterations (with each iteration performing a single GP run on the MSS) can grow quite large, making experimental tests somewhat prohibitive. Nonetheless, even these runs provide a reasonable approximation of how the algorithm will perform on a real-world setting.

The stopping criterion is also based on the size of CS with $v$ being equal or greater than the total inliers, assuming that the MSS may contain up to 50% of outliers, forcing the algorithm to find the most accurate model possible. Table 4 summarizes the number of iterations required by RANSAC-GP to find a model based on the parametrization given in Table 3. Notice that when the contamination is below 50%, RANSAC-GP finds an accurate model in eight or fewer iterations, a nice result that will surely encourage this approach in practice. As the number of outliers increases, it is evident that the number of required iterations

**Table 4.** Iterations RANSAC-GP required to find a CS per level of contamination.

| Problems | % Outliers | | | | | | | | |
|---|---|---|---|---|---|---|---|---|---|
| | 10% | 20% | 30% | 40% | 50% | 60% | 70% | 80% | 90% |
| Benchmark 1 | 1 | 1 | 1 | 2 | 1 | 2 | 138 | 579 | 795 |
| Benchmark 2 | 1 | 1 | 2 | 2 | 6 | 72 | 494 | 456 | 892 |
| Benchmark 3 | 2 | 1 | 4 | 1 | 1 | 13 | 75 | 107 | 690 |
| Benchmark 4 | 1 | 2 | 2 | 3 | 8 | 16 | 563 | 530 | 907 |
| Benchmark 5 | 2 | 2 | 2 | 1 | 4 | 93 | 237 | 249 | 7066 |

(samples) also increases sharply, requiring several hundreds of iterations and in a single extreme case over 7,000. For some this might seem like an excessive price to pay for such scenarios, but in practice there are several important aspects to consider. First, a parallel implementation of RANSAC-GP can be easily derived, since each MSS is taken independently, such that several iterations can be performed in parallel. Moreover, when considering a highly contaminated training set the threshold $t$ for acceptance into the consensus set might reasonably be set much larger than in our experiments, allowing the algorithm to more quickly find a useful model.

Let us now turn to the modeling results on the most extreme case, where 90% of the training data are outliers, presented in Figs. 3, 4, 5, 6 and 7. Each figure shows three plots. First, the contaminated training set, which by visual

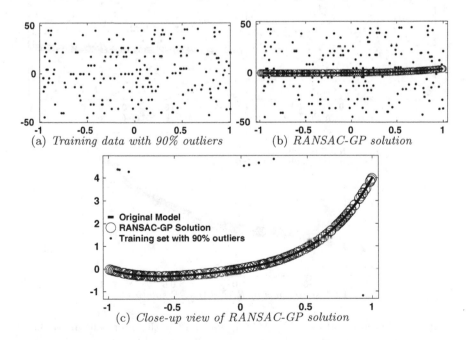

(a) *Training data with 90% outliers*        (b) *RANSAC-GP solution*

(c) *Close-up view of RANSAC-GP solution*

**Fig. 3.** Solution found by RANSAC-GP with 90% outliers for benchmark 1.

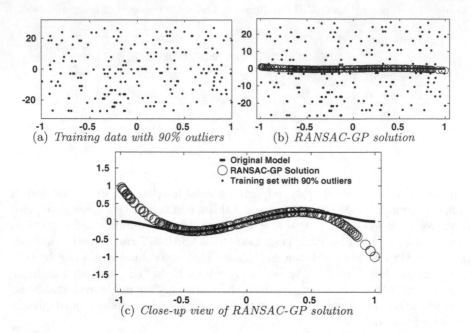

(a) *Training data with 90% outliers*     (b) *RANSAC-GP solution*

(c) *Close-up view of RANSAC-GP solution*

**Fig. 4.** Solution found by RANSAC-GP with 90% outliers for benchmark 2.

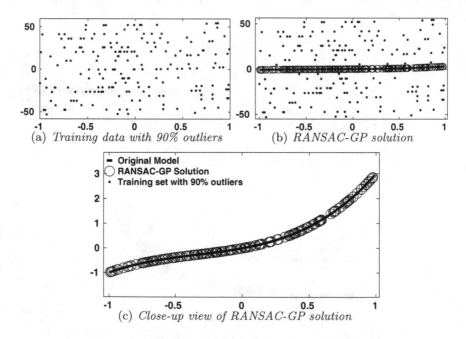

(a) *Training data with 90% outliers*     (b) *RANSAC-GP solution*

(c) *Close-up view of RANSAC-GP solution*

**Fig. 5.** Solution found by RANSAC-GP with 90% outliers for benchmark 3.

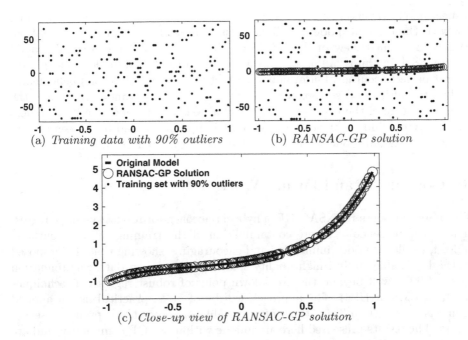

(a) *Training data with 90% outliers*    (b) *RANSAC-GP solution*

(c) *Close-up view of RANSAC-GP solution*

**Fig. 6.** Solution found by RANSAC-GP with 90% outliers for benchmark 4.

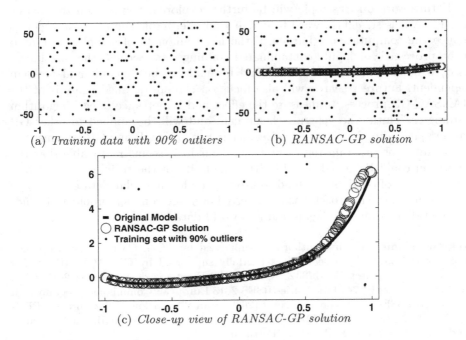

(a) *Training data with 90% outliers*    (b) *RANSAC-GP solution*

(c) *Close-up view of RANSAC-GP solution*

**Fig. 7.** Solution found by RANSAC-GP with 90% outliers for benchmark 5.

inspection clearly shows a very difficult regression problem. Second, the training set with the original model highlighted, as well as the RANSAC-GP solution. Finally, a zoomed view of the RANSAC-GP solution and the original ground truth model. In almost all problems the quality of the found model is clear, with almost perfect accuracy. Performance is slightly worse on the second benchmark, and on the fifth to a lesser extent. Nonetheless, it is important to remember the extreme nature of the training data, where all GP practitioners and researchers will surely conclude that no other GP-based approach would have been able to detect the original model with the level of accuracy reported here.

# 6    Conclusion and Future Work

This paper presents RANSAC-GP, a hybrid approach for robust symbolic regression in the presence of severe contamination of the training set with outliers. Indeed, results are both impressive and encouraging, showing that the proposed method was able to find high quality solutions even when data contamination reached 90%, well beyond the breakdown point of robust regression techniques (such as LMS or LTS). The approach will be of great practical use in applied domains where the observations of a variable of interest are prone to severe errors. The results presented here are unique within the GP community, and are meant to encourage future work on automatic modeling under large amounts of data contamination.

Future work on this topic will be further explored, focusing on the following. First, it is straightforward to derive a parallel implementation, in order to reduce the computational burden of the large amounts of samples that need to be processed for highly contaminated training sets. Second, the threshold $t$ used to build the consensus set should be made to be dynamic and problem dependent. Future research will also focus on a more detailed analysis of the RANSAC-GP process, to measure the effect and importance of each algorithm parameter. Third, the general scheme can, and will, be applied to other GP variants, including bloat-free GP, Geometric Semantic GP and GP systems that utilize different program representations. Fourth, it seems reasonable that the algorithm could be tuned based on the regularity of the training set, where a large amount of outliers will tend to produce highly irregular datasets and vice versa. Finally, the approach has to be tested on more complex problems, including multidimensional problems and real-world datasets.

**Acknowledgments.** First author was supported by CONACYT (México) scholarships No. 573397. This research was partially supported by CONACYT Basic Science Research Project No. 178323, CONACYT Fronteras de la Ciencia 2015-2 No. 944, as well as by FP7- Marie Curie-IRSES 2013 European Commission program with project ACoBSEC with contract No. 612689. Sara Silva acknowledges project PERSEIDS (PTDC/EMS-SIS/0642/2014) and BioISI RD unit, UID/MULTI/04046/2013, funded by FCT/MCTES/PIDDAC, Portugal.

# References

1. Alfons, A., Croux, C., Gelper, S., et al.: Sparse least trimmed squares regression for analyzing high-dimensional large data sets. Annals Appl. Stat. **7**(1), 226–248 (2013)
2. Derpanis, K.G.: Overview of the RANSAC algorithm. Image Rochester NY **4**(1), 2–3 (2010)
3. Fischler, M.A., Bolles, R.C.: Random sample consensus: a paradigm for model fitting with applications to image analysis and automated cartography. Commun. ACM **24**(6), 381–395 (1981)
4. Fortin, F.A., De Rainville, F.M., Gardner, M.A., Parizeau, M., Gagné, C.: DEAP: evolutionary algorithms made easy. J. Mach. Learn. Res. **13**, 2171–2175 (2012)
5. Giloni, A., Padberg, M.: Least trimmed squares regression, least median squares regression, and mathematical programming. Math. Comput. Model. **35**(9), 1043–1060 (2002)
6. Gonçalves, I., Silva, S.: Balancing learning and overfitting in genetic programming with interleaved sampling of training data. In: Krawiec, K., Moraglio, A., Hu, T., Etaner-Uyar, A.Ş., Hu, B. (eds.) EuroGP 2013. LNCS, vol. 7831, pp. 73–84. Springer, Heidelberg (2013). doi:10.1007/978-3-642-37207-0_7
7. Hast, A., Nysjö, J., Marchetti, A.: Optimal RANSAC-towards a repeatable algorithm for finding the optimal set. J. WSCG **21**(1), 21–30 (2013)
8. Kotanchek, M.E., Vladislavleva, E.Y., Smits, G.F.: Symbolic regression via genetic programming as a discovery engine: insights on outliers and prototypes. In: Riolo, R., O'Reilly, U.-M., McConaghy, T. (eds.) Genetic Programming Theory and Practice VII, pp. 55–72. Springer, Heidelberg (2010)
9. La Cava, W., Spector, L., Danai, K.: Epsilon-lexicase selection for regression. In: GECCO 2016 Proceedings of the Genetic and Evolutionary Computation Conference 2016, pp. 741–748. ACM, New York (2016)
10. Lacey, A., Pinitkarn, N., Thacker, N.A.: An evaluation of the performance of RANSAC algorithms for stereo camera calibrarion. In: BMVC, pp. 1–10 (2000)
11. Martínez, Y., Trujillo, L., Naredo, E., Legrand, P.: A comparison of fitness-case sampling methods for symbolic regression with genetic programming. In: Tantar, A.-A. (ed.) EVOLVE - A Bridge between Probability, Set Oriented Numerics, and Evolutionary Computation V. AISC, vol. 288, pp. 201–212. Springer, Heidelberg (2014). doi:10.1007/978-3-319-07494-8_14
12. Martnez, Y., Naredo, E., Trujillo, L., Legrand, P., Lpez, U.: A comparison of fitness-case sampling methods for genetic programming. Journal of Experimental and Theoretical Artificial Intelligence (accepted to appear 2016)
13. McDermott, J., White, D.R., Luke, S., Manzoni, L., Castelli, M., Vanneschi, L., Jaskowski, W., Krawiec, K., Harper, R., De Jong, K., O'Reilly, U.M.: Genetic programming needs better benchmarks. In: GECCO 2012 Proceedings of the Fourteenth International Conference on Genetic and Evolutionary Computation Conference, pp 791–798. ACM, New York (2012)
14. Nunkesser, R., Morell, O.: An evolutionary algorithm for robust regression. Comput. Stat. & Data Anal. **54**(12), 3242–3248 (2010)
15. Pearson, R.K.: Mining imperfect data: dealing with contamination and incomplete records. SIAM (2005)
16. Rousseeuw, P.J.: Least median of squares regression. J. Am. Stat. Assoc. **79**(388), 871–880 (1984)

17. Spector, L.: Assessment of problem modality by differential performance of lexicase selection in genetic programming: a preliminary report. In: GECCO Companion 2012 Proceedings of the Fourteenth International Conference on Genetic and Evolutionary Computation Conference Companion, pp. 401–408. ACM (2012)
18. Tarsha-Kurdi, F., Landes, T., Grussenmeyer, P., et al.: Hough-transform and extended RANSAC algorithms for automatic detection of 3D building roof planes from Lidar data. In: Proceedings of the ISPRS Workshop on Laser Scanning. vol. 36, pp. 407–412 (2007)
19. Torr, P.H., Zisserman, A.: MLESAC: a new robust estimator with application to estimating image geometry. Comput. Vis. Image Underst. **78**(1), 138–156 (2000)
20. Zuliani, M.: RANSAC for Dummies. Vision Research Lab, University of California, Santa Barbara (2009)

# Symbolic Regression on Network Properties

Marcus Märtens$^{(\boxtimes)}$, Fernando Kuipers, and Piet Van Mieghem

Faculty of Electrical Engineering, Mathematics and Computer Science,
Delft University of Technology, PO Box 5031, 2600 GA Delft, The Netherlands
M.Maertens@tudelft.nl

**Abstract.** Networks are continuously growing in complexity, which creates challenges for determining their most important characteristics. While analytical bounds are often too conservative, the computational effort of algorithmic approaches does not scale well with network size. This work uses Cartesian Genetic Programming for symbolic regression to evolve mathematical equations that relate network properties directly to the eigenvalues of network adjacency and Laplacian matrices. In particular, we show that these eigenvalues are powerful features to evolve approximate equations for the network diameter and the isoperimetric number, which are hard to compute algorithmically. Our experiments indicate a good performance of the evolved equations for several real-world networks and we demonstrate how the generalization power can be influenced by the selection of training networks and feature sets.

**Keywords:** Symbolic regression · Complex networks · Cartesian Genetic Programming

## 1 Introduction

One of the first and most important steps for modelling and analyzing complex real-world relationships is to understand their structure. Networks are an effective way to organize our data so that nodes describe certain actors or entities, while relations are expressed as links connecting two nodes with each other. The resulting topological representation (adjacency matrix) provides an abstract model that is amenable for further analysis. For example, algorithms for finding shortest paths, spanning trees or similar structures usually take the topological representation of the network as input. Community detection algorithms can cluster groups of nodes that are more connected within their group than outside. Computing node centrality metrics allows for the identification of important nodes or critical connections. A well-known example is Google's Pagerank algorithm [1], which uses the eigenvector centrality of a node in order to assess the rank of a webpage with respect to Google's search queries.

Eigenvector centrality [2] is interesting from a different perspective as well. It shows that spectral network properties can improve our understanding of such vast aggregations of data like in the world-wide web. Spectral graph theory explicitly seeks to understand the relations between eigenvalues, eigenvectors and characteristic polynomials of various network matrices. Many links to

© Springer International Publishing AG 2017
J. McDermott et al. (Eds.): EuroGP 2017, LNCS 10196, pp. 131–146, 2017.
DOI: 10.1007/978-3-319-55696-3_9

fundamental questions of mathematics and complexity theory arise from spectral graph theory, making this area of research both valuable and intricate. It is possible that many topological network properties are reflected in the spectrum, only waiting to be discovered.

This work proposes symbolic regression as a method to automatically derive insights in the spectral domain and their corresponding topological reflections in the network. Only a minimal number of assumptions are needed, in particular in comparison to the frequently used procedure of curve fitting, which assumes already a pre-knowledge of a certain function like a polynomial, exponential, etc. In contrast, symbolic regression is guided by supervised learning for a regression task that explicitly constructs free-form equations out of numeric features and elementary arithmetic operations.

The topological representation may be a cumbersome feature space for machine learning techniques, if only the binary features of the adjacency matrix are considered. Therefore, we examine the usage of features from the spectral domain of the network. By training the symbolic regression system on a set of carefully constructed networks, we are able to estimate target features. Consequently, symbolic regression may assist researchers to unravel the hidden structures in the spectral domain and to propose first-order approximations for difficult-to-compute properties.

Our work is structured as follows: Sect. 2 introduces the concept of symbolic regression by giving references to previous work where this technique proved useful. Section 3 provides the necessary background in network science by introducing network properties that will be used as features and targets for our experiments. The setup of our experiments is outlined in Sect. 4 and their results are discussed in Sect. 5. We conclude with directions for future research in Sect. 6.

## 2    Related Work

### 2.1    Symbolic Regression

One of the most influential works on symbolic regression is due to Schmidt and Lipson [3], who demonstrated that physical laws can be derived from experimental data (observations of a physical system) by algorithms, rather than physicists. The algorithm is guided by evolutionary principles: a set of (initially random) parameters and constants are used as inputs, which are subsequently combined with arithmetic operators like $\{+, -, \times, \div\}$ to construct building blocks of formulas. Genetic operations like crossover and mutation recombine the building blocks to minimize various error metrics. The algorithm terminates after a certain level of accuracy is reached; the formulas that describe the observed phenomenon best are delivered as output for further analysis.

In the work of Schmidt and Lipson [3], symbolic regression was able to find hidden physical conservation laws, which describe invariants over the observed time of physical systems in motion, like oscillators and pendulums. It is remarkable that symbolic regression was able to evolve the Hamiltonian of the double pendulum, a highly non-linear dynamic system [4] that undergoes complex and chaotic motions. Also, accurate equations of motions were automatically derived for systems of coupled oscillators.

While symbolic regression rarely deduces error-free formulas, the output may deepen our insight in the problem and may help to eventually find exact solutions. One example is the case of solving *iterated functions*, which asks for a function $f(x)$ that fulfills $f(f(x)) = g(x)$ for some given function $g(x)$. Despite the simple description of the problem, there exist difficult cases for which highly non-trivial algebraic techniques seem to be needed to find solutions.

One example is the iterated function $f(f(x)) = x^2 - 2$, for which the best known analytic approach to find $f(x)$ requires the substitution of special function forms and recognizing relations between certain Chebyshev polynomials. Again, Schmidt and Lipson [5] were able to evolve a couple of symbolic expressions that were so close at describing a solution, that a simple proof by basic calculus could be inferred.

Most recently, symbolic regression has been explored in the context of generative network models by Menezes and Roth [6]. They present a stochastic model in which each possible link has a weight computed by an evolved symbolic expression. The weight-computation-rules are executed and the resulting networks are compared by a similarity-metric with some target networks (corresponding to the observations of a physical system), which guides evolution to incrementally improve the underlying generative model.

One particular benefit of symbolic regression and automatic generation of equations is reduction of the bias introduced sometimes unknowingly by human preferences and assumptions. Thus, it is possible for symbolic regression to discover relations that would be deemed counter-intuitive by humans. This makes symbolic regression especially attractive for finding non-linear relationships, for which the human mind often lacks insight and intuition.

With the exception of the deterministic FFX-algorithm by McConaghy [7], most symbolic regression algorithms are based on Genetic Programming [8], where an evolutionary process typically uses grammars [9,10] to evolve expression trees. Our work can be potentially implemented by many of these Genetic Programming variants, but we selected Cartesian Genetic Programming[1] (CGP) for reasons outlined in the following subsection.

## 2.2   Cartesian Genetic Programming (CGP)

CGP was originally developed by Miller [11] to represent electronic circuits on 2d-grids (hence the name *Cartesian*), but it soon became a general purpose tool for genetic programming. It has been used in numerous applications, e.g. to develop Robot Controllers [12], Neural Networks [13], Image Classifiers [14] and Digital Filters [15]. A recent result by Vasicek and Sekanina shows how approximate digital circuits can be efficiently evolved by CGP, giving human-competitive results in the field of approximate computing [16]. Vasicek also shows

---

[1] http://www.cartesiangp.co.uk/resources.html.

how CGP can be scaled to deal with a large number of parameters in order to optimize combinatorial circuits [17].

The reason why CGP is so popular (especially for circuit design) is due to its internal representation of the Genetic program. CGP uses a flexible encoding that represents the wiring of a computational network. Each node in this network is an arithmetic operation that needs a certain amount of inputs to produce an output. A simple $1 + 4$ evolutionary strategy changes the interconnections between those nodes in order to improve a fitness function (minimizing errors). Input parameters and constants are forward-propagated by applying the computational nodes until output nodes are reached. At these output nodes, the chain of mathematical operations on the inputs can be reconstructed as an equation.

A surprising property of CGP is that only a minor fraction of nodes actually contribute to the final computation. Similar to a human genome, only part of it is actively used, while inactive parts are dormant, but subject to genetic drift. This redundancy is often argued to be beneficial for the evolutionary process in CGP (see Miller and Smith [18]). There is also evidence that CGP does not suffer much from bloat [19], a major issue in other genetic programming techniques that tend to produce very large program sizes even for simple tasks.

## 3    Networks

In this section, we formally define some network properties and notation that will be used throughout our experiments.

### 3.1    Network Representations

A network is represented as a graph $G = (\mathcal{N}, \mathcal{L})$, where $\mathcal{N}$ is the set of nodes and $\mathcal{L} \subseteq \mathcal{N} \times \mathcal{N}$ is the set of links. The number of nodes is denoted by $N = |\mathcal{N}|$ and the number of links by $L = |\mathcal{L}|$. The set $\mathcal{L}$ is typically represented by an $N \times N$ adjacency matrix $A$ with elements $a_{ij} = 1$ if node $i$ and $j$ are connected by a link and $a_{ij} = 0$ otherwise. As we restrict ourselves to simple, undirected networks without self-loops in this work, $A$ is always symmetric. We call $A$ the *topological representation* of $G$ as each element of $A$ directly refers to a structural element (a link) of the network. The number of all neighbors of a node $i$ is called its degree $d_i = \sum_{j=1}^{N} a_{ij}$.

The adjacency matrix $A$ is not the only possible representation of a network. Of equal importance is the Laplacian matrix $Q = \Delta - A$, where $\Delta$ is a diagonal matrix consisting of the degrees $d_i$ for each node $i \in \mathcal{N}$.

A different view on the network can be derived by its eigenstructure. Given the adjacency matrix $A$, there exists an eigenvalue decomposition [20]

$$A = X \Lambda X^T \tag{1}$$

such that the columns of $X$ contain the eigenvectors $x_1, x_2, \ldots, x_N$ belonging to the real eigenvalues $\lambda_1 \geq \lambda_2 \geq \ldots \geq \lambda_N$, respectively, contained in the diagonal matrix $\Lambda$.

While obtaining the spectral representation of the network requires computational effort by itself (usually, the network is given in its topological representation for which the eigenvalues still need to be determined), it provides a different perspective on the network's properties. For example, the largest eigenvalue $\lambda_1$ is linked with the vulnerability of a network to epidemic spreading processes [21].

A similar decomposition is possible for the Laplacian matrix, whose eigenvalues are denoted by $\mu_1 \geq \mu_2 \geq \ldots \geq \mu_N$ and whose real eigenvectors are $y_1, y_2, \ldots, y_N$. The second smallest eigenvalue $\mu_{N-1}$ is known as the *algebraic connectivity* [22] and its corresponding eigenvector is known as *Fiedler's vector*. Spectral clustering [20] is a possible application of Fiedler's vector.

Both eigensystems constitute the *spectral representation* of $G$. Our goal is to describe network properties typically computed by algorithms on the *topological representation* of $G$ by simple functions consisting of elements from the *spectral representation* of $G$. An example is the number of triangles $\blacktriangle_G$ in a network. A way of computing $\blacktriangle_G$ is to enumerate all possible triples of nodes in a graph and checking whether they are connected in $A$. However, the number of triangles can also be expressed as

$$\blacktriangle_G = \frac{1}{6} \cdot \sum_{k=1}^{N} \lambda_k^3 \tag{2}$$

and is thus directly computable from the spectral representation without the need of exhaustive enumeration (see Van Mieghem [20], art. 28 for a proof of Eq. (2)).

## 3.2   Network Properties

**Network Diameter.** Many applications of networks are concerned with finding and using shortest-path structures in networks. A *path* between two distinct nodes $i$ and $j$ is a collection of links that can be traversed to reach $i$ from $j$ and vice versa. A *shortest path* is a path with minimal number of links. The *diameter* $\rho$ of a network is defined as the length of the *longest* shortest path in the network, i.e. the maximum over all shortest-path lengths between all node-pairs. Algorithms that solve the all-pairs shortest path problem (like the Floyd-Warshall algorithm) are able to compute the diameter in $\mathcal{O}(N^3)$. While more efficient algorithms exist for sparse networks, an exact computation of the diameter is usually too expensive for very large networks.

There exist multiple upper bounds for the diameter [23,24], but we find the bound of Chung et al. [25] most tight in almost all cases:

$$\rho \leq \left\lfloor \frac{\cosh^{-1}(N-1)}{\cosh^{-1}\left(\frac{\mu_1 + \mu_{N-1}}{\mu_1 - \mu_{N-1}}\right)} \right\rfloor + 1. \tag{3}$$

This bound was independently derived by Van Dam and Haemers [26].

**Isoperimetric Number.** For each subset of nodes $X \subset \mathcal{N}$ we can define the set $\partial X$ as the set of links that have exactly one endpoint in $X$ and the other endpoint in $\mathcal{N} \backslash X$. The *isoperimetric number* $\eta$ of a network is defined as

$$\eta = \min_{\substack{X \subseteq \mathcal{N} \\ |X| \leq \frac{1}{2}N}} \frac{|\partial X|}{|X|}. \tag{4}$$

Essentially, the isoperimetric number is a measure related to *bottlenecks* in networks. Intuitively, a low isoperimetric number indicates that the network can be separated in two reasonably big parts by only cutting a minimum amount of links. While the isoperimetric number is a good descriptor of network robustness, its computation for general networks is intractable, as the computational effort scales with the amount of possible cuts of the network. More information on the isoperimetric number can be found in [20, 27].

## 4   Experiments

This section describes technical details of the symbolic regression process we deployed to infer equations for the network diameter and the isoperimetric number. As symbolic regression is a supervised learning technique, we describe the sets of networks that were used for training and testing, together with the features we extracted for each case.

### 4.1   Network Diameter

In order to find a suitable formula for the network diameter, we trained CGP on 3 different sets of networks:

- augmented path graphs,
- barbell graphs and
- the union of both.

The augmented path graphs were generated by iteratively adding random links to a simple path graph of $N$ nodes. With each additional link, there is a chance to lower the diameter of the network. Following this procedure, it is possible to generate a set of relatively sparse graphs of constant node-size with uniformly distributed diameters.

A barbell graph $B(a, b)$ is generated by taking two cliques of size $a$ and connecting them with a path graph of size $b$. The total number of nodes is $N = 2a + b$. The diameter $\rho(B(a, b))$ is always $b + 3$. Adjusting the length of the path graph allows for generating graphs of different diameters. Changing the size of the cliques allows for creating graphs with the same diameter, but different number of nodes. We sample again such that the network diameter is uniformly distributed within the set of all barbell graphs. Compared with augmented path graphs, barbell graphs are (in general) denser networks.

The set of mixed graphs is the union of the set of augmented path graphs and barbell graphs. See Fig. 1 for examples of these networks and Table 1 for a summary of all sets of networks for the experiments.

One reason why we have chosen these sets of networks instead of, for example, Erdős-Rényi (ER) random graphs [28], is to control the distribution of our target feature, the network diameter. Preliminary experiments have shown that too little variance in our target feature will push CGP to converge to a constant function, which does not include any information about the relation between spectral features and the target that we want to extract. For an ER graph of $N$ nodes and with link probability $p$, Bollobás [28] showed that, for fixed $p$ and $N$ large, $\rho$ can only have one of two possible neighboring values with high probability. Thus, sampling uniform diameters for the random graph model requires careful adjustment of $N$ and $p$, where we found the usage of augmented paths and barbell graphs more convenient.

For the supervised learning of CGP, each set of networks was separated in a 60% training and a 40% test set. Table 2 gives an overview of the various parameters we set for CGP. In preliminary experiments, we changed each of these parameters independently from another and found the settings of Table 2 to provide the most useful results in terms of fitness and formula complexity. A more thorough parameter tuning approach is needed for maximum performance, but is outside the scope of this work. For the meaning of these parameters, see Miller [11]. Effective tuning of CGP was researched by Goldman and Punch [29].

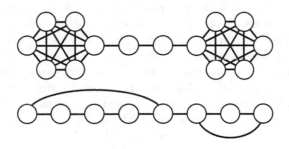

**Fig. 1.** Example of the barbell graph $B(6,2)$ with $\rho = 5$ on top and an augmented path graph with $\rho = 4$ at the bottom.

Table 1. Properties of network sets.

|  | Aug. path | Barbell | Mixed |
|---|---|---|---|
| Networks | 1672 | 1675 | 3347 |
| Nodes | $N = 70$ | $7 \leq N \leq 667$ | $7 \leq N \leq 667$ |
| Diameter | $2 \leq \rho \leq 69$ | $4 \leq \rho \leq 70$ | $2 \leq \rho \leq 70$ |
| Avg. link density | 0.04845 | 0.36985 | 0.20910 |

**Table 2.** Parameterisation of CGP.

| Parameter | Value |
|---|---|
| Fitness function | Sum of absolute errors |
| Evolutionary strategy | $1 + 4$ |
| Mutation type and rate | Probabilistic (0.1) |
| Node layout | 1 row with 200 columns |
| Levels-back | Unrestricted |
| Operators | $+, -, \times, \div, .^2, .^3, \sqrt{\cdot}, \log$ |
| Number of generations | $2 \cdot 10^5$ |

In our experiments, we tried a vast selection of different features to evolve formulas. To keep this section organized, we report only results derived from two of the most generic, but useful, sets of features:

(A) $N, L, \lambda_1, \lambda_2, \lambda_3, \lambda_N$
(B) $N, L, \mu_1, \mu_{N-1}, \mu_{N-2}, \mu_{N-3}$.

Additionally, the natural numbers $1, \ldots, 9$ were provided as network independent constants for CGP to adjust evolved terms appropriately.

The choice of feature sets A and B provides a reasonable trade-off between formula complexity and fitness. While selecting the complete spectrum of eigenvalues as features is possible, we observed that it leads to a high formula complexity without providing considerable improvements in fitness. Additionally, the largest adjacency (smallest Laplacian) eigenvalues are the ones that are suggested to have the strongest influence on network properties [20]. Lastly, since the number of nodes in our network instances is not (in every case) constant, giving the complete spectrum would mean that several features would be missing in networks with low number of nodes. It is unclear, how an appropriate substitution of the missing features should be realized. Thus, some of the discovered formulas could be inapplicable for some networks.

Since the evolutionary procedures of CGP to optimize the fitness of the evolved expressions are stochastic, we deployed multiple runs for each combination of feature and network set. We aggregated those multiple runs into batches, as the test-environment was implemented to run on a computational cluster. Each batch consisted of 20 runs of CGP for a specific set of features. Out of those 20 runs, only the one with the best (lowest) fitness is reported. The fitness is the sum of absolute errors on the test instances of the corresponding set of networks. More formally, if $\hat{\rho}_G$ is the estimate on the diameter $\rho_G$ of network $G$ given by the evolved formula $\hat{\rho}$ and $\mathcal{G}_{test}$ is the set of all networks for testing, the fitness $f(\hat{\rho})$ is defined as:

$$f(\hat{\rho}) = \sum_{G \in \mathcal{G}_{test}} |\rho_G - \hat{\rho}_G|. \tag{5}$$

Furthermore, we define the approximation error $e(\hat{\rho})$ as the average deviation from the diameter over the complete test set:

$$e(\hat{\rho}) = \sum_{G \in \mathcal{G}_{test}} \frac{|\rho_G - \hat{\rho}_G|}{|\mathcal{G}_{test}|}. \tag{6}$$

We present the results over 100 batches for each combination of feature and network test set in Table 3.

**Table 3.** Experimental results for the network diameter.

| Networks | Feature set | Avg. fitness | Min. fitness | Min. approx. error |
|----------|-------------|--------------|--------------|--------------------|
| Aug. path | A | 3694.98750 | 3404.53700 | 5.08899 |
| | B | 842.89691 | 778.98900 | 1.16441 |
| Barbell | A | 1.66654 | 0.00900 | 0.00001 |
| | B | 50.53473 | $<10^{-5}$ | $<10^{-5}$ |
| Mixed | A | 5313.91179 | 4500.68900 | 3.36123 |
| | B | 1462.61943 | 1134.34100 | 0.84716 |

## 4.2  Isoperimetric Number

The training set of networks for the isoperimetric number $\eta$ had to be limited to relatively small networks, since the computation of $\eta$ becomes intractable even for general medium-sized networks. Thus, we decided to exhaustively enumerate all networks of size $N = 7$, for which the computation was still practical. This set consists of 1046 non-isomorphic networks, which we randomly split into a training set of 627 and a test set of 419 networks. We applied the same parameters to CGP as shown in Table 2, with one exception: we created 100 batches for each of the following sets of operators:

1. $+, -, \times, \div, \sqrt{\cdot}, \log$
2. $+, -, \times, \div, .^2, \log$

3. $+, -, \times, \div, \sqrt{\cdot}$
4. $+, -, \times, \div, .^2$.

Since we have only networks of size $N = 7$, we can select the full spectrum as our features, resulting in the following feature sets:

(A') $N, L, \lambda_1, \lambda_2, \lambda_3, \lambda_4, \lambda_5, \lambda_6, \lambda_7$
(B') $N, L, \mu_1, \mu_2, \mu_3, \mu_4, \mu_5, \mu_6$.

Since the smallest Laplacian eigenvalue $\mu_N$ always equals $0, \mu_7 = 0$ and is thus not included as a feature. Additionally, we provided the natural numbers 1, 2 and 3 as constants. Each batch consisted of 5 independent runs from which only the best one is reported.

Feature set A' delivered on average much better results than feature set B'. The best result was found with feature set A' and operator-set $+, -, \times, \div, \cdot^2, \log$, although not all of these operators appear:

$$\hat{\eta}_1 = \frac{L - \lambda_2^2 - 2}{\frac{\lambda_2}{2} + 5} \qquad (7)$$

Although Eq. (7) is short (low complexity), it had still the best fitness (53.215) of all evolved formulas. The approximation of $\eta$ on the test set is shown in Fig. 2.

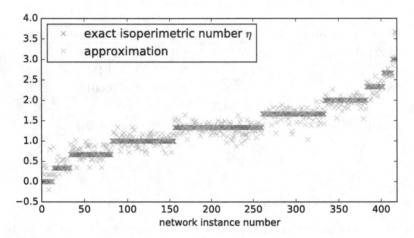

**Fig. 2.** All 419 networks from the test set ordered by their actual isoperimetric number ascending from left to right. The red crosses show the approximation given by Eq. (7). (Color figure online)

## 5    Discussion

In the previous section, we evolved approximate equations for hard to compute network properties. The approximation errors were largely dependent on the used networks for training and testing. For example, the best equations for the diameter found for barbell networks have almost no error, while noticeable errors exist for augmented path and mixed network. This raises two questions:

- how is the quality of the approximate equations influenced by the pre-selected networks and
- how do the approximate equations generalize to other networks?

To give answers, we compute already established analytical equations from the literature as reference points for quality and provide appropriate selections of networks which were not involved in the generation process of the evolved equations.

## 5.1    Network Diameter

As a measure for the quality of our evolved equations, we compare their estimates of diameter $\rho$ to the upper bound given by Eq. (3). An upper bound and an approximation are different: while the bound is always above the real diameter, approximations may be above or below without any guarantees. Yet, we believe the bound can mark a reference point for a qualitative comparison in addition to the exact diameter itself.

As additional networks, we selected 12 real-world data sets available at networkrepository.com [30], where more information, interactive visualization and analytics can be found. While these networks should only be viewed as examples, they might give an idea about the applicability of the presented technique. To eliminate the selection-bias and gather significant results, one would need to sample the network space in a representative and meaningful way, which is notoriously difficult. For example, simple network generators like Erdős-Rényi random graphs or the Barabási-Albert model are not sufficient, as they will only allow to sample certain degree distributions. Consequently, we restrict ourselves to examples.

We expect that the equations with the lowest fitness give the best results. Because feature set B consistently outperformed feature set A in terms of fitness, we analyzed the approximations given by the best equations of feature set B. The explicit equations are:

$$N - \frac{1 - \frac{1}{(L-N)^{\frac{3}{2}}}}{6 - \frac{\frac{6}{(L-N)^{\frac{3}{2}}}}{\sqrt{L-N}} + 4\sqrt{L-N}} - 2\sqrt{L-N} - \frac{1}{\sqrt{L-N}} \tag{8}$$

$$\frac{\log\left(2L\mu_{N-3} + 6\right) + 6}{\log\left(L\mu_{N-3} + \sqrt{5}\sqrt{\frac{1}{\mu_{N-1}}}\right)} + \sqrt{5}\sqrt{\frac{1}{\mu_{N-1}}} + 3\sqrt{82}\sqrt{\frac{1}{729L\mu_{N-2}\mu_{N-3} - 5}} \tag{9}$$

$$\sqrt{\sqrt{N} + \frac{45\mu_{N-3}}{(\mu_{N-1} + \mu_{N-3})^2} + \log\left(\frac{216}{(\mu_{N-1} + \mu_{N-3})^2}\right) - \frac{16}{9\mu_{N-3}} + \frac{8\sqrt[4]{\mu_{N-3}}}{L\mu_{N-1}\mu_{N-2}}} \tag{10}$$

The numerical values are all listed together with some basic properties of our validation networks in Table 4. First, we observe that Eq. (8) performs extremely poorly by giving huge overestimations of $\rho$, despite its fitness of almost 0 for the networks of the original test set. The reason is that Eq. (8) was evolved on barbell graphs only, which have a fixed and symmetric structure. In particular, the difference $N - L$, which is a frequent subterm of the formula, is higher in the dense barbell graphs compared to the rather sparse networks of our validation set. Thus, Eq. (8) seems to be *overfitted* to the class of barbell graphs.

Equation (9) was evolved only on augmented path graphs and provides a much better approximation of $\rho$ for our validation networks. This might be the case since the validation networks are more similar to the sparse augmented path graphs than to the dense barbell graphs. A visual comparison of the approximation of Eq. (9) is given by Fig. 3.

Adding the barbell graphs to the training set, like in the evaluation of Eq. (10), shows that the accuracy of the approximation of $\rho$ increases by roughly 10%, which must be the effect of the barbell graphs adding a selective force towards accuracy on more denser networks. Moreover, it seems that CGP focused on finding a good approximation for the augmented path graphs in the mix rather than considering to find an equation for both classes of networks. In the majority of the cases, $\rho$ is still overestimated a little, but by far not as much as by the upper bound in Eq. (3).

**Table 4.** Diameter on validation networks.

| Name | $N$ | $L$ | $\rho$ | Equation (9) | Equation (8) | Equation (10) | Equation (3) |
|---|---|---|---|---|---|---|---|
| Cca-netscience | 379 | 914 | 17 | 21 | 333 | 24 | 160 |
| Bio-celegans | 453 | 2025 | 7 | 7 | 374 | 8 | 104 |
| Rt-twitter-copen | 761 | 1029 | 14 | 16 | 728 | 18 | 126 |
| Soc-wiki-Vote | 889 | 2914 | 13 | 10 | 799 | 12 | 133 |
| Ia-email-univ | 1133 | 5451 | 8 | 6 | 1002 | 8 | 58 |
| Ia-fb-messages | 1266 | 6451 | 9 | 7 | 1122 | 10 | 96 |
| Web-google | 1299 | 2773 | 14 | 29 | 1222 | 35 | 336 |
| Bio-yeast | 1458 | 1948 | 19 | 19 | 1414 | 22 | 208 |
| Tech-routers-rf | 2113 | 6632 | 12 | 14 | 1979 | 17 | 237 |
| Socfb-nips-ego | 2888 | 2981 | 9 | 52 | 2869 | 61 | 2466 |
| Web-edu | 3031 | 6474 | 11 | 36 | 2914 | 40 | 663 |
| Inf-power | 4941 | 6594 | 46 | 98 | 4860 | 110 | 749 |

## 5.2 Isoperimetric Number

The quality of the equations approximating the isoperimetric number will be related to the Cheeger inequality (see Mohar [27]) that gives us bounds in relation to the algebraic connectivity $\mu_{N-1}$ and the maximum degree $d_{max}$ of the network $G$:

$$\frac{\mu_{N-1}}{2} \le \eta \le \sqrt{\mu_{N-1}(2d_{max} - \mu_{N-1})} \qquad (11)$$

Since our equations were evolved by an exhaustive enumeration of all non-isomorphic networks of $N = 7$, we are interested how their quality of fit will differ with $N$. However, as pointed out before, the computation of the exact value for $\eta$ is in general only feasible for very small networks. Consequently, we cannot use any of the validation-networks from Table 4. Instead we decided to sample random networks of $N = 20$ nodes and links from $22 \le L \le 190$. In total, we generated 4984 non-isomorphic connected networks with roughly uniformly distributed link densities by a variant of the ER random graph model.

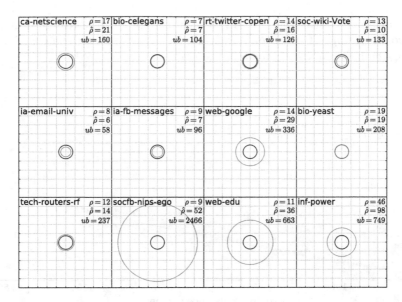

**Fig. 3.** Red circles: approximate diameter $\hat{\rho}$ by Eq. (9) relative to the network diameter $\rho$ as a black circle. All network diameters are scaled in each network to have unit-length in the figure. All values are rounded to the next integer. The upper bound Eq. (3) values are given as $ub$ in blue (too large to plot). (Color figure online)

In these networks, the isoperimetric number $\eta$ ranges from 0.2 to 10.0, while in our training set $\eta$ was between 0 and 4.

Surprisingly, Eq. (7) deduced from all networks with $N = 7$ nodes is performing poorly on the set of random networks, as shown by the green dots in Fig. 4. The estimates are most of the time not even below the bound of the Cheeger inequality, shown in grey. By analyzing the sum of absolute errors on this new set of networks, we found that from all evolved formulas, the following equation for the isoperimetric number gives the best performance:

$$\hat{\eta}_2 = \frac{1}{N^2} \left( L \left( \frac{\mu_1}{\mu_2} + \mu_2 \right) - 1 \right). \tag{12}$$

We observe that for over 98% of the random networks, the estimate of Eq. (12) was within $(1 \pm 0.2) \cdot \eta$. Since this equation incorporates not only spectral features, but also $N$ and $L$, we believe it generalizes better to networks of different size other than those used in the training set. Additionally, our experiments show that a low fitness value does not necessarily correspond to good generalization. Out of the 800 batches used to find a formula for $\eta$, only 259 returned expressions that did not create artifacts (like square roots of negative numbers or divisions by zero, which CGP evaluates to 0 by definition). While Eq. (7) ranked first with a fitness of 53.215, Eq. (12) was one of the unranked expressions, since on some of the unconnected networks of the training set, $\mu_2$ was 0, while $\mu_2 = 0$ did never appear for the connected random networks.

It is also noteworthy that Eq. (12) seems to slightly overestimate $\eta$ as soon as networks with $\eta > 4$ are encountered. This does not seem to be a coincidence, as 4 was the maximum value for $\eta$ in the training set.

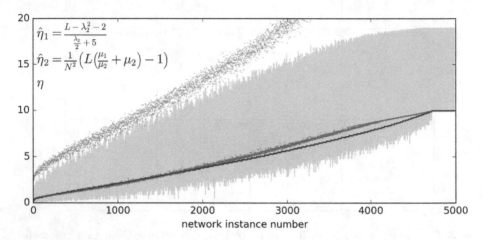

**Fig. 4.** A set of 4984 random networks with $N = 20$ ordered by their isoperimetric number (blue). The grey area corresponds to the Cheeger inequalities given by Eq. (11). The lower bound coincides with $\eta$ and the red approximation formula for the last 100 networks, as they are all fully connected. (Color figure online)

## 6   Conclusion

Our experiments provide a first demonstration that symbolic regression can be applied to analyze networks. For the first time, to the best of our knowledge, an automated system has inferred approximate equations for network properties, which otherwise would have required a high algorithmic effort to be determined. Although these equations are not rigorously proven and might be cumbersome for humans to comprehend, they are able to exploit the hidden relationships between the topological and the spectral representation of networks, which has been elusive to analytical treatment so far.

We do not expect that symbolic regression at the current level will substitute researchers deriving meaningful equations, but we do believe that symbolic regression can be a meaningful tool for these researchers. As the proposed techniques are not biased by human preconceptions, unexpected results might provide inspiration and stimulating starting points for the development of formal proofs or more accurate formulas. While this lack of bias can be an advantage, the proposed system nevertheless allows for the incorporation of *a priori* expert knowledge. If certain features and operators are suspected to be correlated to an unknown target quantity, their usage can be enforced easily.

Understanding which conditions give rise to equations with a high generalization power for networks will be the main challenge for the future. While a good fitness of an equation does not necessarily imply a high generalization power, our experiments indicate that symbolic regression is clearly able to produce equations that are reasonably accurate for unknown networks. In order to prevent overfitting to the training set and to increase this generalization power, selecting a good set of networks for training seems to be the key. This makes symbolic regression especially appealing when dealing with networks that can be characterized by their structural and degree-related properties, like scale-free or small-world networks. As networks with such properties are ubiquitous, discovering explicit relations between their features will pave the way for a deeper insight into our increasingly connected environments.

# References

1. Page, L., Brin, S., Motwani, R., Winograd, T.: The PageRank citation ranking: bringing order to the web. Technical report, Stanford University, Stanford, CA (1998)
2. Van Mieghem, P.: Graph eigenvectors, fundamental weights and centrality metrics for nodes in networks. arXiv preprint arXiv:1401.4580 (2014)
3. Schmidt, M., Lipson, H.: Distilling free-form natural laws from experimental data. Science **324**(5923), 81–85 (2009)
4. Strogatz, S.H.: Nonlinear Dynamics and Chaos: With Applications to Physics, Biology, Chemistry, and Engineering. Studies in Nonlinearity. Westview Press, Boulder (2014)
5. Schmidt, M., Lipson, H.: Solving iterated functions using genetic programming. In: Proceedings of the 11th Annual Conference Companion on Genetic and Evolutionary Computation Conference: Late Breaking Papers, pp. 2149–2154. ACM (2009)
6. Menezes, T., Roth, C.: Symbolic regression of generative network models. Sci. Rep. **4** (2014). Article No. 6284, doi:10.1038/srep06284
7. McConaghy, T.: FFX: fast, scalable, deterministic symbolic regression technology. In: Riolo, R., Vladislavleva, E., Moore, J.H. (eds.) Genetic Programming Theory and Practice IX, pp. 235–260. Springer, New York (2011)
8. Koza, J.R.: Genetic Programming: On the Programming of Computers by Means of Natural Selection, vol. 1. MIT press, Cambridge (1992)
9. McConaghy, T., Eeckelaert, T., Gielen, G.: CAFFEINE: template-free symbolic model generation of analog circuits via canonical form functions and genetic programming. In: Proceedings of the Conference on Design, Automation and Test in Europe, vol. 2, pp. 1082–1087. IEEE Computer Society (2005)
10. Augusto, D.A., Barbosa, H.J.: Symbolic regression via genetic programming. In: 6th Brazilian Symposium on Neural Networks, pp. 173–178. IEEE (2000)
11. Miller, J.F.: Cartesian Genetic Programming. Springer, Heidelberg (2011)
12. Harding, S., Miller, J.F.: Evolution of robot controller using cartesian genetic programming. In: Keijzer, M., Tettamanzi, A., Collet, P., Hemert, J., Tomassini, M. (eds.) EuroGP 2005. LNCS, vol. 3447, pp. 62–73. Springer, Heidelberg (2005). doi:10.1007/978-3-540-31989-4_6
13. Khan, M.M., Khan, G.M.: A novel neuroevolutionary algorithm: cartesian genetic programming evolved artificial neural network (CGPANN). In: Proceedings of the 8th International Conference on Frontiers of Information Technology, p. 48. ACM (2010)

14. Harding, S., Graziano, V., Leitner, J., Schmidhuber, J.: MT-CGP: mixed type cartesian genetic programming. In: Proceedings of the 14th Annual Conference on Genetic and Evolutionary Computation, pp. 751–758. ACM (2012)
15. Miller, J.F.: Evolution of digital filters using a gate array model. In: Poli, R., Voigt, H.-M., Cagnoni, S., Corne, D., Smith, G.D., Fogarty, T.C. (eds.) EvoWorkshops 1999. LNCS, vol. 1596, pp. 17–30. Springer, Heidelberg (1999). doi:10.1007/10704703_2
16. Vasicek, Z., Sekanina, L.: Evolutionary approach to approximate digital circuits design. IEEE Trans. Evol. Comput. **19**(3), 432–444 (2015)
17. Vasicek, Z.: Cartesian GP in optimization of combinational circuits with hundreds of inputs and thousands of gates. In: Machado, P., Heywood, M.I., McDermott, J., Castelli, M., García-Sánchez, P., Burelli, P., Risi, S., Sim, K. (eds.) EuroGP 2015. LNCS, vol. 9025, pp. 139–150. Springer, Heidelberg (2015). doi:10.1007/978-3-319-16501-1_12
18. Miller, J.F., Smith, S.L.: Redundancy and computational efficiency in cartesian genetic programming. IEEE Trans. Evol. Comput. **10**(2), 167–174 (2006)
19. Turner, A.J., Miller, J.F.: Cartesian genetic programming: why no bloat? In: Nicolau, M., Krawiec, K., Heywood, M.I., Castelli, M., García-Sánchez, P., Merelo, J.J., Rivas Santos, V.M., Sim, K. (eds.) EuroGP 2014. LNCS, vol. 8599, pp. 222–233. Springer, Heidelberg (2014). doi:10.1007/978-3-662-44303-3_19
20. Van Mieghem, P.: Graph Spectra for Complex Networks. Cambridge University Press, Cambridge (2011)
21. Van Mieghem, P., Omic, J., Kooij, R.: Virus spread in networks. IEEE/ACM Trans. Netw. **17**(1), 1–14 (2009)
22. Fiedler, M.: Algebraic connectivity of graphs. Czechoslovak Math. J. **23**(2), 298–305 (1973)
23. Alon, N., Milman, V.D.: λ1, isoperimetric inequalities for graphs, and superconcentrators. J. Comb. Theory Ser. B **38**(1), 73–88 (1985)
24. Mohar, B.: Eigenvalues, diameter, and mean distance in graphs. Graphs Comb. **7**(1), 53–64 (1991)
25. Chung, F.R., Faber, V., Manteuffel, T.A.: An upper bound on the diameter of a graph from eigenvalues associated with its Laplacian. SIAM J. Discret. Math. **7**(3), 443–457 (1994)
26. Van Dam, E.R., Haemers, W.H.: Eigenvalues and the diameter of graphs. Linear Multilinear Algebra **39**(1–2), 33–44 (1995)
27. Mohar, B.: Isoperimetric numbers of graphs. J. Comb. Theory Ser. B **47**(3), 274–291 (1989)
28. Bollobás, B.: Random Graphs. Springer, New York (1998)
29. Goldman, B.W., Punch, W.F.: Analysis of cartesian genetic programming's evolutionary mechanisms. IEEE Trans. Evol. Comput. **19**(3), 359–373 (2015)
30. Rossi, R.A., Ahmed, N.K.: An interactive data repository with visual analytics. SIGKDD Explor. **17**(2), 37–41 (2016)

# Evolving Time-Invariant Dispatching Rules in Job Shop Scheduling with Genetic Programming

Yi Mei[1(⊠)], Su Nguyen[1,2], and Mengjie Zhang[1]

[1] Victoria University of Wellington, Wellington, New Zealand
{yi.mei,su.nguyen,mengjie.zhang}@ecs.vuw.ac.nz
[2] Department of Business, Hoa Sen University, Ho Chi Minh City, Vietnam

**Abstract.** Genetic Programming (GP) has achieved success in evolving dispatching rules for job shop scheduling problems, particularly in dynamic environments. However, there is still great potential to improve the performance of GP. One challenge that is yet to be addressed is the huge search space. In this paper, we propose a simple yet effective approach to improve the effectiveness and efficiency of GP. The new approach is based on a newly defined *time-invariance* property of dispatching rules, which is derived from the idea of translational invariance from machine learning. Then, we develop a new terminal selection scheme to guarantee the time-invariance throughout the GP process. The experimental studies show that by considering the time-invariance, GP can achieve much better rules in a much shorter time.

**Keywords:** Job shop scheduling · Genetic programming · Hyper-heuristic

## 1 Introduction

Job Shop Scheduling (JSS) [1] is an important optimisation problem with a wide range of real-world applications in domains such as manufacturing [2,3], project scheduling [4] and cloud computing. Given a set of machines and jobs, JSS is to process the jobs with the machines subject to certain constraints (e.g. each job must follow a specific routing among the machines, and each machine can process only one job at a time) and optimise some criteria such as makespan, flowtime and tardiness.

JSS can be either static or dynamic. In static JSS, all the jobs are available at the beginning of the scheduling horizon, and all the information is known in advance. In dynamic JSS, unpredicted job arrivals can occur in real time, and can affect the subsequent scheduling decisions. In this study, we focus on dynamic JSS, since it is closer to reality and more challenging than static JSS.

Dispatching rules (DRs) are promising decision making heuristics for solving dynamic JSS due to their low complexity, scalability and flexibility. Briefly speaking, a DR is a priority function of the job shop attributes such as the

© Springer International Publishing AG 2017
J. McDermott et al. (Eds.): EuroGP 2017, LNCS 10196, pp. 147–163, 2017.
DOI: 10.1007/978-3-319-55696-3_10

operation processing time and job due date. In each decision situation (e.g. a machine becomes idle and its queue is not empty), the DR is used to calculate the priority for each job/operation waiting in the queue, and the one with the best priority is selected to be processed next.

Recently, automatically designing/evolving DRs using Genetic Programming (GP) has achieved some success [5]. The DRs evolved by GP have shown to be much more effective than the DRs designed by human experts. However, a major challenge for evolving DRs with GP is the huge search space caused by the large number of possible combinations involving the function and terminal sets. It is highly desired to develop intelligent guidances for the GP search process to achieve better effectiveness and efficiency.

### 1.1   Goals

In this paper, we propose a new simple yet effective method to improve the effectiveness of GP. The proposed method is based on a newly defined property called *time-invariance*. A DR is time-invariant if it generates the same schedule for the same *JSS pattern* (defined as a time window in the scheduling horizon, details in Sect. 3) regardless of when such pattern occurs. By restricting the search to only the time-invariant DRs, we expect to reduce the search space and improve the efficiency of the search. The goal of this paper is to investigate the effectiveness of considering time-invariance in GP. Specifically, we have the following objectives:

– Formally define the concept of time-invariance;
– Show that GP may generate DRs that are not time-invariant;
– Develop a new selection scheme of terminals so that GP always generates time-invariant DRs;
– Compare the resultant time-invariance-aware GP with the baseline GP to verify the proposed method.

### 1.2   Organisation

The rest of the paper is organised as follows: Sect. 2 gives the background introduction. Then, the concept of time-invariance is defined in Sect. 3. The new selection scheme of terminal set is developed in Sect. 4. Experimental studies are carried out in Sect. 5. Finally, Sect. 6 concludes the paper.

## 2   Background

### 2.1   Job Shop Scheduling

In JSS, a set of jobs $\mathcal{J} = \{J_1, \ldots, J_n\}$ and machines $\mathcal{M} = \{M_1, \ldots, M_m\}$ are given. Each job $J_j$ has an arrival time $t_0(J_j)$ and a due date $\rho(J_j)$. It consists of a sequence of operations $[O_{1j} \rightarrow \cdots \rightarrow O_{l_j j}]$. Each operation $O_{ij}$ must be processed by machine $\pi_{ij} \in \mathcal{M}$, and its processing time is $\delta_{ij}$. Thus, it can be represented as a tuple $O_{ij} = \langle \pi_{ij}, \delta_{ij} \rangle$. An operation cannot be processed before the completion before its precedent operations. Each machine can process at most one operation at a time. Then, JSS is to find a feasible schedule to optimise some objective(s). The commonly considered JSS objectives include minimising the makespan $(C_{\max})$, total flowtime $(\sum C_j)$, total weighted tardiness $(\sum w_j T_j)$, number of tardy jobs, etc. [1].

## 2.2    Automatic Design of Dispatching Rules

In dynamic JSS, unpredicted job arrivals occur in real time, and thus the scheduling process can be seen as a discrete event simulation. Once a machine becomes idle and its queue is not empty, a *dispatching rule* (DR) [6] is used to prioritise the operations waiting in the queue, and the most prior operation is processed next. For example, the Shortest Processing Time (SPT) rule selects the operation with the shortest processing time to be processed next.

So far, there have been a large number of DRs designed by human experts (e.g. [7–10]), by considering a variety of job shop attributes such as operation processing time, job due date, work remaining and slack. However, the existing manually designed DRs are normally not effective enough, and restricted to the particular job shop scenario they are designed for.

Genetic Programming (GP) has been demonstrated to be powerful for automatically designing/evolving DRs in job shop scheduling. So far, there have been extensive studies [11–16]) in this direction and successfully achieved much better results than the previously man-made rules. A comprehensive review can be found in [5].

For evolving DRs using GP, a major challenge is how to search effectively and efficiently in the huge search space. For example, in the conventional tree-based GP, an individual is represented as a GP tree. Then, the size of the search space depends on the maximal tree depth/size, the function set and the terminal set. The performance of GP can be dramatically improved by incorporating domain knowledge about job shop scheduling. Several works have been done to improve the search efficiency from different perspectives. For example, Nguyen et al. [13] investigated different representations and proposed a grammar-based representation to constrain the search space. Mei et al. [17] proposed to identify a subset of relevant features and remove the other redundant features from the terminal set to reduce the search space. They demonstrated that the more compact terminal set can lead to significantly better rules. Riley et al. [18] proposed a similar terminal selection idea. Durasević et al. [19] developed a dimensionally aware GP that considers the compatibility between the dimensions (e.g. time, weight, counting number) of the terminals and design initialisation and evolutionary operators in such a way that no semantically incorrect rule (e.g. adding time to weight) is generated. This way, the search is focused on only the semantically correct rules.

In this paper, we investigate a new way of incorporating domain knowledge by borrowing the idea of invariant features in machine learning.

# 3    Time-Invariant Dispatching Rule

*Invariance* is an important property that has been extensively studied in machine learning, particularly classification (e.g. [20]). Using invariant features tends to improve the generalisability and robustness of the classifier.

Evolving DRs using GP is similar to the training process of a classifier. In the GP process, the fitness of a DR is evaluated on a set of training JSS instances/simulations. By providing a sufficient number of diversely distributed decision situations [21] in the training set, the learned DRs can make proper decisions for the decision situations in both the training instances and the unseen test instances.

As introduced in [21], a decision situation is represented as a feature table, where each row stands for an operation in the queue, and each column indicates a feature that may be considered by the DR. Therefore, how to select the features in the decision situation is an important issue. If there are too few features, some important information may be missed. If there are too many features, the limited training set may not be able to cover all the possible feature values, and the evolved DRs may not generalise well on the unseen test set which can be quite different from the training set. To achieve a good tradeoff between the loss of information and generalisation, we define a new property of time-invariance.

It is natural to assume that JSS instances contain some key patterns that are invariant under translation, i.e. shift on the time horizon. Here, a *JSS pattern* is defined as a time window of the entire scheduling horizon, including the *initial job shop state* (next idle time of each machine and the uncompleted jobs) when the time window starts and the *new job arrivals* during the time window.

**Definition 1 (JSS pattern).** *A JSS pattern for a window $[t_1, t_2]$ is defined as a tuple $\vartheta_{[t_1,t_2]} = \langle t_\vartheta^{idle}, \mathcal{J}_\vartheta^{init}, \mathcal{J}_\vartheta^{new} \rangle$, where $t_\vartheta^{idle}$ is the vector of the next idle time of the machines at $t_1$, $\mathcal{J}_\vartheta^{init}$ is the set of uncompleted jobs, and $\mathcal{J}_\vartheta^{new}$ is a sequence of new job arrivals.*

An example of a JSS pattern $\vartheta_{[0,8]}$ is given in Table 1 (top half). In this pattern, the two machines are both idle at time 0, and there is no uncompleted job. During the time window $[0, 8]$, there are two new job arrivals. The first job arrives at time 0 with a due date of 18, and the second arrives at time 2 with a due date of 20. Both jobs have two operations, as shown in the table. The job and operation ids are ignored.

**Table 1.** An example of a JSS pattern $\vartheta_{[0,8]}$ and the shifted pattern $h(\vartheta_{[0,8]}, 10)$. The pattern contains two jobs, each with two operations.

|  | $t^{idle}$ | $\mathcal{J}^{init}$ | $\mathcal{J}^{new}$ | | |
|---|---|---|---|---|---|
|  |  |  | $t_0$ | $\rho$ | Sequence of operations |
| $\vartheta_{[0,8]}$ | $(0, 0)$ | $\emptyset$ | 0 | 18 | $[\langle M_1, 2 \rangle \rightarrow \langle M_2, 4 \rangle]$ |
|  |  |  | 2 | 20 | $[\langle M_2, 2 \rangle \rightarrow \langle M_1, 4 \rangle]$ |
| $h(\vartheta_{[0,8]}, 10)$ | $(10, 10)$ | $\emptyset$ | 10 | 28 | $[\langle M_1, 2 \rangle \rightarrow \langle M_2, 4 \rangle]$ |
|  |  |  | 12 | 30 | $[\langle M_2, 2 \rangle \rightarrow \langle M_1, 4 \rangle]$ |

Obviously, a decision situation $\vartheta_t^{ds} = \vartheta_{[t,t]}$ is a special instantaneous JSS pattern. Then, we define the time-shift transformation for JSS patterns as follows.

**Definition 2 (Time-shift transformation for JSS patterns).** *Given a JSS pattern $\vartheta$ and a shift value $\Delta t$, the time-shift transformation $h(\cdot)$ generates a new pattern $h(\vartheta, \Delta t)$ by increasing the next idle times of the machines, and the arrival times and due dates of all the jobs by $\Delta t$, i.e. $t_\vartheta^{idle} \to t_\vartheta^{idle} + \Delta t$, $t_0(J) \to t_0(J) + \Delta t$, $\rho(J) \to \rho(J) + \Delta t$, $\forall J \in \mathcal{J}_\vartheta^{init} \cup \mathcal{J}_\vartheta^{new}$.*

Table 1 gives an example of a shifted pattern $h(\vartheta_{[0,8]}, 10)$ at the bottom. The changed parts are highlighted in bold.

Two patterns $\vartheta_1$ and $\vartheta_2$ are said to be *equivalent under time-shift* if there exists a shift value $\Delta t$, so that $\vartheta_1 = h(\vartheta_2, \Delta t)$.

By applying a DR to a pattern, the output is a *schedule*, which can be represented as sequences of operations, each for a machine. For example, a schedule $\Phi$ (shown in Fig. 1) for the pattern $\vartheta$ given in Table 1 can be represented as follows:

$$\Phi = \begin{bmatrix} M_1 : O_{11} \to O_{22} \\ M_2 : O_{12} \to O_{21} \end{bmatrix} \tag{1}$$

where $O_{ij}$ stands for the $i^{th}$ operation of the job in the $j^{th}$ row in the pattern. Note that the starting times of the operations are ignored. Since no delay is allowed, each operation will start as soon as it becomes ready (all its precedent operations in the same job and the same machine have been completed, and the machine is idle).

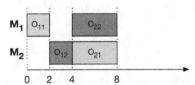

**Fig. 1.** A schedule for the JSS instance $\vartheta$ given in Table 1.

Then, the time-invariance property for DRs is defined as follows.

**Definition 3 (Time-invariant dispatching rule).** *A rule $\Upsilon$ is time-invariant if its generated schedule is invariant under the time-shift transformation, i.e. $\Phi(\Upsilon, \vartheta) = \Phi(\Upsilon, g(\vartheta, \Delta t)), \forall \vartheta, \Delta t \in \mathbb{R}$.*

For example, the SPT (shortest processing time) rule is time-invariant, as its decision making does not depend on time.

Time-invariant DRs have the following promising properties.

- Time-invariant DRs generate the same schedule for the patterns that are equivalent under time-shift. Thus, they tend to show repetitive behaviours throughout the simulation horizon, which are also likely to generalise to the same patterns occur in the unseen test instances no matter when the patterns occur. Therefore, time-invariant DRs tend to have better generalisability.

– Time-invariant DRs are good at dealing with periodic job arrivals, e.g. the same order arrivals for weekly scheduling periods. In this case, time-invariant DRs generate the same schedule for each period (equivalent patterns under time-shift) without explicitly splitting the scheduling horizon into periods. This is particularly useful when the periodic pattern is not obvious.

### 3.1  An Example: Time-Invariance v.s. Time-Dependence

In a JSS instance, the same pattern may occur at different times. An example of a JSS instance is given in Table 2. When applying the SPT rule to this instance, the schedule for the first 6 jobs is shown in Fig. 2. Obviously, the two patterns $\vartheta_{[0,8]}$ and $\vartheta_{[16,24]}$ (described in Table 3) are equivalent under time-shift. It can be seen that $\vartheta_{[16,24]} = h(\vartheta_{[0,8]}, 16)$. As a result, the time-invariant SPT rule generates the same schedule in the two time windows.

**Table 2.** An example of a JSS instance.

| $j$ | $t_0(J_j)$ | $\rho(J_j)$ | Sequence of operations |
|---|---|---|---|
| 1 | 0 | 18 | $[\langle M_1, 2 \rangle \to \langle M_2, 4 \rangle]$ |
| 2 | 2 | 20 | $[\langle M_2, 2 \rangle \to \langle M_1, 4 \rangle]$ |
| 3 | 9 | 21 | $[\langle M_1, 2 \rangle \to \langle M_2, 2 \rangle]$ |
| 4 | 10 | 28 | $[\langle M_2, 4 \rangle \to \langle M_1, 2 \rangle]$ |
| 5 | 16 | 34 | $[\langle M_1, 2 \rangle \to \langle M_2, 4 \rangle]$ |
| 6 | 18 | 36 | $[\langle M_2, 2 \rangle \to \langle M_1, 4 \rangle]$ |
| $\cdots$ | | | |

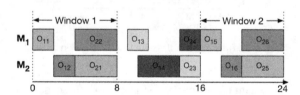

**Fig. 2.** A schedule obtained by the SPT rule for the JSS instance given in Table 2.

**Table 3.** The patterns occurring in time windows $[0, 8]$ and $[16, 24]$ of the instance shown in Table 2.

| Pattern | $t^{\mathrm{idle}}$ | $\mathcal{J}^{\mathrm{init}}$ | $\mathcal{J}^{\mathrm{new}}$ | | |
|---|---|---|---|---|---|
| | | | $t_0$ | $\rho$ | Sequence of operations |
| $\vartheta_{[0,8]}$ | $(0,0)$ | $\emptyset$ | 0 | 18 | $[\langle M_1, 2 \rangle \to \langle M_2, 4 \rangle]$ |
| | | | 2 | 20 | $[\langle M_2, 2 \rangle \to \langle M_1, 4 \rangle]$ |
| $\vartheta_{[16,24]}$ | $(16,16)$ | $\emptyset$ | 16 | 34 | $[\langle M_1, 2 \rangle \to \langle M_2, 4 \rangle]$ |
| | | | 18 | 36 | $[\langle M_2, 2 \rangle \to \langle M_1, 4 \rangle]$ |

However, GP cannot guarantee to always generate time-invariant DRs. For example, GP may generate a rule $PT \times (DD/t - PT)$ during the evolutionary process (where PT and DD stand for the process time and due date). To show that the above rule is not time-invariant (or time-dependent), we examine its behaviour in the two decision situations at time 2 and time 18. In the former decision situation, $O_{12}$ and $O_{21}$ are waiting in the queue of $M_2$. In the latter one, $O_{16}$ and $O_{25}$ are waiting in the queue of $M_2$.

**Decision situation 1.** $\text{priority}(O_{12}) = 2 \times (20/2 - 2) = 16$,
$\text{priority}(O_{21}) = 4 \times (18/2 - 4) = 20$, select $O_{12}$;
**Decision situation 2.** $\text{priority}(O_{16}) = 2 \times (36/18 - 2) = 0$,
$\text{priority}(O_{25}) = 4 \times (34/18 - 4) = -8.44$, select $O_{25}$.

As a result, the schedule obtained by the rule $PT \times (DD/t - PT)$ is shown in Fig. 3. It can be seen that although the patterns $\vartheta_{[0,8]}$ and $\vartheta_{[16,24]}$ are equivalent under time-shift, i.e. $\vartheta_{[16,24]} = h(\vartheta_{[0,8]}, 16)$, the rule $PT \times (DD/t - PT)$ generates different schedules in the two time windows.

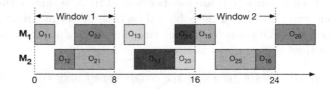

**Fig. 3.** A schedule obtained by the rule $PT \times (DD/t - PT)$ for the JSS instance given in Table 2. It generates different schedules

### 3.2 Relationship Between Existing Rule Classifications

In [1], the DRs are classified into *static* and *dynamic* rules, or *local* and *global* rules. Static rules are time-independent, i.e. the priority values do not depend on time. In dynamic rules, the priority of jobs/operations change over time. Local rules only uses the local information pertaining to either the queue where the job is waiting or to the machine where the job is queued. Global rules may use information of other machines such as the work in the next queue.

The classification of *time-invariant* and *time-dependent* rules is different from both rule classifications in the following aspects.

- Static rules are time-invariant since they make time-independent decisions. Some dynamic rules can be time-invariant as well. For example, the dynamic minimum slack rule $SL = \max\{DD - WKR - t, 0\}$ (WKR is the work remaining) is time-invariant, since SL do not change under the time-shift transformation. For any operation $O$ and decision situation $\vartheta^{ds}$, $SL(O|h(\vartheta^{ds}, \Delta t)) = \max\{DD(O|\vartheta^{ds}) + \Delta t - WKR(O|\vartheta^{ds}) - (t + \Delta t), 0\} = SL(O|\vartheta^{ds})$.
- Both local and global rules can be time-invariant. For example, the global rule $PT + WINQ$ is time-invariant, since both PT and WINQ do not depend on time.

## 4    Selection of Terminals for Time-Invariance

In Sect. 3.1, we have shown that GP cannot guarantee to evolve time-invariant DRs. In this section, we propose a new selection scheme for terminals so that GP can always generate time-invariant DRs. Recalling that *a decision situation is special instantaneous JSS pattern*. Therefore, the time-shift transformation of decision situations is the same as that of JSS patterns.

Then, we define a new concept called *time-invariant feature* as follows.

**Definition 4 (Time-invariant feature).** *A feature $\alpha$ is time-invariant if given any decision situation $\vartheta^{ds}$ and any operation $O$ waiting in the queue, the value $a(O)$ is invariant under the time-shift transformation on $\vartheta^{ds}$. That is,*

$$\alpha(O|\vartheta^{ds}) = \alpha(O|h(\vartheta^{ds}, \Delta t)), \forall \Delta t \in \mathbb{R}, \tag{2}$$

*where $\alpha(O|\vartheta^{ds})$ is the value of feature $\alpha$ of operation $O$ in $\vartheta^{ds}$.*

Based on the definition of the time-shift transformation (Definition 2), it is obvious that all the features that are independent of the idle time of the machine, the arrival time and due date of the jobs will be time-invariant. In addition, as a feature, the current time is obviously not time-invariant since its value changes along with the time-shift transformation.

**Theorem 1.** *If GP uses a terminal set consisting of only time-invariant features, then GP always generates time-invariant DRs.*

*Proof.* Given a terminal set $\mathcal{T}$, any DR $\Upsilon$ generated by GP is a priority function $f(\mathcal{T}(\cdot))$ of the terminals. For any decision situation $\vartheta^{ds}$ and operation $O$, the priority value of $O$ is $f(\mathcal{T}(O|\vartheta^{ds}))$. Since all the terminals are time-invariant features, we have

$$\alpha(O|\vartheta^{ds}) = \alpha(O|h(\vartheta^{ds}, \Delta t)), \forall \alpha \in \mathcal{T}, \Delta t \in \mathbb{R}.$$

Therefore, $f(\mathcal{T}(O|\vartheta^{ds})) = f(\mathcal{T}(O|h(\vartheta^{ds}, \Delta t)), \forall \Delta t \in \mathbb{R}$. That is, the priority values calculated by $\Upsilon$ do not change under the time-shift transformation. In other words, $\Upsilon$ is a time-invariant DR.

Based on Theorem 1, we can guarantee the time-invariance of the GP-evolved rules simply by only including time-invariant features in the terminal set. To this end, we examined the time-invariance of the commonly used features in literature [17,22], using the definition of the features and Definitions 2 and 4. If a feature is not time-invariant, then it is either removed from the terminal set, or replaced by a time-invariant counterpart. The detail is given in Table 4. It is easy to ensure the time-invariance of the time-invariant counterparts. For example, for any decision situation $\vartheta^{ds}$ and operation $O$, TIS$(O|h(\vartheta^{ds}, \Delta t)) = (t + \Delta t) - (\text{AT}(O|\vartheta^{ds}) + \Delta t) = \text{TIS}(O|\vartheta^{ds})$.

# 5    Experimental Studies

To evaluate the effectiveness of considering time-invariance, we compare between the baseline GP (BaselinGP) with the original terminals (the left part of Table 4) and the time-invariant GP (TivGP) with the time-invariant terminals (right part of Table 4).

In the experiments, we consider three objectives: (1) the maximal tardiness (Tmax), (2) the mean tardiness (Tmean) and (3) the total weighted tardiness (TWT). For each objective, we consider utilisation levels of 0.85 and 0.95. Therefore, the experiments consist of $3 \times 2 = 6$ different job shop scenarios. The configuration parameters of the simulation model are given in Table 5. This simulation configuration has been used in previous studies [21, 23].

For each scenario, we use the standard GP process to train DRs. The parameter setting of the GP is given in Table 6. Note that the only difference between BaselineGP and TivGP is the terminal set. This way, one can analyse how the use of time-invariant terminals affects the performance of GP.

During the training process, an individual is evaluated using a randomly generated simulation. To improve generalisation, the random seed for generating the training simulation changes per generation. In addition, the fitness is normalised by the objective value of the reference rule. The reference rule is set to EDD, ATC and WATC for Tmax, Tmean and TWT, respectively. Finally, the best individual in the last generation is selected as the best individual of the GP run.

For testing, a test set of 50 simulation replications is randomly generated for each scenario. The test fitness of a rule $x$ is defined as the normalised total

**Table 4.** The terminals used in [17, 22], and the time-invariant counterpart.

| Original | | Time-invariant counterpart | |
|---|---|---|---|
| Notation | Description | Notation | Description |
| $t$ | Current time | - | - |
| NIQ | Number of operations In Queue | same | same |
| WIQ | Work In Queue | same | same |
| MRT | Machine Ready Time | MWT ($t$-MRT) | Machine Waiting Time |
| PT | Processing Time | same | same |
| NPT | Next Processing Time | same | same |
| ORT | Operation Ready Time | OWT ($t$-ORT) | Operation Waiting Time |
| NRT | Next Machine Ready Time | NWT (NRT-$t$) | Next Machine Waiting Time |
| WKR | Work Remaining | same | same |
| NOR | Number of Operations Remaining | same | same |
| WINQ | Work In Next Queue | same | same |
| NINQ | Number of operations In Next Queue | same | same |
| FDD | Flow Due Date | rFDD (FDD-$t$) | Relative FDD |
| DD | Due Date | rDD (DD-$t$) | Relative DD |
| W | Weight | same | same |
| AT | Arrival Time | TIS ($t$-AT) | Time In System |
| SL | Slack | same | same |

**Table 5.** The Dynamic JSS simulation system configuration.

| Parameter | Value |
|---|---|
| #machines | 10 |
| #jobs | 5000 |
| #warmup jobs | 1000 |
| #operations per job | Random from 2 to 10 |
| Job arrival process | Poisson process |
| Utilisation level | {0.85, 0.95} |
| Due date | 4× total processing time |
| Eligible machine | Uniform discrete distribution |
| Processing time | Uniform discrete distribution between 1 and 99 |

**Table 6.** The parameter setting of GP.

| Parameter | Value |
|---|---|
| Terminal set | The original ones in Table 4 for BaselineGP |
| | The time-invariant ones in Table 4 for TivGP |
| Function set | $\{+, -, *, /, \min, \max\}$ |
| Population size | 1024 |
| Maximal depth | 8 |
| Crossover rate | 80% |
| Mutation rate | 15% |
| Reproduction rate | 5% |
| Parent selection | Tournament selection with size 7 |
| Elitism | 10 best individuals |
| Number of generations | 51 |

objective value over the test replications, i.e. $\Gamma(x, \Pi, F) = \frac{\sum_{\pi \in \Pi} F(x, \pi)}{\sum_{\pi \in \Pi} F(\text{RefRule(Obj)}, \pi)}$, where $F \in \{\text{Tmax}, \text{Tmean}, \text{TWT}\}$.

In the experiments, both BaselineGP and TivGP were implemented in Java using the ECJ library [24]. The experiments were run on desktops with Intel(R) Core(TM) i7 CPU @3.60 GHz. Both algorithms were run 30 times independently for each scenario.

## 5.1 Results and Discussions

Figure 4 shows the curves of the test fitness of the compared algorithms in 30 runs. The ribbon around each curve is the standard error of the mean. For each scenario, we conducted Wilcoxon's rank sum test between the two algorithms with the significance level of 0.05. From the figure, one can see

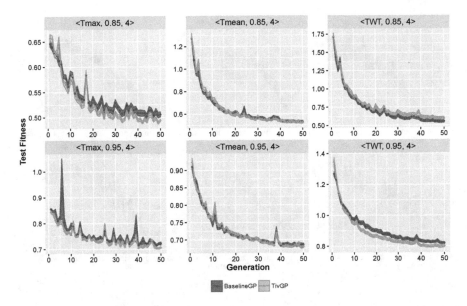

**Fig. 4.** The curves of the test fitness of the 30 runs of BaselineGP and TivGP.

that TivGP significantly outperformed BaselinGP in scenarios $\langle$Tmax, 0.85, 4$\rangle$, $\langle$Tmax, 0.95, 4$\rangle$ and $\langle$TWT, 0.95, 4$\rangle$. For the two scenarios with objective Tmean, the two algorithms performed almost the same. TivGP was defeated by BaselineGP in only the scenario $\langle$TWT, 0.85, 4$\rangle$. Overall, TivGP performed much better than BaselineGP (3 wins, 2 draws and 1 lose) over the 6 tested scenarios.

From Fig. 4, it can also be seen that in general, the curves of the test fitness is smoothly improving as the search continues. This indicates that it is reasonable to simply select the best rule in the last generation as the best rule of the run.

In addition to the test fitness, we investigate the growth of the tree size (number of nodes) of the best rules during the GP runs, which is shown in Fig. 5. From the figure, it is clear that TivGP led to significantly smaller tree sizes than BaselineGP in most of the scenarios. This implies that using the time-invariant features in the terminal set can help evolving smaller (and possibly simpler) rules.

In order to analyse the usefulness of the terminals in GP, we investigate the growth of the number of unique terminals used in the GP tree over generations, which is shown in Fig. 6. From the figure, one can clearly see that TivGP uses significantly fewer unique terminals than BaselineGP, On average, the number of unique terminals in TivGP is one or two smaller than that in BaselineGP. This is partially because TivGP has a smaller terminal set than BaselineGP (the terminal $t$ is included in BaselineGP but not in TivGP). This shows that the current time $t$ can be safely removed from the terminal set, if all the other terminals are time-invariant features.

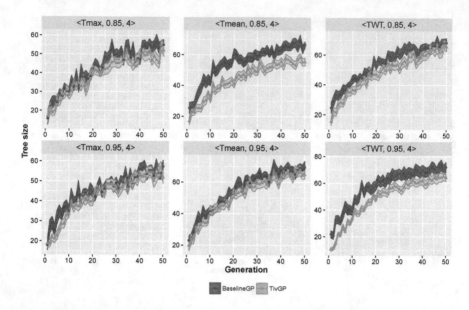

**Fig. 5.** The curves of the program size of the 30 runs of BaselineGP and TivGP.

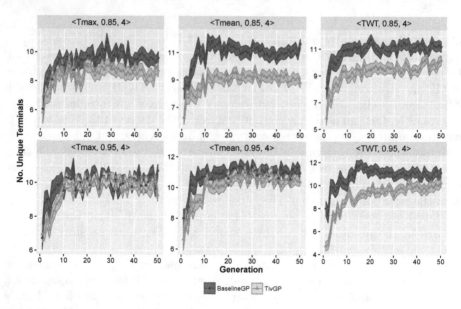

**Fig. 6.** The curves of the number of unique terminals of the 30 runs of BaselineGP and TivGP.

Figure 7 shows the generational running time of the compared algorithms. One can clearly see that TivGP is much faster than BaselineGP in all the 6 scenarios. Since the computational effort of fitness evaluation largely depends on the tree size of the individuals, Fig. 7 indicates that the individuals in TivGP are generally much smaller than the individuals in BaselineGP.

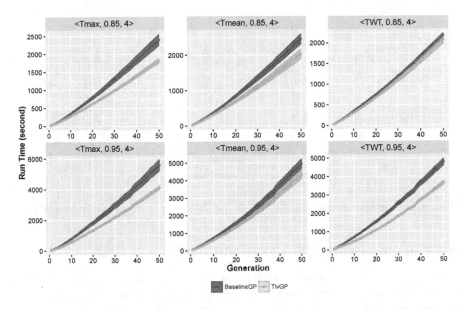

**Fig. 7.** The generational running time of the 30 runs of BaselineGP and TivGP.

## 5.2 Further Analysis

Since the test fitness depends on both the training fitness and generalisation, it is interesting to know the relationship between the training and test fitnesses to show the generalisation of the rules. To this end, we show the scatter plot of the training and test fitnesses of the best rules obtained by BaselineGP and TivGP for all the scenarios in Fig. 8.

From the figure, one can see that the two algorithms have similar generalisations in all the scenarios. Briefly speaking, the training and test fitnesses are strongly correlated. If a rule has a better training fitness, then it is very likely to have a better test fitness, no matter which algorithm it is from. For the three scenarios where TivGP outperformed BaselineGP ($\langle \text{Tmax}, 0.85, 4 \rangle$, $\langle \text{Tmax}, 0.95, 4 \rangle$ and $\langle \text{TWT}, 0.95, 4 \rangle$), the rules obtained by TivGP tend to have both better training and test fitnesses (more towards the bottom-left corner). In the scenario $\langle \text{Tmean}, 0.85, 4 \rangle$, the two algorithms have similar distributions. In the scenario $\langle \text{Tmax}, 0.95, 4 \rangle$, TivGP tends to have better generalisation (better test fitness for the same training fitness), but suffered from the two outliers in the top-right corner. In the scenario $\langle \text{TWT}, 0.85, 4 \rangle$ where TivGP was outperformed by BaselineGP, the two algorithms seem to have very similar train-vs-test distributions. The only difference is that BaselineGP has more points towards the bottom-left corner (around the region $(0.42, 0.45)$).

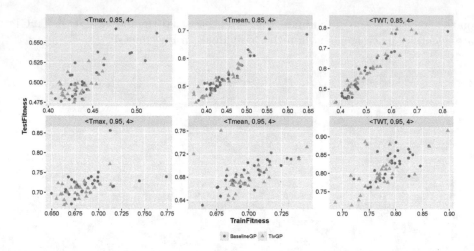

**Fig. 8.** The generational running time of the 30 runs of BaselineGP and TivGP.

## 5.3 Time-Invariance of the Evolved Rules

Obviously, all the rules evolved by TivGP are time-invariant. Then, we are interested in examining the time-invariance of the rules evolved by BaselineGP. To this end, we conduct a case study on the best rule obtained by BaselineGP for the scenario $\langle \text{TWT}, 0.85, 4 \rangle$, where BaselineGP showed better performance than TivGP. The rule is shown below:

$$((B_1 + B_2) \times B_3 + \text{WIQ}) \times B_4 \times \min\{\text{PT}, \text{MRT}\}/B_5,$$
$$\text{where } B_1 = \max\{\text{NINQ}, \text{SL}\} \times \text{W}/(\text{WKR} - \text{NIQ}),$$
$$B_2 = \max\{\text{NINQ}, \text{SL}\} + \text{NRT}/\text{PT},$$
$$B_3 = (t/\text{AT}) \times \text{WINQ} \times \text{PT}/\text{NRT},$$
$$B_4 = \max\{(\text{WKR} - \text{NIQ})/\text{W}, \text{SL}\},$$
$$B_5 = \text{WKR} - (\text{WKR} - \text{NIQ}) \times \text{PT}/\text{NRT}.$$

The rule consists of four features that are not time-invariant, i.e. NRT, $t$, AT and MRT. NRT occurs in the sub-trees NRT/PT and PT/NRT, which occur in three places. $t$ and AT occur together as a pattern $t/\text{AT}$. MRT occurs in only one pattern $\min\{\text{PT}, \text{MRT}\}$. Therefore, the rule is not time-invariant.

However, based on the feature definitions, we know that the values of these features increase as the simulation continues, and thus tend to be large in most time of the simulation. Therefore, we can transform the above rule into a time-invariant rule as follows:

– Replace NRT with a large value (1000000 in this case);
– Replace $t/\text{AT}$ with 1, which is the converged value when both $t$ and AT approach infinity.

– Replace min{PT, MRT} with PT, since PT is almost always smaller than MRT.

After the transformation, we compare the test fitness of the resultant time-invariant rule to that of the original rule, and found that their test fitnesses are very close (0.476 versus 0.448).

We have examined several other rules evolved by BaselineGP, and found that they are either time-invariant, or can be easily transformed to time-variant rules with very close test fitnesses. This also validates the motivation of evolving time-invariant rules.

In summary, we have the following findings from the experimental studies:

– By using the time-invariant features as terminals, TivGP outperformed the BaselineGP counterpart in terms of test fitness, program size and efficiency.
– BaselineGP and TivGP have similar generalisability, and the test fitnesses of the evolved rules are strongly correlated with their training fitnesses.
– Although the rules evolved by BaselineGP are not necessarily time-invariant, they can be transformed into time-invariant rules with similar test performances.

## 6    Conclusions and Future Work

In this paper, to improve the effectiveness and efficiency of GP to search in the huge space, we proposed to consider only *time-invariant* DRs during the search process. The concept of time-invariance is borrowed from the translational invariance in machine learning. To this end, we defined the concepts of *JSS pattern*, which is a time window of the scheduling horizon, and the *time-shift transformation* for JSS patterns. To guarantee that GP always generates time-invariant DRs, we proposed to select only the *time-invariant* features in the terminal set, and the resultant GP is called the time-invariant GP (TivGP). The experimental studies showed that TivGP can achieve DRs with significantly better test performances and smaller sizes than the rules obtained by the baseline GP. TivGP also has a faster convergence than the baseline GP. Furthermore, although the baseline GP sometimes outperform TivGP, the resultant time-dependent DRs can be transformed into time-invariant DRs with similar performance. This demonstrates the efficacy of considering time-invariant DRs.

In the future, we will investigate more schemes that take the time-invariance into account, such as developing new fitness functions and search operators.

## References

1. Pinedo, M.L.: Scheduling: Theory, Algorithms, and Systems. Springer Science & Business Media, New York (2012)
2. Ceberio, J., Irurozki, E., Mendiburu, A., Lozano, J.A.: A distance-based ranking model estimation of distribution algorithm for the flowshop scheduling problem. IEEE Trans. Evol. Comput. **18**(2), 286–300 (2014)

3.  Marichelvam, M.K., Prabaharan, T., Yang, X.S.: A discrete firefly algorithm for the multi-objective hybrid flowshop scheduling problems. IEEE Trans. Evol. Comput. **18**(2), 301–305 (2014)
4.  Xiong, J., Liu, J., Chen, Y., Abbass, H.A.: A knowledge-based evolutionary multi-objective approach for stochastic extended resource investment project scheduling problems. IEEE Trans. Evol. Comput. **18**(5), 742–763 (2014)
5.  Branke, J., Nguyen, S., Pickardt, C., Zhang, M.: Automated design of production scheduling heuristics: a review. IEEE Trans. Evol. Comput. **20**(1), 110–124 (2016)
6.  Blackstone, J.H., Phillips, D.T., Hogg, G.L.: A state-of-the-art survey of dispatching rules for manufacturing job shop operations. Int. J. Prod. Res. **20**(1), 27–45 (1982)
7.  Sels, V., Gheysen, N., Vanhoucke, M.: A comparison of priority rules for the job shop scheduling problem under different flow time-and tardiness-related objective functions. Int. J. Prod. Res. **50**(15), 4255–4270 (2012)
8.  Holthaus, O., Rajendran, C.: Efficient dispatching rules for scheduling in a job shop. Int. J. Prod. Econ. **48**(1), 87–105 (1997)
9.  Jayamohan, M., Rajendran, C.: New dispatching rules for shop scheduling: a step forward. Int. J. Prod. Res. **38**(3), 563–586 (2000)
10. Rajendran, C., Holthaus, O.: A comparative study of dispatching rules in dynamic flowshops and jobshops. Eur. J. Oper. Res. **116**(1), 156–170 (1999)
11. Hildebrandt, T., Heger, J., Scholz-Reiter, B.: Towards improved dispatching rules for complex shop floor scenarios: a genetic programming approach. In: Proceedings of Genetic and Evolutionary Computation Conference, pp. 257–264. ACM (2010)
12. Jakobović, D., Budin, L.: Dynamic scheduling with genetic programming. In: Collet, P., Tomassini, M., Ebner, M., Gustafson, S., Ekárt, A. (eds.) EuroGP 2006. LNCS, vol. 3905, pp. 73–84. Springer, Heidelberg (2006). doi:10.1007/11729976_7
13. Nguyen, S., Zhang, M., Johnston, M., Tan, K.: A computational study of representations in genetic programming to evolve dispatching rules for the job shop scheduling problem. IEEE Trans. Evol. Comput. **17**(5), 621–639 (2013)
14. Ho, N., Tay, J.: Evolving dispatching rules for solving the flexible job-shop problem. In: IEEE Congress on Evolutionary Computation, vol. 3, pp. 2848–2855. IEEE (2005)
15. Pickardt, C., Hildebrandt, T., Branke, J., Heger, J., Scholz-Reiter, B.: Evolutionary generation of dispatching rule sets for complex dynamic scheduling problems. Int. J. Prod. Econ. **145**(1), 67–77 (2013)
16. Hunt, R., Johnston, M., Zhang, M.: Evolving less-myopic scheduling rules for dynamic job shop scheduling with genetic programming. In: Proceedings of the 2014 Conference on Genetic and Evolutionary Computation, pp. 927–934. ACM (2014)
17. Mei, Y., Zhang, M., Nyugen, S.: Feature selection in evolving job shop dispatching rules with genetic programming. In: Proceedings of Genetic and Evolutionary Computation Conference, pp. 365–372. ACM (2016)
18. Riley, M., Mei, Y., Zhang, M.: Feature selection in evolving job shop dispatching rules with genetic programming. In: IEEE Congress on Evolutionary Computation. IEEE (2016)
19. Durasević, M., Jakobović, D., Knežević, K.: Adaptive scheduling on unrelated machines with genetic programming. Appl. Soft Comput. **48**, 419–430 (2016)
20. Ojala, T., Pietikainen, M., Maenpaa, T.: Multiresolution gray-scale and rotation invariant texture classification with local binary patterns. IEEE Trans. Pattern Anal. Mach. Intell. **24**(7), 971–987 (2002)

21. Hildebrandt, T., Branke, J.: On using surrogates with genetic programming. Evol. Comput. **23**(3), 343–367 (2015)
22. Nguyen, S., Zhang, M., Tan, K.C.: Surrogate-assisted genetic programming with simplified models for automated design of dispatching rules. IEEE Trans. Cybern. 1–15 (2016). doi:10.1109/TCYB.2016.2562674
23. Nguyen, S., Zhang, M., Johnston, M., Tan, K.C.: Dynamic multi-objective job shop scheduling: a genetic programming approach. In: Uyar, A.S., Ozcan, E., Urquhart, N. (eds.) Automated Scheduling and Planning. SCI, vol. 505, pp. 251–282. Springer, Heidelberg (2013). doi:10.1007/978-3-642-39304-4_10
24. Luke, S., et al.: A Java-based evolutionary computation research system. https:// cs.gmu.edu/~eclab/projects/ecj/

# Strategies for Improving the Distribution of Random Function Outputs in GSGP

Luiz Otavio V.B. Oliveira$^{(\boxtimes)}$, Felipe Casadei, and Gisele L. Pappa

Universidade Federal de Minas Gerais, Belo Horizonte, Brazil
{luizvbo,glpappa}@dcc.ufmg.br, felipe.casadei@gmail.com

**Abstract.** In the last years, different approaches have been proposed to introduce semantic information to genetic programming. In particular, the geometric semantic genetic programming (GSGP) and the interesting properties of its evolutionary operators have gotten the attention of the community. This paper is interested in the use of GSGP to solve symbolic regression problems, where semantics is defined by the output set generated by a given individual when applied to the training cases. In this scenario, both mutation and crossover operators defined with fitness function based on Manhattan distance use randomly built functions to generate offspring. However, the outputs of these random functions are not guaranteed to be uniformly distributed in the semantic space, as the functions are generated considering the syntactic space. We hypothesize that the non-uniformity of the semantics of these functions may bias the search, and propose three different standard normalization techniques to improve the distribution of the outputs of these random functions over the semantic space. The results are compared with a popular strategy that uses a logistic function as a wrapper to the outputs, and show that the strategies tested can improve the results of the previous method. The experimental analysis also indicates that a more uniform distribution of the semantics of these functions does not necessarily imply in better results in terms of test error.

**Keywords:** Geometric semantic genetic programming · Normalization · Logistic function

## 1 Introduction

In the last years, different approaches have been proposed to introduce semantic information to genetic programming (GP) [1,2]. In particular, geometric semantic GP (GSGP) [3] introduced geometric semantic crossover and mutation operators—for Boolean, program and arithmetic domains—capable of searching the underlying semantics of the programs. When applied to symbolic regression problems, given the set of input-output training cases, the semantics of an individual is represented by the vector of outputs the individual generates when applied to the training set. This vector is defined in a $n$-dimensional space, called semantic space.

© Springer International Publishing AG 2017
J. McDermott et al. (Eds.): EuroGP 2017, LNCS 10196, pp. 164–177, 2017.
DOI: 10.1007/978-3-319-55696-3_11

The geometric semantic crossover operator combines two parents such that the resulting offspring is located on the metric segment connecting the semantics of both parents in the semantic space. The geometric semantic mutation operator, in turn, applies a perturbation to a given individual, generating an offspring inside a hypersphere centred at the parent.

In symbolic regression problems, the aforementioned operators employ real functions, randomly generated, to combine or modify the parent individuals. These functions are sampled from the syntactic domain, which does not reflect the semantic distribution [4,5] and may lead to unexpected behaviours. In order to control the range of the semantics of these randomly generated functions, different approaches have been proposed. They include applying a logistic wrapper to the output of the functions [6] and employing interval arithmetic during the initialization of the trees in order to compute the expected input of the non-terminal nodes [7]. However, these strategies may lead to random functions with non-uniformly distributed semantics.

Given the properties of the geometric semantic operators, we hypothesize that the unevenly semantic distribution of these randomly generated functions has a negative impact on the way the geometric semantic operators explore the semantic space. Based on this hypothesis, this paper presents an analysis of the distribution of the semantics of randomly generated functions used by the crossover and mutation operators. As this analysis shows, most functions, even when they have their outputs corrected by the logistic methods, still remain non-uniformly distributed. Hence, we propose to use three different normalization strategies to restrain the output of these functions and distribute them more uniformly in the semantic space.

The results show that generating more uniformly distributed functions do not always ensure better performance in terms of test error. From the three strategies proposed, min-max$_\alpha$ is the only one capable of improving search in terms of test error when compared to the well-known logistic normalization method employed in GSGP, although the logistic z-score normalization also generates random functions with better uniform semantic distribution.

The remainder of this paper is organized as follows. Section 2 presents some background and explains the motivations for the work. Section 3 reviews related work, while Sect. 4 presents the strategies employed for function output normalization. Finally, Sect. 5 presents experimental results, and Sect. 6 draws conclusions and presents directions of future work.

## 2   Background and Motivation

This section presents the geometric semantic operators, their geometric representation in the semantic space, and shows the impact of the distribution of the semantics of the random functions employed by these operators on the resulting offspring, which serves as the main motivation for this work.

## 2.1   Geometric Semantic Genetic Programming

As previously mentioned, this paper focuses on symbolic regression problems. In this scenario, the semantics of a program $p$ represented by an individual evolved by GSGP, denoted by $s(p)$, is the $n$-element vector of outputs it produces when applied to the training cases defined by the problem, where $n$ is the size of the training set. The semantics of an individual can be represented as a point in the $n$-dimensional semantic space $\mathcal{S}$.

The geometric semantic operators presented by GSGP act on the syntax of the programs to induce a geometric behaviour on the semantic level. Given two parent individuals, $p_1$ and $p_2$, the geometric semantic crossover (GSX) operator is defined as

$$GSX(p_1, p_1) = r \cdot p_1 + (1 - r) \cdot p_2, \tag{1}$$

where $r$ is a random real constant in $[0, 1]$ (for fitness function based on Euclidean distance) or a random real function with codomain $[0, 1]$ (for fitness function based on Manhattan distance).

In the same fashion, given a parent individual $p$, the geometric semantic mutation (GSM) operator is given by

$$GSM(p, ms) = p + ms \cdot (r_1 - r_2), \tag{2}$$

where $ms \in \mathbb{R}^+$ is the mutation step and $r_1$ and $r_2$ are real functions randomly generated.

These operators generate individuals with a known geometric behaviour in $\mathcal{S}$. The offspring generated by $GSX(p_1, p_2)$ is located on the metric segment connecting $s(p_1)$ to $s(p_2)$—corresponding to a line connecting $s(p_1)$ and $s(p_2)$ when the Euclidean metric is adopted in the semantic space, or to a hyperrectangle[1] with vertices in $s(p_1)$ and $s(p_2)$ when the metric corresponds to the Manhattan distance. The offspring generated by $GSM(p, ms)$ belong to a closed ball centred in $s(p)$ with radius $\varepsilon \in \mathbb{R}^+$, where $\varepsilon$ is proportional to $ms$ [3,5].

Figure 1 presents geometric representations, in two-dimensional semantic spaces, of the region where the offspring resulting from the geometric semantic crossover and mutation operators are generated when using a Manhattan based fitness function. Given that the geometric semantic crossover operator for fitness functions based on Euclidean distance does not employ random functions in its construction, it will not be considered in this paper.

## 2.2   The Impact of the Random Functions

Figure 1 shows the regions of the semantic space where the offspring generated by the geometric semantic operators can be placed. However, the distribution of the resulting offspring within these regions is determined by the distribution of the semantics of the random functions generated by these operators (represented

---

[1] The hyperrectangle in a semantic space under the Euclidean metric is the equivalent to a segment in a semantic space under the Manhattan distance.

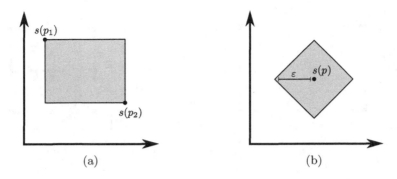

**Fig. 1.** Geometric representation of the geometric semantic (a) crossover and (b) mutation operators in two-dimensional semantic spaces defined by the Manhattan metric.

by $r$, $r_1$ and $r_2$ in Eqs. 1 and 2, respectively). If the semantics of these functions is non-uniformly distributed, the resulting offspring is prone to concentrate in some areas.

Previous empirical studies suggest that functions randomly generated—similarly to the functions generated by the geometric semantic operators—present semantics unevenly distributed. Pawlak and collaborators [4] present an empirical analysis of the distribution of behaviours of all possible trees that can be generated up to height 4 from the function set $\{+, -, \times, /, sin, cos, exp, log\}$. The results indicated that the semantics of the generated programs is highly non-uniformly distributed. Jackson [8], on the other hand, analyse the behavioural variability of individuals generated by the ramped half-and-half (RHH) method [9] in problems from different domains, including one symbolic regression problem. The results pointed out that the tree sampling performed by RHH is likely to produce individuals with the same behaviour—the number of semantically duplicated tress corresponds to about 60% of the population for the symbolic regression problem.

Another source of non-uniformity in the semantic distribution of the functions used by geometric semantic operators comes from "wrapper" functions used to restrain the output of the generated functions to the range $[0, 1]$. Moraglio et al. [3] state that the function used by GSX ($r$ in Eq. 1) has codomain $[0, 1]$. In addition, Castelli et al. [6] affirm that GSGP is able to produce better results when the output of the functions used by the geometric semantic mutation operator ($r_1$ and $r_2$ in Eq. 2) are also limited to $[0, 1]$. The solution presented in [6,10] to restrain the outputs to $[0, 1]$ consists in "wrapping" the randomly generated functions employed by GSGP operators within the standard logistic or sigmoid function

$$logis(x) = \frac{1}{1 + e^{-x}}. \tag{3}$$

However, as shown in [7], $logis(x)$ approximates to 0 and 1 for $x < -5$ and $x > 5$, respectively, biasing the outputs and impacting directly on the distribution of the offspring resulting from geometric semantic operators. To illustrate

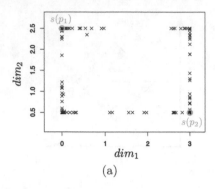 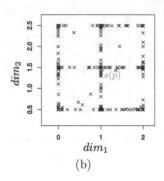

(a)                               (b)

**Fig. 2.** Distribution of the offspring resulting from (a) $GSX(p_1, p_2)$ and (b) $GSM(p, 1)$ in a two-dimensional semantic space.

this statement, assume that the functions randomly generated are uniformly distributed in $[-100, 100]^2$ in a two-dimensional semantic space—assume that the only source of non-uniformity comes from the logistic function. Although at first this range might sound odd, as it is two orders of magnitude larger than the logistic output range, these values were chosen because we are interested in simulating the behaviour of the random functions normalized by the logistic function in the GSGP literature—as observed in Sect. 5, in our testbed composed of datasets from the GP and GSGP literature, data ranges in intervals even larger than $[-100, 100]$. We generate 1,000 points uniformly distributed in $[-100, 100]^2$ for each of the functions used by the geometric semantic operators—i.e., $r$, $r_1$ and $r_2$—representing their semantics and normalize to $[0, 1]^2$ through the logistic function. Figure 2 presents the resulting offspring from $GSX(p_1, p_2)$ and $GSM(p, 1)$, respectively, where $s(p_1) = [0, 2.5]$, $s(p_2) = [3, 0.5]$ and $s(p) = [1, 1.5]$. As observed in the figure, for crossover the resulting offspring concentrates on the edges of the hyperrectangle with vertices in $s(p_1)$ and $s(p_2)$. For mutation, the resulting offspring concentrates on the edges of a hypercube and two new axes, both centred in $s(p)$.

As shown in Fig. 2, the results normalized by the logistic function are far from uniform. This behaviour makes the geometric semantic operators avoid regions of the semantic space which could lead to potentially good solutions.

## 3   Related Work

Although Moraglio et al. [3] do not present any strategy to restrain the output of the random functions employed by GSX and state nothing about the codomain of the functions used within the GSM, the logistic function proposed in [6] is widely adopted to restrain the output of these functions to $[0, 1]$—e.g., see [6, 11–13].

Given its frequent usage, Gonçalves et al. [14] investigate the impact on the search of using the logistic function to restrain the random functions generated within the GSM. The experimental analysis in two datasets indicated that the GSGP performs better in terms of test error when the random functions generated by the GSM are "wrapped" with the logistic function than when they are not.

However, Dick [7] points out limitations of the logistic function, which can negatively impact GSGP search: (1) the logistic function biases input values less than $-5$ and greater than 5, as presented in Sect. 2.2; and (2) the logistic function implicitly assumes the semantics of the randomly generated functions are centred around zero. The author suggests an alternative approach based on safe initialisation using interval arithmetic, such that the logistic function is not used to restrain the outputs. The proposed approach employs interval arithmetic during the initialization of the trees representing the random functions in order to compute the expected input of the non-terminal nodes. When the input interval of a non-terminal node disrespects some property—e.g., divisor equal to zero in a division node—the tree is penalised with a very poor fitness. Notice, however, that this strategy does not restrain the output of the initialized trees to any interval.

## 4   Strategies for Normalizing Outputs of Random Functions

As previously shown, normalizing the output of the random functions used by the crossover and mutation operators, to restrain the semantics of the functions generated to $[0, 1]^n$, leads to non-uniform distributions. Although the methods might benefit from the normalization, the problems with the logistic function and its concentration of points in certain regions of the search space (see Fig. 2) may be detrimental to the search.

In order to better distribute the outputs of the random functions employed by GSX and GSM in the same interval, we test other normalization strategies as alternatives to the logistic function. They are:

- **Min-max** normalization performs a linear transformation on the original data, preserving the relationships among the original data values [15]. In order to ensure the normalization generates values always in $[0, 1]$, even for unseen data—present in the test set—we employ a bounded version of the min-max, as presented by Eq. 4, where $x_{min}$ and $x_{max}$ are the minimum and maximum possible value of $x$, respectively.

$$minmax(x, x_{min}, x_{max}) = min\left(1, max\left(0, \frac{x - x_{min}}{x_{max} - x_{min}}\right)\right) \quad (4)$$

- **Min-max$_\alpha$** normalization is the same as the min-max, but assign the value of the $(\alpha/2)$th percentile to $x_{min}$ and the value of the $(100 - (\alpha/2))$th percentile to $x_{max}$ in Eq. 4. The rationale behind this strategy is to deal with outliers by removing data from the defined intervals.

- **Logistic z-score** normalization applies the logistic function (Eq. 3) to the data transformed by the z-score normalization, scaling the data to the range $[0, 1]$ as given by:

$$logis_z(x) = logis\left(\frac{x - \bar{x}}{s_x}\right),$$ (5)

where $\bar{x}$ and $s_x$ are the mean and standard deviation of $x$ estimated over the set of known values of $x$.

Given a finite set of input-output pairs representing the training cases, defined as $T = \{(\mathbf{x_i}, y_i)\}_{i=1}^n$—where $(\mathbf{x}_i, y_i) \in \mathbb{R}^d \times \mathbb{R}$ $(i = 1, 2, \ldots, n)$—and a function $r : \mathbb{R}^d \rightarrow \mathbb{R}$, generated by a geometric semantic operator, the normalization strategies presented above consider the set of outputs generated by $r$ when applied to $T$—i.e., its semantics—defined by the vector $s(r) = [r(\mathbf{x_1}), r(\mathbf{x_2}), \ldots, r(\mathbf{x_n})]$. Thus, the values $x_{min}$, $x_{max}$, $\bar{x}$ and $s_x$, presented in Eqs. 4 and 5, are computed over the vector $s(r)$. The resulting function, after normalization, corresponds to plugging $r$ in the right-hand side of the normalization function, replacing $x$.

The min-max is based on normalization techniques to equalize the weight of the attributes of the dataset, employed mainly in classification algorithms involving neural networks or distance measurements, such as nearest-neighbour classification and clustering [15]. The logistic z-score and min-max$_\alpha$ approaches take into account the presence of outliers which can skew the distribution of the normalized values.

## 5    Experimental Analysis

In this section we present an experimental analysis of the effect of different normalization strategies on the distribution of the semantics of randomly generated functions used by the geometric semantic operators. We analyse two different aspects: (i) the impact on the output distribution and (ii) on GSGP performance, measured by the root mean squared error (RMSE) in a testbed composed of eight datasets from real and synthetic domains, presented in Table 1.

For each real-world dataset, we performed a 5-fold cross-validation with 6 replications, resulting in 30 executions. For synthetic datasets, we adopted different approaches according to the sampling presented in Table 2. For Keijzer-5 and Vladislavleva-1 we generated 5 samples—training and test—and, for each sample, applied the algorithms 6 times, resulting again in 30 executions. For Keijzer-6 and Keijzer-7, the training and test sets were fixed, and hence we performed 30 executions.

The GSGP method adopted in our experiments employed the geometric semantic crossover for fitness function based on Manhattan distance and mutation operators, as presented in [6], both with probability 0.5. The mutation step required by the mutation operator was defined as 10% of the standard deviation of the desired outputs given by the training data. The grow method

**Table 1.** Main characteristics of the datasets used in the experiments.

| Dataset | No. of input variables | No. of instances | | Nature | Source |
|---|---|---|---|---|---|
| | | Train. | Test | | |
| Airfoil Self-noise | 5 | 1503 | | Real | [17] |
| Concrete Comp. Strength | 8 | 1030 | | Real | [18] |
| Wine Quality (Red Wine) | 11 | 1599 | | Real | [17] |
| Yacht Hydrodynamics | 6 | 768 | | Real | [17] |
| Keijzer-5 | 3 | 1000 | 10000 | Synt. | [19] |
| Keijzer-7 | 1 | 100 | 991 | Synt. | [19] |
| Keijzer-8 | 1 | 101 | 1001 | Synt. | [19] |
| Vladislavleva-1 | 2 | 100 | 2025 | Synt. | [19] |

**Table 2.** Input range and sampling strategy of the synthetic datasets. $U[a, b]$ indicates values uniformly selected from $[a, b]$ and $E[a', b']$ indicates values evenly spaced in a grid with range $[a', b']$.

| Dataset | Training set | Test set |
|---|---|---|
| Keijzer-5 | $x_1, x_3 : U[-1, 1]$ | $x_1, x_3 : U[-1, 1]$ |
| | $x_2 : U[1, 2]$ | $x_2 : U[1, 2]$ |
| Keijzer-7 | $x_1 : E[1, 100]$ | $x_1 : E[1, 100]$ |
| Keijzer-8 | $x_1 : E[0, 100]$ | $x_1 : E[0, 100]$ |
| Vladislavleva-1 | $x_1, x_2 : U[0.3, 4]$ | $x_1, x_2 : E[-0.2, 4.2]^*$ |

* Values sampled from a two-dimensional grid

[9] was adopted to generate the random functions inside the geometric semantic crossover and mutation operators, and the ramped half-and-half method [9] used to generate the initial population, both with maximum individual depth equal to 6. The terminal set included constant values randomly picked from the interval $[-1, 1]$ and the variables of the problem. Notice that the terminal set has direct impact on the outputs—normalized or not—of the randomly generated functions employed by the geometric semantic operators. The function set was composed by the functions $\{+, -, \times, AQ\}$, where AQ stands for analytic quotient [16], an alternative to the arithmetic division[2] with similar properties, but without discontinuity, defined as:

$$AQ(a, b) = \frac{a}{\sqrt{1 + b^2}}. \tag{6}$$

---

[2] The use of AQ instead of division—or protected division—makes the approach proposed by Dick [7] redundant. For this reason, we do not present the method in our experimental analysis.

## 5.1   Normalization Impact on the Distribution of the Semantics of Random Functions

The first aspect we investigated is the impact of different normalization strategies on the resulting semantics of randomly generated functions, both in real-domain and synthetic datasets adopted by GP and GSGP literature [13, 19]. In this stage, only the input variables of the dataset are considered—Tables 2 and 3 present their respective ranges.

In order to perform this study, we randomly generated 1,000 function trees using the grow method [9] with maximum depth equal to 6. Then, we applied the functions represented by the trees to the training sets of our testbed and normalized the outputs using the logistic, min-max and logistic z-score approaches. The function trees were applied to all training set samples[3]. Then, the relative frequency of the outputs is measured over the training set(s). The relative frequencies for our testbed are presented in Figs. 3 and 4. The shaded regions represent the standard deviations of the functions in a given interval.

The curves for the results of the original data and of the logistic function are overlapped in these figures. However, this behaviour is expected, since the intervals of the $x$ axis regarding the original outputs are adjusted to reflect the intervals of the logistic function.

Overall, as expected, the values normalized by the logistic function—and consequently the original data—are likely to concentrate in the middle and in the extremes (mainly the right one). However, when the data feeding the logistic function is normalized by the z-score—through the logistic z-score normalization—the data spread better in the interval $[0.2, 0.8)$. The min-max, on the other hand, is inclined to concentrate in the middle—and in some cases in the left extreme. Notice that the results for the Keijzer-5 dataset presents the better distribution, for any of the normalization strategies analysed. Also, we can observe that the curves for the datasets Keijzer-7 and Keijzer-8 are very similar given that their input ranges are almost identical—as shown in Table 2.

**Table 3.** Range of each input variable of the real-domain datasets. The range is presented by the lower and upper bounds.

| Dataset | Bound | Range of the input variables | | | | | | | | | | |
|---|---|---|---|---|---|---|---|---|---|---|---|---|
| | | $x_1$ | $x_2$ | $x_3$ | $x_4$ | $x_5$ | $x_6$ | $x_7$ | $x_8$ | $x_9$ | $x_{10}$ | $x_{11}$ |
| Airfoil | Lower | 200.00 | 0.00 | 0.03 | 31.70 | 0.00 | | | | | | |
| Self-noise | Upper | 20000.00 | 22.20 | 0.30 | 71.30 | 0.06 | | | | | | |
| Concrete | Lower | 102.00 | 0.00 | 0.00 | 121.80 | 0.00 | 801.00 | 594.00 | 1.00 | | | |
| Comp. Str | Upper | 540.00 | 359.40 | 200.10 | 247.00 | 32.20 | 1145.00 | 992.60 | 365.00 | | | |
| Wine Qual | Lower | 4.60 | 0.12 | 0.00 | 0.90 | 0.01 | 1.00 | 6.00 | 0.99 | 2.74 | 0.33 | 8.40 |
| (Red Wine) | Upper | 15.90 | 1.58 | 1.00 | 15.50 | 0.61 | 72.00 | 289.00 | 1.00 | 4.01 | 2.00 | 14.90 |
| Yacht | Lower | −5.00 | 0.53 | 4.34 | 2.81 | 2.73 | 0.13 | | | | | |
| Hydrodyn | Upper | 0.00 | 0.60 | 5.14 | 5.35 | 3.64 | 0.45 | | | | | |

---

[3] Five training partitions of the 5-fold cross-validation, 5 samples generated for Keijzer-5 and Vladislavleva-1 and a single set generated for Keijzer-6 and Keijzer-7.

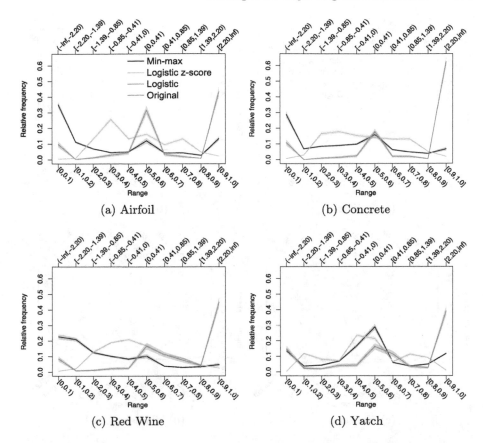

**Fig. 3.** Relative frequency of the outputs of randomly generated functions in different datasets. The bottom and top $x$ axes refer to the intervals regarding the normalized and original outputs, respectively. The intervals adopted in the top axis approximately reflect the logistic function output intervals, making the original data overlap the logistic function.

## 5.2    The Impact on the GSGP Performance

The next stage of our experimental analysis consists in comparing the performance of the GSGP—in terms of test RMSE—with different normalization strategies to restrain the output of the functions generated by the geometric semantic operators. All the experiments performed in this section used a population of 1,000 individuals evolved for 2,000 generations with tournament selection of size 10.

Tables 4 and 5 present the median and IQR (Interquartile Range) of the training and test RMSE, respectively, according to 30 executions employing each normalization method. We adopted different values of $\alpha$ for the min-max$_\alpha$, in addition to the min-max—represented in the table by $\alpha = 0\%$.

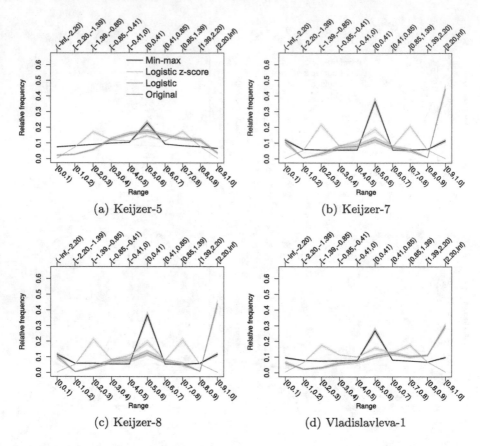

(a) Keijzer-5           (b) Keijzer-7

(c) Keijzer-8           (d) Vladislavleva-1

**Fig. 4.** Relative frequency of the outputs of randomly generated functions in different datasets (continued).

In order to analyse the statistical difference of the results we adopted a less conservative variant of the Friedman test [20], here called adjusted Friedman test. We performed an adjusted Friedman test under the null hypothesis that the performances of the methods—measured by their median test RMSE—are equal.

The resulting *p-value*, equal to 0.0132, implies discarding the null hypothesis with a confidence level of 95%. Thus, a Nemenyi post-hoc test [21] was performed to verify statistical differences among the normalization methods analysed in this paper in relation to the test RMSE. The *p-values* resulting from this test are presented in Table 6. Highlighted values indicate the method in the row is statistically better than the method in the column with 95% confidence.

The results show that the min-max$_{10\%}$ and min-max$_{20\%}$ normalizations, present significant RMSE improvement when compared to the logistic function, which indicates these methods can be used as replacement for this function, representing new alternatives to the normalization of the outputs of random func-

**Table 4.** Training RMSE (median and IQR) obtained by the normalization methods for each dataset.

| | $\alpha$ | Airfoil med. | IQR | Concrete med. | IQR | Keijzer-5 med. | IQR | Keijzer-7 med. | IQR | Keijzer-8 med. | IQR | Vladis-1 med. | IQR | Yacht med. | IQR | RedWine med. | IQR |
|---|---|---|---|---|---|---|---|---|---|---|---|---|---|---|---|---|---|
| Min-max | 0% | 1.998 | 0.195 | 4.107 | 0.111 | 0.023 | 0.003 | 0.004 | 0.005 | 0.006 | 0.009 | 0.014 | 0.003 | 1.279 | 0.130 | 0.493 | 0.011 |
| | 2% | 1.955 | 0.138 | 4.109 | 0.120 | 0.023 | 0.003 | 0.004 | 0.007 | 0.006 | 0.007 | 0.014 | 0.003 | 1.207 | 0.154 | 0.524 | 0.011 |
| | 5% | 1.886 | 0.117 | 4.114 | 0.076 | 0.023 | 0.003 | 0.004 | 0.003 | 0.008 | 0.009 | 0.013 | 0.002 | 1.107 | 0.114 | 0.519 | 0.010 |
| | 10% | 1.807 | 0.146 | 3.999 | 0.100 | 0.023 | 0.003 | 0.004 | 0.005 | 0.006 | 0.006 | 0.012 | 0.002 | 1.009 | 0.135 | 0.518 | 0.013 |
| | 20% | 1.757 | 0.171 | 3.863 | 0.146 | 0.023 | 0.003 | 0.004 | 0.004 | 0.006 | 0.012 | 0.011 | 0.003 | 1.174 | 0.166 | 0.513 | 0.012 |
| Logistic | | 7.931 | 0.532 | 3.623 | 0.088 | 0.044 | 0.003 | 0.016 | 0.008 | 0.031 | 0.018 | 0.012 | 0.001 | 2.096 | 0.226 | 0.514 | 0.012 |
| Log. z-score | | 2.214 | 0.193 | 4.104 | 0.085 | 0.025 | 0.002 | 0.003 | 0.002 | 0.005 | 0.003 | 0.010 | 0.003 | 1.634 | 0.100 | 0.504 | 0.011 |

**Table 5.** Test RMSE (median and IQR) obtained by the normalization methods for each dataset.

| | $\alpha$ | Airfoil med. | IQR | Concrete med. | IQR | Keijzer-5 med. | IQR | Keijzer-7 med. | IQR | Keijzer-8 med. | IQR | Vladis-1 med. | IQR | Yacht med. | IQR | RedWine med. | IQR |
|---|---|---|---|---|---|---|---|---|---|---|---|---|---|---|---|---|---|
| Min-max | 0% | 2.285 | 0.348 | 5.294 | 0.356 | 0.027 | 0.004 | 0.006 | 0.009 | 0.015 | 0.009 | 0.052 | 0.019 | 1.774 | 0.576 | 0.616 | 0.039 |
| | 2% | 2.245 | 0.266 | 5.269 | 0.434 | 0.027 | 0.004 | 0.006 | 0.008 | 0.014 | 0.008 | 0.063 | 0.037 | 1.736 | 0.580 | 0.605 | 0.032 |
| | 5% | 2.162 | 0.265 | 5.298 | 0.558 | 0.027 | 0.005 | 0.008 | 0.006 | 0.018 | 0.009 | 0.052 | 0.018 | 1.521 | 0.657 | 0.608 | 0.029 |
| | 10% | 2.145 | 0.381 | 5.221 | 0.440 | 0.028 | 0.004 | 0.005 | 0.007 | 0.014 | 0.006 | 0.053 | 0.022 | 1.480 | 0.435 | 0.609 | 0.026 |
| | 20% | 2.091 | 0.292 | 5.046 | 0.535 | 0.028 | 0.007 | 0.006 | 0.008 | 0.014 | 0.012 | 0.047 | 0.014 | 2.088 | 0.885 | 0.605 | 0.034 |
| Logistic | | 8.437 | 0.879 | 5.319 | 0.662 | 0.048 | 0.005 | 0.016 | 0.008 | 0.035 | 0.020 | 0.042 | 0.029 | 2.334 | 0.602 | 0.606 | 0.032 |
| Log. z-score | | 2.392 | 0.321 | 5.217 | 0.531 | 0.029 | 0.005 | 0.004 | 0.003 | 0.018 | 0.008 | 0.048 | 0.015 | 2.220 | 0.572 | 0.600 | 0.027 |

**Table 6.** P-values obtained by the Nemenyi test regarding the median test RMSE of the normalization methods.

| | | Logistic | Min-max | | | | |
|---|---|---|---|---|---|---|---|
| | $\alpha$ | – | 0% | 2% | 5% | 10% | 20% |
| Min-max | 0% | 0.24 | – | – | – | – | – |
| | 2% | 0.24 | 1.000 | – | – | – | – |
| | 5% | 0.59 | 1.000 | 1.000 | – | – | – |
| | 10% | **0.02** | 0.97 | 0.97 | 0.74 | – | – |
| | 20% | **0.02** | 0.97 | 0.97 | 0.74 | 1.00 | – |
| Log. z-score | | 0.59 | 1.00 | 0.74 | 0.74 | 1.00 | 1.00 |

tions. The logistic z-score normalization, on the other hand, does not present significant improvement regarding the other methods, although being the normalization method that generated outputs more uniformly distributed (Sect. 5.1). This fact indicates that our initial hypothesis does not always hold—i.e., random functions more uniformly distributed in the semantic space do not necessarily imply in better performance.

# 6    Conclusions and Future Work

This work presented and analysed the use of three different data normalization strategies to improve the distribution of the outputs produced by randomly generated functions used by GSGP crossover and mutation operators for fitness function based on Manhattan distance. Experiments showed how the distribution of the outputs of the functions changes with different strategies, being the distributions presented by the logistic z-score normalization one of the best. When looking at the results regarding the error obtained by GSGP with these different strategies, we observed that the min-max$_{10\%}$ and min-max$_{20\%}$ normalization strategies can improve the performance regarding median test error when compared to the logistic function. The logistic z-score normalization, in turn, besides the potential to improve the distribution of the outputs, presented no statistical difference when compared to other normalization functions, which indicates that a more uniform semantic distribution does not necessarily imply in better performance.

As future work, we intend to analyse the impact of other distributions, alternatives to the uniform, on GSGP performance, measured by the test error. We also intend to look at the impact of generating semantically distinct random functions for the geometric semantic operators.

**Acknowledgements.** The authors would like to thank the anonymous reviewers for their valuable comments and suggestions. This work was partially supported by the following Brazilian Research Support Agencies: CNPq, FAPEMIG, and CAPES.

# References

1. Vanneschi, L., Castelli, M., Silva, S.: A survey of semantic methods in genetic programming. Genet. Program. Evolvable Mach. **15**(2), 195–214 (2014)
2. Oliveira, L.: Improving search in geometric semantic genetic programming. Ph.D. thesis, Universidade Federal de Minas Gerais, September 2016
3. Moraglio, A., Krawiec, K., Johnson, C.G.: Geometric semantic genetic programming. In: Coello, C.A.C., Cutello, V., Deb, K., Forrest, S., Nicosia, G., Pavone, M. (eds.) PPSN 2012. LNCS, vol. 7491, pp. 21–31. Springer, Heidelberg (2012). doi:10.1007/978-3-642-32937-1_3
4. Pawlak, T., Wieloch, B., Krawiec, K.: Semantic backpropagation for designing search operators in genetic programming. IEEE Trans. Evol. Comput. **19**(3), 326–340 (2014)
5. Pawlak, T.P.: Competent algorithms for geometric semantic genetic programming. Ph.D. thesis, Poznan University of Technology, Pozna'n, Poland (2015)
6. Castelli, M., Silva, S., Vanneschi, L.: A C++ framework for geometric semantic genetic programming. Genet. Program. Evolvable Mach. **16**(1), 73–81 (2015)
7. Dick, G.: Improving geometric semantic genetic programming with safe tree initialisation. In: Machado, P., Heywood, M.I., McDermott, J., Castelli, M., García-Sánchez, P., Burelli, P., Risi, S., Sim, K. (eds.) EuroGP 2015. LNCS, vol. 9025, pp. 28–40. Springer, Heidelberg (2015). doi:10.1007/978-3-319-16501-1_3
8. Jackson, D.: Phenotypic diversity in initial genetic programming populations. In: Esparcia-Alcázar, A.I., Ekárt, A., Silva, S., Dignum, S., Uyar, A.Ş. (eds.) EuroGP 2010. LNCS, vol. 6021, pp. 98–109. Springer, Heidelberg (2010). doi:10.1007/978-3-642-12148-7_9
9. Koza, J.R.: Genetic Programming: On the Programming of Computers by Means of Natural Selection, vol. 1. MIT Press, Cambridge (1992)

10. Vanneschi, L., Castelli, M., Manzoni, L., Silva, S.: A new implementation of geometric semantic gp and its application to problems in pharmacokinetics. In: Krawiec, K., Moraglio, A., Hu, T., Etaner-Uyar, A.Ş., Hu, B. (eds.) EuroGP 2013. LNCS, vol. 7831, pp. 205–216. Springer, Heidelberg (2013). doi:10.1007/978-3-642-37207-0_18

11. Castelli, M., Trujillo, L., Vanneschi, L., Silva, S., Z-Flores, E., Legrand, P.: Geometric semantic genetic programming with local search. In: Proceedings of the 2015 Genetic and Evolutionary Computation Conference, GECCO 2015, pp. 999–1006. ACM, New York (2015)

12. Oliveira, L.O.V.B., Miranda, L.F., Pappa, G.L., Otero, F.E.B., Takahashi, R.H.C.: Reducing dimensionality to improve search in semantic genetic programming. In: Handl, J., Hart, E., Lewis, P.R., López-Ibáñez, M., Ochoa, G., Paechter, B. (eds.) PPSN 2016. LNCS, vol. 9921, pp. 375–385. Springer, Heidelberg (2016). doi:10.1007/978-3-319-45823-6_35

13. Oliveira, L., Otero, F.E.B., Pappa, G.L.: A dispersion operator for geometric semantic genetic programming. In: Proceedings of the Genetic and Evolutionary Computation Conference 2016, pp. 773–780. ACM (2016)

14. Gonçalves, I., Silva, S., Fonseca, C.M.: On the generalization ability of geometric semantic genetic programming. In: Machado, P., Heywood, M.I., McDermott, J., Castelli, M., García-Sánchez, P., Burelli, P., Risi, S., Sim, K. (eds.) EuroGP 2015. LNCS, vol. 9025, pp. 41–52. Springer, Heidelberg (2015). doi:10.1007/978-3-319-16501-1_4

15. Han, J., Pei, J., Kamber, M.: Data Mining: Concepts and Techniques. The Morgan Kaufmann Series in Data Management Systems. Elsevier Science, Amsterdam (2011)

16. Ni, J., Drieberg, R.H., Rockett, P.I.: The use of an analytic quotient operator in genetic programming. IEEE Trans. Evol. Comput. 17(1), 146–152 (2013)

17. Bache, K., Lichman, M.: UCI machine learning repository (2014). http://archive.ics.uci.edu/ml

18. Castelli, M., Vanneschi, L., Silva, S.: Prediction of high performance concrete strength using genetic programming with geometric semantic genetic operators. Expert Syst. Appl. 40(17), 6856–6862 (2013)

19. McDermott, J., White, D.R., Luke, S., Manzoni, L., Castelli, M., Vanneschi, L., Jaskowski, W., Krawiec, K., Harper, R., De Jong, K., O'Reilly, U.M.: Genetic programming needs better benchmarks. In: Proceedings of the 14th Annual Conference on Genetic and Evolutionary Computation, pp. 791–798. ACM (2012)

20. Iman, R.L., Davenport, J.M.: Approximations of the critical region of the Friedman statistic. Commun. Stat. - Theory Methods 9(6), 571–595 (1980)

21. Demšar, J.: Statistical comparisons of classifiers over multiple data sets. J. Mach. Learn. Res. 7, 1–30 (2006)

# Synthesis of Mathematical Programming Constraints with Genetic Programming

Tomasz P. Pawlak$^{(\boxtimes)}$ and Krzysztof Krawiec

Institute of Computing Science, Poznan University of Technology, Poznań, Poland
{tpawlak,krawiec}@cs.put.poznan.pl

**Abstract.** We identify a novel application of Genetic Programming to automatic synthesis of mathematical programming (MP) models for business processes. Given a set of examples of states of a business process, the proposed *Genetic Constraint Synthesis* (GENETICS) method constructs well-formed constraints for an MP model. The form of synthesized constraints (e.g., linear or polynomial) can be chosen accordingly to the nature of the process and the desired type of MP problem. In experimental part, we verify syntactic and semantic fidelity of the synthesized models to the actual benchmark models of varying complexity. The obtained symbolic models of constraints can be combined with an objective function of choice, fed into an off-shelf MP solver, and optimized.

**Keywords:** Linear programming · Constraint acquisition · Business process

## 1 Introduction

A business process is a sequence of steps to achieve a business goal, e.g., manufacturing of a product or delivery of goods. Once identified, a *model* of a process can be built and used to analyze the performance and capabilities of the process. A common representation of process models is that of *Mathematical Programming* (MP) [22], i.e., a mathematical structure built from variables defining the business process, constraints delineating the feasible operating conditions, and an objective function for assessing individual *candidate solutions*. MP problems are suitable for simulating business processes under different operating conditions and for optimization, i.e., determining the operating conditions that maximize performance as defined by the objective function.

A common practice is to build MP models manually by experts familiar with the process. Unfortunately, manual model building for real-world processes is laborious for several reasons. First, the relationships between the variables can be unknown and/or implicit, hence mistakenly not included in the model. Second, the true relationships between the variables may be incompatible with the requirements for the model (e.g., linearity) or difficult to represent using arithmetic formulas (e.g., logical implication), so requiring advanced modeling techniques. Third, experts that are both knowledgeable in a given business process

© Springer International Publishing AG 2017
J. McDermott et al. (Eds.): EuroGP 2017, LNCS 10196, pp. 178–193, 2017.
DOI: 10.1007/978-3-319-55696-3_12

and competent in formal modeling techniques do not come by the dozen. In effect, the models are often inconsistent with reality, and need to be revised, sometimes multiple times. However, once an accurate enough model is built, its further use is largely automated, i.e., specific instances of the modeled problem are being solved by *solvers* – tools that simulate and optimize the model.

The practical upshot of the above is that manual model *preparation* requires often more effort than its *use*. An alternative to manual model building is *learning* models from process behavior. Given a sample of process states, each composed of input variables and the associated outputs, machine learning (ML) methods can synthesize regression and classification models. However, such models represent only dependencies between variables, while ignoring the *constraints*, i.e., logical clauses that declare as feasible only some combinations of variable values. A model of a process that ignores constraints is incomplete, and in some cases even misleading: it can be queried for arbitrary inputs, also those out of range of feasible operating conditions of the process, for which the real process fails. Missing constraints are also detrimental in optimization, where an optimized solution may be in practice infeasible. Constraints are thus indispensable for comprehensive model synthesis.

This work aims at automation of constraint synthesis using *Genetic Programming* (GP) [14] and proposes *Genetic Constraint Synthesis* (GENETICS). The input to GENETICS is a set of examples of process states. The *feasible* examples represent the valid states of the process (i.e., normal, operating conditions), and the *infeasible* ones correspond to invalid (erroneous, failing) states. For instance of a real-world manufacturing process, such data can be acquired by monitoring the production volume and the corresponding parameters of production plant over time. GENETICS synthesizes a set of constraints, represented using abstract syntax trees and directly applicable in an MP model. Five search operators inspired by the tree-swapping crossover and mutation [14] are proposed and evaluated on synthetic benchmarks. We assess fidelity of the synthesized models to the true ones using a number of syntactic and semantic measures, and discuss the results.

## 2    Related Work

A related method [1] builds multidimensional convex hull that encloses the feasible examples in a bounded region using correlated constraints and clusters this region's facets to minimize their number. The authors apply the method to synthesizing constraints in robust optimization in supply chain management.

In our other work [18], we formulate constraint synthesis problem as a Mixed-Integer Linear Programming (MILP) model and solve it with the guarantee of achieving a set of minimal constraints w.r.t. the adopted measure. However, solving a MILP model is NP-hard, and the proposed method becomes infeasible for large instances of synthesis problem (i.e., such that involve many examples).

Constraint synthesis as posed in this paper resembles *constraint acquisition* considered in the constraint programming community. For instance, in [13] inductive logic programming is used to learn a set of first-order clauses from examples and preference learning to estimate the weights of those clauses. The *Model Seeker* system [2] synthesizes Prolog constraints that satisfy a set of feasible examples. The *Conacq* system [5] uses version space learning to build from examples constraints in the conjunctive normal form of terms, each performing a binary comparison of variables. The *QuAcq* interactive constraint synthesis system [4] constructs constraints of predefined types from examples with missing values. However, none of these systems can synthesize models that are compatible with the paradigm of linear and nonlinear programming.

A less related topic in MP is *cutting plane generation*, which refers to techniques used to facilitate problem solving. A solver equipped with a cutting plane generation starts with an underconstrained variant of a given problem, and then iteratively generates new cuts (i.e., constraints) to reduce the solution space. The cutting plane methods originate in the work by Gomory [7], who proposed an algorithm for mixed-integer linear programming that generates cutting planes, relaxes the requirement for the variables to be integer-valued and then solves the problem using the simplex algorithm.

A recent proposal of probabilistic nature adopts the Bayesian approach to modeling of optimization problems [9]. The authors focus on the problems in which the objective function and the constraints can be calculated independently. They come up with an *acquisition function* that estimates the desirability of a given candidate solution being evaluated, and apply their approach to benchmark problems of various difficulty, with promising outcomes.

## 3    Constraint Synthesis

Below, we describe the constraint synthesis problem and the GENETICS method.

### 3.1    Constraint Synthesis Problem

Let $x_1, x_2, ..., x_n$ be variables, a tuple $h = (x_1, x_2, ..., x_n)$ be an *example*, and $H$ be the set of examples. If an example corresponds to a feasible state of the process, i.e., represent its normal, operating conditions, we call it *feasible*, otherwise we call it *infeasible*. A *constraint* $\phi = (p, \triangle, q)$ is a logical clause of the form $p(h) \triangle q(h)$, where $\triangle$ is a comparison operator, $\triangle \in \{\leq, =, \geq\}$, $p(h) \equiv p(x_1, x_2, ..., x_n)$, $q(h) \equiv q(x_1, x_2, ..., x_n)$ are functions of $x_i$s, and $\Phi$ is a set of constraints. A *constraint synthesis problem* (*synthesis problem* for short) is defined by a set of examples $H$ partitioned into two sets $H^+$ and $H^-$ of feasible and infeasible examples, respectively. Solving a constraint synthesis problem consists in finding a set of constraints $\Phi^*$, such that:

1. All constraints $\phi_i \in \Phi^*$ are met for each feasible example (*positive requirements*), i.e., $\forall_{h \in H^+} \forall_{(p, \triangle, q) \in \Phi^*} p(h) \triangle q(h)$,
2. At least one constraint $\phi_i \in \Phi^*$ is not met for each infeasible example (*negative requirements*), i.e., $\forall_{h \in H^-} \exists_{(p, \triangle, q) \in \Phi^*} \neg(p(h) \triangle q(h))$.

In the following, we use the terms 'set of constraints' and 'model' interchangeably, even though a complete model includes also an objective function.

## 3.2   Genetic Constraint Synthesis (GENETICS)

We divide the description of GENETICS into four parts: solution representation, fitness function, search operators and post-processing.

**Representation.** We represent constraints as abstract syntax trees (ASTs [14]), with leaf nodes returning values of variables $x_i$ or constants, and inner tree nodes implementing arithmetic operations. A constraint $\phi = (p, \triangle, q)$ is represented as an AST with a comparison operator $\triangle$ in its root node and the subtrees $p$ and $q$ as the left and the right child of $\triangle$, respectively. A model $\Phi$ comprising of one or more ASTs forms an *individual*.

A class of considered constraints in $\Phi$ is determined by a set of operators used in $p$ and $q$. If they are limited to $\{+, -\}$, ASTs represent linear constraints suitable for linear programming (LP) models [22]. Since we assume $\Phi$ to be a set, it cannot contain duplicated constraints. To ensure this technically, each time a new $\Phi$ is created, we remove the syntactically duplicate ASTs.

**Fitness Function.** We assess a model $\Phi$ by calculating for each example $h_i$, $i = 1..z$ the number of violated positive and negative requirements formulated in Sect. 3.1:

$$g_i(\Phi) = \begin{cases} \sum_{(p,\triangle,q)\in\Phi} \mathbf{1}(\neg(p(h_i) \triangle q(h_i))) & \text{for } h_i \in H^+ \\ \prod_{(p,\triangle,q)\in\Phi} \mathbf{1}(p(h_i) \triangle q(h_i)) & \text{for } h_i \in H^-, \end{cases} \tag{1}$$

where $\mathbf{1}()$ is an indicator function: $\mathbf{1}(true) = 1$, $\mathbf{1}(false) = 0$. For feasible $h_i$s, $g_i(\Phi)$ calculates the number of violated constraints (violation of the positive requirement); for infeasible $h_i$s, $g_i(\Phi)$ returns 1 if all constraints $\phi \in \Phi$ are met (violation of the negative requirement). We employ *parsimony pressure* [17] by defining two extra functions:

$$g_{z+1}(\Phi) = 0.1|\Phi|$$

$$g_{z+2}(\Phi) = 0.01 \sum_{\phi\in\Phi} \text{NODES}(\phi)$$

that count respectively, the number of ASTs in $\Phi$ and the number of nodes in all ASTs in $\Phi$. The constants 0.1 and 0.01 were chosen based on model sizes observed in preliminary experiments. The contributions of all $g_i$s are then summed to form the minimized fitness function:

$$f(\Phi) = \sum_{i=1}^{z+2} g_i(\Phi). \tag{2}$$

However, particular $g_i$s are considered separately by the Lexicase selection operator, detailed in the following.

**Algorithm 1.** FULL and GROW initialization operators. $MinConstr$ and $MaxConstr$ are minimum and maximum numbers of constraints, respectively, $MaxDepth$ is maximum depth of initialized ASTs, $U(a, b)$ is an integer drawn uniformly from range $[a, b]$.

| | |
|---|---|
| 1: **function** FULL( ) | 1: **function** GROW( ) |
| 2: | 2:     $c \leftarrow U(MinConstr, MaxConstr)$ |
| 3:     $\Phi \leftarrow \emptyset$ | 3:     $\Phi \leftarrow \emptyset$ |
| 4:     **while** $|\Phi| < MaxConstr$ **do** | 4:     **while** $|\Phi| < c$ **do** |
| 5:         $\Phi \leftarrow \Phi \cup \{\text{FULLCONSTR}(1)\}$ | 5:         $\Phi \leftarrow \Phi \cup \{\text{GROWCONSTR}(1)\}$ |
| 6:     **return** $\Phi$ | 6:     **return** $\Phi$ |
| 7: **function** FULLCONSTR($d$) | 7: **function** GROWCONSTR($d$) |
| 8:     **if** $d < MaxDepth$ **then** | 8:     **if** $d < MaxDepth$ **then** |
| 9:         $r \leftarrow$ PICKNONTERMINAL( ) | 9:         $r \leftarrow$ PICKINSTRUCTION( ) |
| 10:         **for** $i = 1..\text{ARITY}(r)$ **do** | 10:         **for** $i = 1..\text{ARITY}(r)$ **do** |
| 11:             $r_i \leftarrow$ FULLCONSTR($d + 1$) | 11:             $r_i \leftarrow$ GROWCONSTR($d + 1$) |
| 12:     **else** | 12:     **else** |
| 13:         $r \leftarrow$ PICKTERMINAL( ) | 13:         $r \leftarrow$ PICKTERMINAL( ) |
| 14:     **return** $r$ | 14:     **return** $r$ |

**GP Operators.** We employ two groups of initialization and search operators: four of them being adaptations of operators for strongly typed tree-based GP [14,16], and three new search operators we designed specifically for our representation of candidate solutions.

We use FULL and GROW initialization methods presented in Algorithm 1. The main difference w.r.t. [14] is that they build a *set* of ASTs, instead of just one. FULL initializes new trees in a loop, until the assumed maximal number $MaxConstr$ of them is reached. FULL starts with the root node and picks a random terminal for the current node $r$ if the maximum tree depth $MaxDepth$ is reached; otherwise, a nonterminal is picked and FULL recursively repeats the same steps for each child node $r_i$. GROW draws a number of constraints $c$ uniformly from $[MinConstr, MaxConstr]$, and then produces $c$ ASTs by randomly picking an instruction from the set of all instructions until $MaxDepth$ depth is reached. When $MaxDepth$ is reached, GROW draws from the terminal set instead. When drawing, we consider only the instructions of the type required by the current context (the parent AST node). We produce a population using the Ramped Half-and-Half (RHH) scheme [14], calling FULL with 50% probability and GROW otherwise.

Algorithm 2 presents Constraint Tree Swapping Crossover (CTX) and Constraint Tree Swapping Mutation (CTM). Given two parents $\Phi_1$ and $\Phi_2$, CTX removes two randomly selected ASTs $\phi_1$ and $\phi_2$ respectively from each parent. Next, the strongly-typed tree swapping crossover [14,16] draws one subtree in $\phi_1$ and one in $\phi_2$, and swaps them, producing so new ASTs $\phi_1'$ and $\phi_2'$. They are added to offspring $\Phi_1'$ and $\Phi_2'$, respectively. These steps repeat until either $\Phi_1$ or $\Phi_2$ is emptied. The remaining constraints in the non-emptied parent are copied intact to the respective offspring.

**Algorithm 2.** Constraint Tree Crossover (CTX) and Constraint Tree Mutation (CTM) operators. $\text{POP}(X)$ picks uniformly and removes an item from $X$, $\text{TREESWAPPINGXOVER}(\cdot, \cdot)$ is the strongly-typed tree swapping crossover [16,14].

| | |
|---|---|
| 1: **function** $\text{CTX}(\Phi_1, \Phi_2)$ | 1: **function** $\text{CTM}(\Phi)$ |
| 2:     $\Phi_1', \Phi_2' \leftarrow \emptyset$ | 2:     $\Phi_r \leftarrow \text{RHH}(\ )$ |
| 3:     **while** $\Phi_1 \neq \emptyset \wedge \Phi_2 \neq \emptyset$ **do** | 3:     $\{\Phi', \Phi_r'\} \leftarrow \text{CTX}(\Phi, \Phi_r)$ |
| 4:         $\phi_1 \leftarrow \text{POP}(\Phi_1)$ | 4:     **return** $\Phi'$ |
| 5:         $\phi_2 \leftarrow \text{POP}(\Phi_2)$ | |
| 6:         $\{\phi_1', \phi_2'\} \leftarrow \text{TREESWAPPINGXOVER}(\phi_1, \phi_2)$ | |
| 7:         $\Phi_1' \leftarrow \Phi_1' \cup \{\phi_1'\}$ | |
| 8:         $\Phi_2' \leftarrow \Phi_2' \cup \{\phi_2'\}$ | |
| 9:     $\Phi_1' \leftarrow \Phi_1' \cup \Phi_1$ | |
| 10:    $\Phi_2' \leftarrow \Phi_2' \cup \Phi_2$ | |
| 11:    **return** $\{\Phi_1', \Phi_2'\}$ | |

**Algorithm 3.** Constraint Swapping Crossover (CSX) and Constraint Swapping Mutation (CSM) operators. $U(a, b)$ is an integer drawn uniformly from the range $[a, b]$.

| | |
|---|---|
| 1: **function** $\text{CSX}(\Phi_1, \Phi_2)$ | 1: **function** $\text{CSM}(\Phi)$ |
| 2:     $\Phi_1', \Phi_2' \leftarrow \emptyset$ | 2:     $\Phi_r \leftarrow \text{RHH}(\ )$ |
| 3:     **for** $\phi \in \Phi_1 \cup \Phi_2$ **do** ▷ Duplicates are allowed | 3:     $\{\Phi', \Phi_r'\} \leftarrow \text{CSX}(\Phi, \Phi_r)$ |
| 4:         $i \leftarrow U(1, 2)$ | 4:     **return** $\Phi'$ |
| 5:         $\Phi_i' \leftarrow \Phi_i' \cup \{\phi\}$ | |
| 6:     **return** $\{\Phi_1', \Phi_2'\}$ | |

**Algorithm 4.** Gaussian Constant Mutation (GCM) operator. $N(\mu, \sigma)$ is a number drawn from normal distribution of $\mu$ mean and $\sigma$ standard deviation.

1: **function** $\text{GCM}(\Phi)$
2:     $\Phi' \leftarrow \emptyset$
3:     **for** $\phi \in \Phi$ **do**
4:         $c \leftarrow \text{PICKCONSTANT}(\phi)$
5:         $c' \leftarrow N(c, 1)$
6:         $\phi' \leftarrow \text{REPLACE}(\phi, c, c')$
7:         $\Phi' \leftarrow \Phi' \cup \{\phi'\}$
8:     **return** $\Phi'$

**Algorithm 5.** Lexicase parent selection operator. $\text{SHUFFLE}(X)$ randomly orders elements in $X$, $\text{PICK}(X)$ uniformly picks an element from $X$.

1. **function** $\text{LEXICASE}(P)$
2:     $P' \leftarrow P$
3:     **for** $i \in \text{SHUFFLE}(\{1, 2, .., z + 2\})$ **do**
4:         $P' \leftarrow \{\Phi \in P' : g_i(\Phi) = \min_{\Phi \in P'} g_i(\Phi)\}$
5:         **if** $|P'| = 1$ **then**
6:             **break**
7:     **return** $\text{PICK}(P')$

CTM takes one parent $\Phi$, randomly initializes another individual $\Phi_r$ using RHH, and applies CTX to $\Phi$ and $\Phi_r$, producing so a pair of offspring, of which it returns the first one (i.e., the one corresponding to $\Phi$).

Algorithm 3 presents the Constraint Swapping Crossover (CSX) and Constraint Swapping Mutation (CSM) operators. CSX mixes the constraints of the given parents $\Phi_1$ and $\Phi_2$ without modifying the constraint trees. It iterates over the list composed of constraints in both parents and adds each constraint to the randomly chosen offspring: $\Phi_1'$ or $\Phi_2'$.

CSM, given a parent $\Phi$, initializes another individual $\Phi_r$ using RHH and applies CSX to $\Phi$ and $\Phi_r$, producing so the offspring $\Phi'$ and $\Phi_r'$, of which the first one is returned.

The expected cardinalities of the offspring $\Phi_1'$ and $\Phi_2'$ produced by CSX are equal to the mean cardinalities of $\Phi_1$ and $\Phi_2$, i.e., $E(|\Phi_1'|) = E(|\Phi_2'|) = \frac{|\Phi_1|+|\Phi_2|}{2}$. However, in an extreme case, one of the offspring may inherit all constraints of both parents, and the other one none. Inheriting all constraints can be beneficial when synthesizing a model for a highly constrained process. The empty offspring, if crossed-over with a non-empty individual $\Phi$ in the next generation, causes $\Phi$'s constraints to be split into two offspring, each one being simpler than $\Phi$.

To provide for fine-tuning of coefficients in evolving constraints, we introduce also the Gaussian Constant Mutation (GCM) presented in Algorithm 4. Given a parent $\Phi$, for each AST $\phi \in \Phi$ GCM picks uniformly a constant terminal tree node $c$, replaces it with a constant drawn from the Gaussian distribution $N(c, 1)$, and adds the produced constraint $\phi'$ to the offspring $\Phi'$. If $\phi$ does not contain any constants, it is copied to $\Phi'$ intact.

For selection, we use the Lexicase operator [8] presented in Algorithm 5. Given a population $P$, LEXICASE copies all individuals to the selection pool $P'$. Then, it iterates over a set of fitness cases in random order, and for each fitness case $i$ preserves in $P'$ the individuals $\Phi$ that have the best assessment on $i$ w.r.t. $g_i(\Phi)$. If $P'$ shrinks to one individual, LEXICASE terminates and returns that individual. After iterating over all fitness cases, if $P'$ contains more than one individual, LEXICASE returns an individual randomly drawn from $P'$.

**Post-processing.** The constraints synthesized using the above operators may become complex, which is common phenomenon observed in GP [15]. To address this problem, we employ three post-processing mechanisms that simplify constraints and keep their size and number at bay.

In every generation, after offspring are bred, we remove from them the constraints $(p, \triangle, q)$ in which $p$ and $q$ are constant functions, as they are either always met or always violated. This simplifies offspring and simultaneously repairs an offspring that would be otherwise violated for all feasible examples.

When GP terminates, we remove from the best found model $\Phi$ the constraints not met for any feasible example, or not violated for any infeasible example, since they bring nothing to the separation of examples. We also symbolically simplify[1] the constraints in $\Phi$ to reduce the size of expressions and improve readability.

---

[1] Technically, we use the MathNet.Symbolics library [20].

# 4   Experiment

## 4.1   Setup

We carry out two experiments. In the first one, we determine which operators from Sect. 3.2 lead to best fitness, i.e., (2) calculated for the training set of examples. In the second experiment, we employ the best GP setup determined in the first experiment and assess several properties of the resulting models, including syntactic fidelity to the actual benchmark model (the agreement between the symbolic representations), as well as semantic accuracy (the agreement of model's feasible region with the actual one).

In both experiments, we use two families of synthetic benchmarks: $n$-dimensional balls and $n$-dimensional cubes, for $n \in [3, 7]$ (Table 1). A Ball$n$ benchmark uses one quadratic constraint with $n + 1$ variables, while Cube$n$ uses $2n$ linear constraints, each involving a single variable. These benchmarks are to verify how well GENETICS performs in extreme requirements for the synthesized model: Ball$n$ one requires synthesizing one quite complex constraint, while Cube$n$ many simple constraints. For each benchmark, we draw $z$ examples of which $z/2$ are feasible, drawn uniformly from the feasible region, and $z/2$ are infeasible and drawn uniformly from the infeasible region, i.e., from the hypercube formed by the Cartesian product of variables' domains minus the feasible region. These examples serve as input data to the method.

The parameters of evolution are the same in both experiments (Table 2). The *Linear* instruction set enables synthesis of linear constraints only, via the

**Table 1.** Benchmark problems (MP models). $b$ is a parameter that determines the domains of variables, set to the non-integer value of 2.7 to make the ranges resemble arbitrary ranges occurring in real world problems.

| Ball$n$ | Cube$n$ |
|---|---|
| $r^2 \geq \sum_{i=1}^{n}(x_i - i)^2$ | $\forall_{i=1}^{n} : x_i \geq i$ |
| $r \in [0.1b, 10b]$ | $\forall_{i=1}^{n} : x_i \leq i + b$ |
| $\forall_{i=1}^{n} : x_i \in [0.1i, 10i]$ | $\forall_{i=1}^{n} : x_i \in [i - b, i + 2b]$ |

**Table 2.** Parameters of evolution.

| Parameter | Value |
|---|---|
| Population size | 2000 |
| Max generations | 100 |
| $MinConstr$, $MaxConstr$ | 1, 14 |
| $MaxDepth$ | 6 |
| Linear instruction set | $\leq, =, \geq, +, -, \times, x_i, 1, \text{ERC}$ |
| Polynomial instruction set | $\leq, =, \geq, +, -, \times, *, x^2, x_i, 1, \text{ERC}$ |
| Initialization method | Ramped Half-and-Half |
| Parent selection method | Lexicase selection |
| Runs per benchmark | 30 |

use of strongly typed multiplication × that accepts as the left argument only subtrees formed of instructions from $\{+, -\}$ and constants (and thus evaluate to a constant value). The *Polynomial* instruction set is a superset of *Linear* and includes also the square function and multiplication ∗ that accepts any subtrees as arguments. ERCs are drawn from $N(0, 1)$. Comparison operators may appear in root nodes only and other instructions must not occur there.

## 4.2   Evaluation of GP Setups

We assess GENETICS on five GP setups, corresponding to the search operators from Sect. 3.2. Each setup employs the corresponding operator with 60% probability and the remaining ones with 10% probability each, providing the former one with dominating impact:

```
CTX setup: CTX 60%, CTM 10%, CSX: 10%, CSM: 10%, GCM: 10%,
CTM setup: CTX 10%, CTM 60%, CSX: 10%, CSM: 10%, GCM: 10%,
CSX setup: CTX 10%, CTM 10%, CSX: 60%, CSM: 10%, GCM: 10%,
CSM setup: CTX 10%, CTM 10%, CSX: 10%, CSM: 60%, GCM: 10%,
GCM setup: CTX 10%, CTM 10%, CSX: 10%, CSM: 10%, GCM: 60%.
```

We set the number of examples $z = 500$ throughout this experiment (other $z$s not shown due to space limit, however they lead to similar conclusions).

Figure 1 presents the mean and .95-confidence interval of the best-so-far fitness over generations, for linear and polynomial models and – due to limited

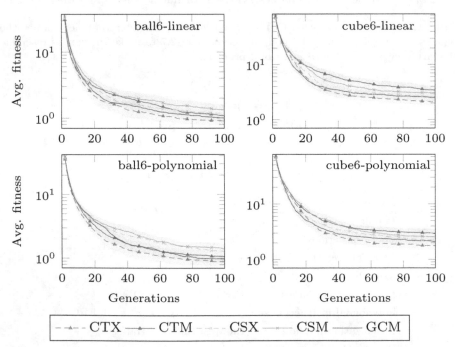

**Fig. 1.** Mean and .95-confidence interval of the best-so-far fitness over generations.

**Table 3.** Mean and .95-confidence interval of the best-of-run fitness; best in bold.

*Linear models*

| Problem | CTX | CTM | CSX | CSM | GCM |
|---|---|---|---|---|---|
| ball3 | **1.74** ± 0.27 | 2.54 ± 0.42 | 1.99 ± 0.40 | 2.54 ± 0.38 | 1.82 ± 0.28 |
| ball4 | 2.41 ± 0.37 | 3.22 ± 0.40 | 2.81 ± 0.50 | 3.07 ± 0.49 | **2.36** ± 0.42 |
| ball5 | **1.64** ± 0.25 | 2.60 ± 0.56 | 2.27 ± 0.35 | 2.09 ± 0.36 | 1.86 ± 0.32 |
| ball6 | **0.90** ± 0.13 | 1.09 ± 0.14 | 1.18 ± 0.21 | 1.34 ± 0.26 | 1.01 ± 0.12 |
| ball7 | **0.59** ± 0.07 | 0.68 ± 0.11 | 0.77 ± 0.12 | 0.75 ± 0.14 | 0.63 ± 0.09 |
| cube3 | **1.10** ± 0.11 | 2.00 ± 0.36 | 1.30 ± 0.32 | 1.49 ± 0.22 | 1.19 ± 0.20 |
| cube4 | **1.49** ± 0.14 | 3.27 ± 0.52 | 1.95 ± 0.33 | 2.63 ± 0.52 | 1.87 ± 0.31 |
| cube5 | 2.25 ± 0.32 | 3.22 ± 0.46 | 2.14 ± 0.25 | 2.89 ± 0.50 | **2.06** ± 0.33 |
| cube6 | **2.08** ± 0.29 | 3.44 ± 0.63 | 2.58 ± 0.44 | 3.06 ± 0.62 | 2.50 ± 0.40 |
| cube7 | **2.01** ± 0.27 | 2.76 ± 0.54 | 2.48 ± 0.35 | 3.03 ± 0.42 | 2.25 ± 0.52 |
| Rank: | 1.30 | 4.40 | 3.30 | 4.20 | 1.80 |

*Polynomial models*

| Problem | CTX | CTM | CSX | CSM | GCM |
|---|---|---|---|---|---|
| ball3 | 1.52 ± 0.33 | **1.48** ± 0.36 | 1.86 ± 0.37 | 2.31 ± 0.53 | 1.63 ± 0.31 |
| ball4 | **1.66** ± 0.33 | 2.73 ± 0.55 | 2.12 ± 0.47 | 2.60 ± 0.58 | 1.78 ± 0.28 |
| ball5 | **1.50** ± 0.33 | 1.83 ± 0.38 | 1.84 ± 0.36 | 2.15 ± 0.47 | 1.79 ± 0.56 |
| ball6 | **0.79** ± 0.08 | 1.13 ± 0.24 | 0.98 ± 0.14 | 1.15 ± 0.28 | 0.96 ± 0.14 |
| ball7 | **0.58** ± 0.10 | 0.68 ± 0.16 | 0.63 ± 0.10 | 0.71 ± 0.13 | 0.63 ± 0.11 |
| cube3 | 1.55 ± 0.25 | 3.29 ± 0.71 | **1.38** ± 0.29 | 2.18 ± 0.45 | 1.41 ± 0.24 |
| cube4 | **2.40** ± 0.57 | 4.30 ± 0.77 | 3.27 ± 0.63 | 3.04 ± 0.64 | 2.41 ± 0.48 |
| cube5 | **2.08** ± 0.33 | 5.69 ± 0.76 | 2.66 ± 0.44 | 3.68 ± 0.69 | 2.24 ± 0.39 |
| cube6 | **2.36** ± 0.42 | 4.30 ± 0.78 | 3.52 ± 0.77 | 3.73 ± 0.78 | 2.83 ± 0.48 |
| cube7 | **2.50** ± 0.40 | 5.67 ± 0.92 | 3.30 ± 0.52 | 3.83 ± 0.70 | 2.86 ± 0.52 |
| Rank: | 1.30 | 4.20 | 3.00 | 4.30 | 2.20 |

space – for a representative selection of problems. Table 3 presents the same statistics for the best-of-run fitness for all considered problems. For linear models, the CTX setup achieves the top best-of-run fitness in eight out of ten problems and maintains its top position for almost entire runs in Fig. 1. The runner-up is GCM, which achieves the best average fitness on two problems.

Among the setups that use the *Polynomial* instruction set, CTX holds the top best-of-run fitness in eight out of ten problems and maintains top fitness throughout the entire runs. The GCM setup is runner-up with the second-best aggregated rank. The remaining setups are worse for both instruction sets.

To assess the significance of the differences between setups, we conduct Friedman's test for multiple achievements of multiple subjects [12], separately for linear and polynomial models, resulting in the conclusive p-values of $1.08 \times 10^{-4}$

**Table 4.** Post-hoc analysis of Friedman's test on data from Table 3. The p-values of incorrectly judging a setup in a row as worse than a setup in a column; p-values ≤ 0.05 are in bold and visualized as arcs in the outranking graphs.

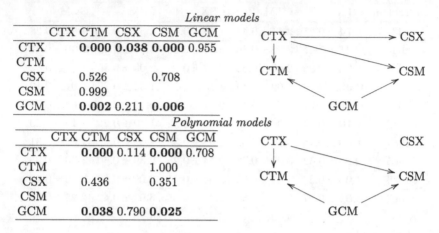

*Linear models*

|      | CTX   | CTM   | CSX   | CSM   | GCM   |
|------|-------|-------|-------|-------|-------|
| CTX  |       | **0.000** | **0.038** | **0.000** | 0.955 |
| CTM  |       |       |       |       |       |
| CSX  |       | 0.526 |       | 0.708 |       |
| CSM  |       | 0.999 |       |       |       |
| GCM  |       | **0.002** | 0.211 | **0.006** |       |

*Polynomial models*

|      | CTX   | CTM   | CSX   | CSM   | GCM   |
|------|-------|-------|-------|-------|-------|
| CTX  |       | **0.000** | 0.114 | **0.000** | 0.708 |
| CTM  |       |       |       | 1.000 |       |
| CSX  |       | 0.436 |       | 0.351 |       |
| CSM  |       |       |       |       |       |
| GCM  |       | **0.038** | 0.790 | **0.025** |       |

and $2.07 \times 10^{-4}$, respectively. To find the pairs of significantly different setups, we carry out post-hoc analysis using symmetry test [10] and present the p-values and the outranking graphs in Table 4. CTX is the best setup that outranks CTM and CSM for both types of models and CSX for linear models.

### 4.3   Evaluation of Synthesized Models

In this experiment our goal is to quantitatively assess the syntactic and semantic properties of the synthesized models. To verify the impact of training set size, we conduct a series of experiments for $z \in \{20, 30, 40, 50, 100, 200, 300, 400, 500, 1000\}$. Given the dominating performance of CTX in the previous experiment, we use that setup to evolve models throughout the rest of this paper.

Table 5 shows the mean and .95-confidence interval of the difference between the number of constraints in the synthesized model and the number of constraints in the benchmark. The number of synthesized constraints clearly increases with the number of examples $z$. This indicates increasing difficulty of separating the examples, since the greater $z$, the smaller the mean distance between the feasible and infeasible examples. The number of constraints also decreases with dimensionality $n$ of the problem. By the *curse of dimensionality* [3], the spatial density of examples decreases exponentially with $n$, and it becomes easier for GP to separate them using a low number of constraints. However, according to two-tailed $t$-test at $\alpha = 0.05$, for Ball$n$ the mean number of constraints is equal to the actual number if $z \leq 40$ and $n \geq 5$ for linear models, and $z \leq 40$ and $n \geq 5$ for polynomial models. The observed pattern of such cases (the underlined values) might suggest that GENETICS performs *better* for larger $n$ and same number of examples $z$, which would contradict the curse of dimensionality. The explanation to this puzzle is the fact that Ball$n$ features only one actual constraint,

**Table 5.** Mean and .95-confidence interval of difference of the number of the synthesized and the actual constraints; positive values stand for the former being greater, negative for the latter; underlining marks indifference to 0 w.r.t. two-tailed $t$-test at $\alpha = 0.05$.

|  | Balln | | | | | Cuben | | | | |
|---|---|---|---|---|---|---|---|---|---|---|
| | *Linear models* | | | | | | | | | |
| $z\backslash n$ | 3 | 4 | 5 | 6 | 7 | $n$: 3 | 4 | 5 | 6 | 7 |
| 20 | 0.10 ±0.11 | 0.10 ±0.11 | 0.00 ±0.00 | 0.03 ±0.06 | 0.00 ±0.00 | −3.93 ±0.16 | −5.93 ±0.16 | −7.97 ±0.11 | −10.10 ±0.14 | −12.07 ±0.13 |
| 30 | 0.30 ±0.16 | 0.17 ±0.13 | 0.07 ±0.09 | 0.03 ±0.06 | 0.00 ±0.00 | −3.57 ±0.27 | −5.57 ±0.26 | −7.43 ±0.20 | −9.53 ±0.20 | −11.77 ±0.18 |
| 40 | 0.37 ±0.17 | 0.20 ±0.14 | 0.03 ±0.06 | 0.03 ±0.06 | 0.00 ±0.00 | −3.07 ±0.23 | −5.30 ±0.21 | −7.17 ±0.26 | −9.23 ±0.24 | −11.20 ±0.25 |
| 50 | 0.57 ±0.20 | 0.33 ±0.17 | 0.23 ±0.18 | 0.07 ±0.09 | 0.03 ±0.06 | −2.83 ±0.31 | −4.63 ±0.22 | −6.87 ±0.29 | −8.80 ±0.25 | −10.90 ±0.34 |
| 100 | 1.03 ±0.24 | 1.07 ±0.26 | 0.77 ±0.27 | 0.23 ±0.15 | 0.03 ±0.06 | −1.47 ±0.27 | −3.27 ±0.29 | −5.33 ±0.27 | −7.37 ±0.31 | −9.20 ±0.30 |
| 200 | 1.97 ±0.30 | 2.17 ±0.26 | 1.83 ±0.28 | 0.93 ±0.24 | 0.40 ±0.22 | −0.33 ±0.17 | −1.60 ±0.22 | −3.70 ±0.23 | −5.27 ±0.31 | −7.20 ±0.28 |
| 300 | 2.63 ±0.25 | 2.80 ±0.35 | 2.63 ±0.40 | 1.53 ±0.32 | 0.77 ±0.26 | −0.07 ±0.09 | −1.07 ±0.26 | −2.67 ±0.30 | −4.47 ±0.32 | −6.43 ±0.33 |
| 400 | 3.23 ±0.34 | 3.53 ±0.30 | 3.00 ±0.33 | 2.03 ±0.35 | 1.03 ±0.31 | 0.00 ±0.00 | −0.60 ±0.24 | −1.93 ±0.28 | −3.90 ±0.31 | −5.40 ±0.30 |
| 500 | 3.80 ±0.40 | 4.40 ±0.38 | 3.80 ±0.52 | 2.53 ±0.47 | 1.27 ±0.29 | 0.07 ±0.09 | −0.03 ±0.17 | −1.27 ±0.29 | −3.30 ±0.26 | −4.90 ±0.27 |
| 1000 | 5.83 ±0.43 | 6.67 ±0.50 | 6.63 ±0.65 | 4.63 ±0.50 | 2.57 ±0.38 | 0.30 ±0.23 | 0.30 ±0.19 | −0.43 ±0.26 | −1.87 ±0.27 | −3.43 ±0.39 |
| | *Polynomial models* | | | | | | | | | |
| $z\backslash n$ | 3 | 4 | 5 | 6 | 7 | $n$: 3 | 4 | 5 | 6 | 7 |
| 20 | 0.08 ±0.10 | 0.04 ±0.07 | 0.04 ±0.07 | 0.04 ±0.07 | 0.00 ±0.00 | −4.81 ±0.15 | −6.92 ±0.11 | −8.92 ±0.11 | −10.96 ±0.07 | −12.96 ±0.07 |
| 30 | 0.12 ±0.12 | 0.15 ±0.14 | 0.04 ±0.07 | 0.04 ±0.07 | 0.00 ±0.00 | −4.58 ±0.24 | −6.64 ±0.19 | −8.56 ±0.22 | −10.54 ±0.24 | −12.85 ±0.14 |
| 40 | 0.31 ±0.21 | 0.08 ±0.10 | 0.12 ±0.13 | 0.04 ±0.07 | 0.00 ±0.00 | −4.19 ±0.32 | −6.48 ±0.25 | −8.08 ±0.37 | −10.54 ±0.22 | −12.38 ±0.28 |
| 50 | 0.35 ±0.18 | 0.27 ±0.17 | 0.27 ±0.17 | 0.08 ±0.10 | 0.04 ±0.07 | −4.08 ±0.38 | −6.28 ±0.30 | −7.64 ±0.38 | −9.92 ±0.40 | −12.08 ±0.35 |
| 100 | 0.54 ±0.27 | 0.73 ±0.23 | 0.50 ±0.22 | 0.23 ±0.16 | 0.08 ±0.10 | −2.42 ±0.48 | −4.64 ±0.53 | −6.56 ±0.46 | −8.58 ±0.39 | −10.50 ±0.42 |
| 200 | 1.00 ±0.38 | 1.69 ±0.37 | 1.65 ±0.28 | 1.00 ±0.28 | 0.23 ±0.19 | −1.38 ±0.51 | −2.96 ±0.46 | −4.60 ±0.41 | −6.46 ±0.34 | −8.04 ±0.52 |
| 300 | 2.04 ±0.43 | 2.12 ±0.44 | 2.12 ±0.40 | 1.42 ±0.34 | 0.69 ±0.32 | −0.77 ±0.34 | −1.96 ±0.44 | −3.76 ±0.39 | −5.65 ±0.41 | −7.23 ±0.42 |
| 400 | 2.15 ±0.45 | 3.19 ±0.49 | 2.96 ±0.38 | 1.92 ±0.38 | 0.85 ±0.30 | −0.31 ±0.33 | −1.04 ±0.44 | −2.44 ±0.52 | −4.77 ±0.33 | −6.31 ±0.42 |
| 500 | 3.35 ±0.64 | 4.38 ±0.60 | 3.58 ±0.70 | 2.31 ±0.40 | 1.15 ±0.31 | 0.19 ±0.19 | −0.84 ±0.44 | −2.52 ±0.42 | −4.04 ±0.49 | −5.73 ±0.42 |
| 1000 | 4.62 ±0.75 | 7.38 ±0.73 | 6.65 ±0.87 | 4.46 ±0.52 | 2.35 ±0.40 | 0.88 ±0.49 | 1.08 ±0.43 | −0.76 ±0.47 | −2.38 ±0.44 | −4.00 ±0.53 |

and the candidate models need at least one constraint to separate examples, so it is unlikely in this benchmark to have fewer constraints than needed. For Cuben, with $2n$ actual constraints, this characteristics does not hold anymore: the $t$-test concludes equality only for few cases where $z \geq 300$ and most models have fewer constraints than the benchmark. The numbers of constraints seem to be independent from instruction set.

To analyze the number of terms in the synthesized models, we transform constraints' expressions into expanded polynomial form (weighted sum of products of variables),[1] for both synthesized models and benchmarks. Then, we count the number of terms (summands) in the constraints transformed in this way. Table 6 shows the mean and .95-confidence interval of the difference between those numbers for the synthesized constraints and for the actual constraints in the benchmarks. For linear models, most values are close to zero, i.e., the number of terms is similar to the actual one. The differences diminish with increasing $z$. The two-tailed $t$-test concludes no difference for Balln for $z \geq 200$ and $n \leq 3$, and for Cuben for $z \geq 50$ and $n \leq 4$. Better results for Cuben probably stem from the linear representation of constraints in that benchmark. On the other hand, the synthesized polynomial models are clearly overgrown. They contain more terms than the benchmarks for about half of combinations of $z$ and $n$ for Balln, and for all but four combinations for Cuben. Nevertheless, for some of them, particularly for the Balln benchmarks, the two-tailed $t$-test renders mean difference statistically indifferent from 0.

Finally, we assess how well the feasible regions of the synthesized models match the actual feasible regions in the benchmarks. Table 7 presents the mean and the .95-confidence interval of *Jaccard index* [11] that measures the overlap of those regions, and ranges from 0 (no overlap) to 1 (perfect overlap). To estimate

**Table 6.** Mean and .95-confidence interval of difference of the number of the synthesized and the actual terms; positive values stand for the former being greater, negative for the latter; underlining marks indifference to 0 w.r.t. two-tailed $t$-test at $\alpha = 0.05$.

Ball$n$ / Cube$n$

Linear models

| $z\backslash n$ | 3 | 4 | 5 | 6 | 7 | $n$: | 3 | 4 | 5 | 6 | 7 |
|---|---|---|---|---|---|---|---|---|---|---|---|
| 20 | $-0.97$ ±0.28 | $-1.87$ ±0.32 | $-2.90$ ±0.32 | $-3.90$ ±0.28 | $-4.73$ ±0.23 | | $-0.93$ ±0.24 | $-1.30$ ±0.38 | $-2.53$ ±0.29 | $-2.90$ ±0.51 | $-3.87$ ±0.35 |
| 30 | $-0.77$ ±0.18 | $-1.60$ ±0.22 | $-2.20$ ±0.28 | $-3.50$ ±0.27 | $-4.73$ ±0.21 | | $-0.17$ ±0.52 | $-0.97$ ±0.35 | $-2.30$ ±0.29 | $-2.73$ ±0.31 | $-3.40$ ±0.39 |
| 40 | $-0.63$ ±0.17 | $-1.17$ ±0.23 | $-1.63$ ±0.27 | $-3.07$ ±0.26 | $-4.43$ ±0.26 | | $\underline{-0.13}$ ±0.46 | $-0.83$ ±0.39 | $-1.37$ ±0.52 | $-2.03$ ±0.63 | $-2.93$ ±0.58 |
| 50 | $-0.63$ ±0.17 | $-1.07$ ±0.23 | $-1.60$ ±0.25 | $-2.73$ ±0.26 | $-4.30$ ±0.26 | | $\underline{0.83}$ ±1.51 | $-0.57$ ±0.80 | $-0.53$ ±1.66 | $-2.03$ ±0.59 | $-3.03$ ±0.31 |
| 100 | $-0.37$ ±0.17 | $-0.77$ ±0.20 | $-1.03$ ±0.24 | $-2.00$ ±0.32 | $-3.10$ ±0.32 | | $\underline{0.10}$ ±0.30 | $-0.23$ ±0.24 | $-1.07$ ±0.28 | $-1.13$ ±0.95 | $-1.93$ ±0.44 |
| 200 | $\underline{-0.03}$ ±0.06 | $-0.47$ ±0.18 | $-0.70$ ±0.25 | $-1.30$ ±0.26 | $-2.20$ ±0.35 | | $0.00$ ±0.00 | $-0.07$ ±0.13 | $-0.53$ ±0.20 | $-0.80$ ±0.34 | $-0.93$ ±0.32 |
| 300 | $\underline{-0.03}$ ±0.06 | $-0.20$ ±0.14 | $-0.67$ ±0.17 | $-1.03$ ±0.24 | $-1.83$ ±0.28 | | $\underline{0.07}$ ±0.13 | $\underline{-0.03}$ ±0.06 | $-0.33$ ±0.17 | $-0.87$ ±0.29 | $-0.83$ ±0.29 |
| 400 | $\underline{0.07}$ ±0.13 | $-0.17$ ±0.16 | $-0.47$ ±0.18 | $-0.77$ ±0.26 | $-1.43$ ±0.33 | | $\underline{0.00}$ ±0.00 | $\underline{0.00}$ ±0.00 | $-0.13$ ±0.12 | $-0.43$ ±0.22 | $-0.57$ ±0.22 |
| 500 | $\underline{0.10}$ ±0.19 | $\underline{-0.20}$ ±0.14 | $-0.30$ ±0.16 | $-0.73$ ±0.21 | $-1.20$ ±0.27 | | $\underline{0.00}$ ±0.00 | $\underline{0.00}$ ±0.00 | $-0.07$ ±0.09 | $-0.33$ ±0.17 | $-0.47$ ±0.22 |
| 1000 | $\underline{0.00}$ ±0.00 | $\underline{-0.03}$ ±0.06 | $-0.27$ ±0.16 | $-0.30$ ±0.16 | $-0.67$ ±0.27 | | $\underline{0.00}$ ±0.00 | $\underline{0.00}$ ±0.00 | $\underline{0.03}$ ±0.06 | $-0.10$ ±0.11 | $-0.27$ ±0.21 |

Polynomial models

| $z\backslash n$ | 3 | 4 | 5 | 6 | 7 | $n$: | 3 | 4 | 5 | 6 | 7 |
|---|---|---|---|---|---|---|---|---|---|---|---|
| 20 | $-0.88$ ±0.47 | $-1.58$ ±0.40 | $-2.88$ ±0.36 | $-3.81$ ±0.30 | $-4.73$ ±0.25 | | $0.77$ ±0.79 | $\underline{0.64}$ ±0.97 | $-0.76$ ±1.08 | $-2.38$ ±0.95 | $-3.19$ ±0.68 |
| 30 | $-0.65$ ±0.30 | $-1.31$ ±0.33 | $-1.62$ ±0.97 | $-3.27$ ±0.33 | $-4.42$ ±0.39 | | $2.58$ ±1.31 | $0.84$ ±0.81 | $\underline{0.28}$ ±0.73 | $-1.46$ ±0.62 | $-2.04$ ±0.87 |
| 40 | $\underline{-0.19}$ ±0.38 | $-0.92$ ±0.26 | $-1.76$ ±0.45 | $-3.04$ ±0.27 | $-4.19$ ±0.32 | | $2.65$ ±1.48 | $2.04$ ±1.62 | $\underline{2.40}$ ±2.50 | $4.73$ ±3.87 | $\underline{0.08}$ ±1.62 |
| 50 | $\underline{-0.08}$ ±0.71 | $\underline{-0.54}$ ±0.57 | $-1.27$ ±0.74 | $-2.73$ ±0.36 | $-3.88$ ±0.33 | | $3.23$ ±1.49 | $4.00$ ±2.10 | $4.00$ ±3.14 | $5.19$ ±4.43 | $6.77$ ±5.05 |
| 100 | $1.19$ ±1.06 | $\underline{2.50}$ ±2.35 | $-0.38$ ±0.65 | $-1.92$ ±0.24 | $-3.00$ ±0.48 | | $3.69$ ±2.26 | $5.40$ ±3.57 | $2.88$ ±2.47 | $11.42$ ±9.05 | $\underline{8.46}$ ±14.38 |
| 200 | $3.88$ ±1.72 | $1.08$ ±0.80 | $\underline{-0.46}$ ±0.62 | $-1.04$ ±0.38 | $-2.00$ ±0.32 | | $2.54$ ±1.27 | $6.92$ ±3.76 | $9.48$ ±6.27 | $6.69$ ±5.21 | $26.65$ ±40.02 |
| 300 | $5.65$ ±4.10 | $5.73$ ±4.82 | $\underline{3.77}$ ±4.48 | $\underline{0.69}$ ±2.33 | $-1.23$ ±0.64 | | $3.54$ ±1.47 | $11.88$ ±4.86 | $20.98$ ±15.51 | $10.69$ ±11.31 | $10.35$ ±5.34 |
| 400 | $6.77$ ±2.72 | $\underline{13.73}$ ±14.11 | $2.27$ ±3.16 | $\underline{1.54}$ ±3.28 | $-0.96$ ±0.73 | | $5.04$ ±2.74 | $18.04$ ±7.14 | $16.08$ ±6.88 | $23.88$ ±16.12 | $11.85$ ±10.58 |
| 500 | $8.62$ ±5.25 | $6.54$ ±4.83 | $4.08$ ±2.80 | $\underline{2.73}$ ±5.65 | $-0.88$ ±0.42 | | $4.92$ ±2.34 | $14.72$ ±7.47 | $17.76$ ±11.13 | $21.23$ ±12.32 | $44.81$ ±31.25 |
| 1000 | $11.27$ ±4.93 | $11.85$ ±11.72 | $3.42$ ±3.21 | $0.50$ ±1.17 | $2.15$ ±2.59 | | $8.42$ ±3.50 | $12.24$ ±6.17 | $19.96$ ±8.46 | $46.15$ ±33.44 | $29.35$ ±16.10 |

**Table 7.** Mean and .95-confidence interval of Jaccard index of feasible regions of the synthesized and the actual models.

Ball$n$ / Cube$n$

Linear models

| $z\backslash n$ | 3 | 4 | 5 | 6 | 7 | $n$: | 3 | 4 | 5 | 6 | 7 |
|---|---|---|---|---|---|---|---|---|---|---|---|
| 20 | 0.75 ±0.02 | 0.54 ±0.05 | 0.31 ±0.04 | 0.10 ±0.01 | 0.05 ±0.01 | | 0.10 ±0.01 | 0.03 ±0.00 | 0.01 ±0.00 | 0.00 ±0.00 | 0.00 ±0.00 |
| 30 | 0.80 ±0.03 | 0.59 ±0.04 | 0.40 ±0.05 | 0.14 ±0.01 | 0.05 ±0.01 | | 0.15 ±0.02 | 0.04 ±0.00 | 0.01 ±0.00 | 0.00 ±0.00 | 0.00 ±0.00 |
| 40 | 0.82 ±0.02 | 0.64 ±0.03 | 0.44 ±0.04 | 0.22 ±0.04 | 0.06 ±0.01 | | 0.17 ±0.02 | 0.05 ±0.01 | 0.02 ±0.00 | 0.00 ±0.00 | 0.00 ±0.00 |
| 50 | 0.83 ±0.02 | 0.64 ±0.03 | 0.46 ±0.04 | 0.24 ±0.04 | 0.07 ±0.01 | | 0.19 ±0.03 | 0.06 ±0.01 | 0.02 ±0.00 | 0.01 ±0.00 | 0.00 ±0.00 |
| 100 | 0.87 ±0.01 | 0.73 ±0.02 | 0.52 ±0.02 | 0.27 ±0.03 | 0.13 ±0.02 | | 0.41 ±0.05 | 0.13 ±0.02 | 0.04 ±0.00 | 0.01 ±0.00 | 0.00 ±0.00 |
| 200 | 0.92 ±0.01 | 0.80 ±0.02 | 0.62 ±0.02 | 0.35 ±0.03 | 0.17 ±0.02 | | 0.65 ±0.07 | 0.24 ±0.03 | 0.08 ±0.01 | 0.02 ±0.00 | 0.01 ±0.00 |
| 300 | 0.93 ±0.00 | 0.82 ±0.01 | 0.66 ±0.02 | 0.39 ±0.03 | 0.20 ±0.01 | | 0.74 ±0.05 | 0.31 ±0.03 | 0.11 ±0.02 | 0.04 ±0.01 | 0.01 ±0.00 |
| 400 | 0.94 ±0.00 | 0.85 ±0.01 | 0.67 ±0.02 | 0.43 ±0.02 | 0.21 ±0.02 | | 0.77 ±0.03 | 0.38 ±0.05 | 0.15 ±0.02 | 0.04 ±0.01 | 0.01 ±0.00 |
| 500 | 0.94 ±0.00 | 0.86 ±0.01 | 0.69 ±0.01 | 0.45 ±0.02 | 0.24 ±0.02 | | 0.81 ±0.03 | 0.48 ±0.05 | 0.18 ±0.02 | 0.05 ±0.01 | 0.02 ±0.00 |
| 1000 | 0.96 ±0.00 | 0.89 ±0.01 | 0.76 ±0.01 | 0.53 ±0.02 | 0.29 ±0.02 | | 0.89 ±0.02 | 0.59 ±0.03 | 0.32 ±0.04 | 0.09 ±0.01 | 0.03 ±0.01 |

Polynomial models

| $z\backslash n$ | 3 | 4 | 5 | 6 | 7 | $n$: | 3 | 4 | 5 | 6 | 7 |
|---|---|---|---|---|---|---|---|---|---|---|---|
| 20 | 0.73 ±0.03 | 0.50 ±0.05 | 0.30 ±0.04 | 0.11 ±0.02 | 0.04 ±0.01 | | 0.07 ±0.01 | 0.02 ±0.00 | 0.01 ±0.00 | 0.00 ±0.00 | 0.00 ±0.00 |
| 30 | 0.78 ±0.03 | 0.59 ±0.05 | 0.36 ±0.04 | 0.15 ±0.03 | 0.05 ±0.01 | | 0.10 ±0.02 | 0.03 ±0.00 | 0.01 ±0.00 | 0.00 ±0.00 | 0.00 ±0.00 |
| 40 | 0.79 ±0.03 | 0.66 ±0.03 | 0.41 ±0.04 | 0.22 ±0.04 | 0.06 ±0.01 | | 0.12 ±0.02 | 0.03 ±0.01 | 0.01 ±0.00 | 0.00 ±0.00 | 0.00 ±0.00 |
| 50 | 0.79 ±0.05 | 0.64 ±0.05 | 0.47 ±0.04 | 0.22 ±0.04 | 0.09 ±0.02 | | 0.14 ±0.03 | 0.04 ±0.01 | 0.02 ±0.00 | 0.00 ±0.00 | 0.00 ±0.00 |
| 100 | 0.86 ±0.03 | 0.72 ±0.04 | 0.50 ±0.03 | 0.29 ±0.03 | 0.12 ±0.02 | | 0.31 ±0.06 | 0.09 ±0.02 | 0.03 ±0.00 | 0.01 ±0.00 | 0.00 ±0.00 |
| 200 | 0.80 ±0.09 | 0.77 ±0.04 | 0.59 ±0.02 | 0.35 ±0.04 | 0.18 ±0.03 | | 0.49 ±0.08 | 0.17 ±0.03 | 0.06 ±0.01 | 0.02 ±0.00 | 0.01 ±0.00 |
| 300 | 0.87 ±0.05 | 0.74 ±0.07 | 0.60 ±0.06 | 0.38 ±0.03 | 0.20 ±0.02 | | 0.60 ±0.06 | 0.21 ±0.03 | 0.08 ±0.01 | 0.03 ±0.00 | 0.01 ±0.00 |
| 400 | 0.85 ±0.08 | 0.83 ±0.03 | 0.65 ±0.03 | 0.43 ±0.03 | 0.22 ±0.02 | | 0.60 ±0.08 | 0.26 ±0.04 | 0.10 ±0.02 | 0.03 ±0.01 | 0.01 ±0.00 |
| 500 | 0.86 ±0.06 | 0.81 ±0.07 | 0.66 ±0.04 | 0.42 ±0.04 | 0.22 ±0.03 | | 0.64 ±0.08 | 0.33 ±0.06 | 0.12 ±0.02 | 0.04 ±0.01 | 0.01 ±0.00 |
| 1000 | 0.90 ±0.06 | 0.84 ±0.06 | 0.71 ±0.05 | 0.54 ±0.02 | 0.25 ±0.03 | | 0.65 ±0.09 | 0.48 ±0.06 | 0.20 ±0.04 | 0.07 ±0.02 | 0.02 ±0.00 |

the volumes of feasible regions, we use Monte Carlo sampling with 50,000 examples drawn uniformly from the Cartesian product of domains of all variables (Table 1).

The results for all benchmarks are consistent in that Jaccard index decreases with $n$, which is again due to curse of dimensionality. The Jaccard index increases with $z$, which helps reproducing the true feasible region. The observed indexes are higher for linear models than the corresponding polynomial ones, which may suggest some degree of overfitting in the latter. The one-tailed $t$-test concludes that all means are significantly smaller than 1 at $\alpha = 0.05$.

**Table 8.** Mean and .95-confidence interval of test-set accuracy of the synthesized models.

| | Balln | | | | | | Cuben | | | | |
|---|---|---|---|---|---|---|---|---|---|---|---|
| | *Linear models* | | | | | | | | | | |
| $z\backslash n$ | 3 | 4 | 5 | 6 | 7 | $n$: | 3 | 4 | 5 | 6 | 7 |
| 20 | 0.89 ±0.03 | 0.84 ±0.03 | 0.88 ±0.04 | 0.88 ±0.03 | 0.88 ±0.02 | | 0.79 ±0.03 | 0.77 ±0.04 | 0.73 ±0.03 | 0.73 ±0.04 | 0.71 ±0.04 |
| 30 | 0.91 ±0.02 | 0.89 ±0.03 | 0.90 ±0.03 | 0.91 ±0.02 | 0.90 ±0.02 | | 0.80 ±0.03 | 0.83 ±0.03 | 0.79 ±0.03 | 0.79 ±0.03 | 0.74 ±0.03 |
| 40 | 0.92 ±0.02 | 0.91 ±0.02 | 0.92 ±0.02 | 0.93 ±0.02 | 0.93 ±0.02 | | 0.87 ±0.02 | 0.85 ±0.03 | 0.79 ±0.04 | 0.80 ±0.03 | 0.79 ±0.02 |
| 50 | 0.92 ±0.02 | 0.91 ±0.02 | 0.93 ±0.02 | 0.94 ±0.01 | 0.93 ±0.02 | | 0.89 ±0.02 | 0.87 ±0.02 | 0.84 ±0.02 | 0.85 ±0.03 | 0.85 ±0.02 |
| 100 | 0.95 ±0.01 | 0.94 ±0.01 | 0.96 ±0.01 | 0.96 ±0.01 | 0.97 ±0.01 | | 0.95 ±0.01 | 0.92 ±0.01 | 0.92 ±0.01 | 0.90 ±0.01 | 0.93 ±0.01 |
| 200 | 0.97 ±0.01 | 0.96 ±0.01 | 0.97 ±0.01 | 0.97 ±0.01 | 0.98 ±0.00 | | 0.98 ±0.01 | 0.97 ±0.01 | 0.96 ±0.01 | 0.96 ±0.01 | 0.97 ±0.01 |
| 300 | 0.97 ±0.00 | 0.96 ±0.00 | 0.97 ±0.00 | 0.97 ±0.00 | 0.98 ±0.00 | | 0.99 ±0.00 | 0.98 ±0.00 | 0.97 ±0.00 | 0.98 ±0.00 | 0.98 ±0.00 |
| 400 | 0.97 ±0.00 | 0.97 ±0.00 | 0.97 ±0.00 | 0.98 ±0.00 | 0.98 ±0.00 | | 0.99 ±0.00 | 0.98 ±0.00 | 0.98 ±0.00 | 0.98 ±0.00 | 0.98 ±0.00 |
| 500 | 0.98 ±0.00 | 0.97 ±0.00 | 0.97 ±0.00 | 0.98 ±0.00 | 0.99 ±0.00 | | 0.99 ±0.00 | 0.99 ±0.00 | 0.99 ±0.00 | 0.99 ±0.00 | 0.99 ±0.00 |
| 1000 | 0.99 ±0.00 | 0.98 ±0.00 | 0.98 ±0.00 | 0.99 ±0.00 | 0.99 ±0.00 | | 1.00 ±0.00 | 0.99 ±0.00 | 1.00 ±0.00 | 0.99 ±0.00 | 0.99 ±0.00 |
| | *Polynomial models* | | | | | | | | | | |
| $z\backslash n$ | 3 | 4 | 5 | 6 | 7 | $n$: | 3 | 4 | 5 | 6 | 7 |
| 20 | 0.90 ±0.02 | 0.83 ±0.04 | 0.87 ±0.04 | 0.84 ±0.04 | 0.88 ±0.02 | | 0.75 ±0.04 | 0.77 ±0.03 | 0.70 ±0.04 | 0.71 ±0.03 | 0.67 ±0.03 |
| 30 | 0.91 ±0.02 | 0.87 ±0.04 | 0.90 ±0.03 | 0.92 ±0.02 | 0.91 ±0.02 | | 0.81 ±0.03 | 0.79 ±0.03 | 0.77 ±0.03 | 0.76 ±0.04 | 0.73 ±0.03 |
| 40 | 0.92 ±0.02 | 0.92 ±0.02 | 0.92 ±0.02 | 0.93 ±0.02 | 0.93 ±0.02 | | 0.84 ±0.02 | 0.82 ±0.03 | 0.77 ±0.03 | 0.76 ±0.03 | 0.74 ±0.03 |
| 50 | 0.92 ±0.02 | 0.91 ±0.02 | 0.94 ±0.02 | 0.95 ±0.01 | 0.94 ±0.02 | | 0.88 ±0.02 | 0.82 ±0.03 | 0.84 ±0.03 | 0.81 ±0.03 | 0.80 ±0.02 |
| 100 | 0.95 ±0.01 | 0.94 ±0.01 | 0.95 ±0.01 | 0.97 ±0.01 | 0.97 ±0.01 | | 0.95 ±0.01 | 0.91 ±0.01 | 0.91 ±0.01 | 0.88 ±0.02 | 0.91 ±0.02 |
| 200 | 0.97 ±0.01 | 0.95 ±0.01 | 0.97 ±0.01 | 0.97 ±0.01 | 0.98 ±0.00 | | 0.98 ±0.00 | 0.95 ±0.01 | 0.95 ±0.01 | 0.95 ±0.01 | 0.96 ±0.01 |
| 300 | 0.98 ±0.00 | 0.97 ±0.00 | 0.97 ±0.01 | 0.98 ±0.00 | 0.98 ±0.00 | | 0.98 ±0.00 | 0.97 ±0.01 | 0.96 ±0.01 | 0.97 ±0.00 | 0.97 ±0.01 |
| 400 | 0.98 ±0.00 | 0.97 ±0.00 | 0.97 ±0.00 | 0.98 ±0.00 | 0.99 ±0.00 | | 0.99 ±0.00 | 0.98 ±0.00 | 0.97 ±0.01 | 0.98 ±0.00 | 0.98 ±0.00 |
| 500 | 0.98 ±0.00 | 0.98 ±0.00 | 0.97 ±0.00 | 0.98 ±0.00 | 0.99 ±0.00 | | 0.99 ±0.00 | 0.98 ±0.00 | 0.98 ±0.00 | 0.98 ±0.00 | 0.98 ±0.00 |
| 1000 | 0.99 ±0.00 | 0.98 ±0.00 | 0.98 ±0.00 | 0.99 ±0.00 | 0.99 ±0.00 | | 0.99 ±0.00 | 0.99 ±0.00 | 0.99 ±0.00 | 0.99 ±0.00 | 0.99 ±0.00 |

Table 8 shows the mean and the .95-confidence interval of test-set *accuracy* [6], i.e., the empirical probability that a point is correctly included in a feasible or infeasible region. The test-set of $z$ examples for this measure is drawn the same way like the training set (i.e., differently than for Jaccard index).

Overall, the accuracy is high, most often well above 0.9, even though in all cases it is significantly smaller than 1 according to one-tailed $t$-test. It clearly increases with $z$, i.e., the models synthesized from larger training sets reproduce the feasible region more accurately. Accuracy seems also to be largely independent from $n$. This is important from practical view, since real-world problems are often highly dimensional. When the number of examples is low ($\leq 40$), accuracy is significantly higher for Balln than for Cuben. We hypothesize that a low number of actual constraints facilitates achieving high accuracy.

## 5   Conclusion

The conducted experiments indicate the overall capability of GENETICS to synthesize good approximations of constraints, especially in semantic terms (cf. the results for Jaccard index and accuracy of classification). The syntactic similarity of synthesized models to the actual constraints, as measured in terms of the number of constraints and terms, is not entirely satisfying: the models tend to be more complex than the actual constraints. However, such redundancy will only be troublesome in usage scenarios related to knowledge discovery, where human insight is required. When understanding of the underlying model is not essential, having a few more constraints than the actual ones is not necessarily very problematic for contemporary solvers (it is typically the number of variables, not constraints, that determines the computational overhead).

In a complete usage scenario, one would apply GENETICS to a (preferably large) sample of feasible and infeasible examples of behavior of a real-world business process, obtain a symbolic model of constraints, combine it with an objective function of choice, feed them together into an off-shelf MP solver, and obtain an optimized solution. As the performance in terms of semantic indicators suggest, such a solution would be likely feasible w.r.t. the actual process, and as such implementable in practice. The benchmarks used in this paper represent the extremes on the number of required constraints and their complexity; real-world problems are often intermediate in those respects, so GENETICS should be able to cope with them too. In this way, GENETICS addresses the problem signaled in Introduction, i.e., it can significantly reduce the time required to prepare a model or automate its synthesis altogether. However, it goes without saying that the accuracy of the synthesized constraints (and thus also the actual feasibility of solutions optimized with them) highly depends on the quality and amount of training data.

The curse of dimensionality remains a challenge for GENETICS: most performance indicators observe deterioration with increasing dimensionality. We hypothesize that this holds also for other problems (both synthetic and taken from the real world), independently on the nature of the problem, in particular whether the volume of the feasible region relative to the volume of the domain tends to decrease or increase with growing dimensionality. This weakness needs to be addressed in further work, as real-world problems tend to be highly-dimensional. On the other hand, it is quite obvious that it is not entirely the number, but the distribution of examples that is critical to GENETICS's performance. A relatively small sample of 'critical' examples, located close to the feasibility boundary, can be sufficient for complete separation of feasible and infeasible regions by the synthesized constraints. Thus, one of the promising further research directions is combining GENETICS with appropriate sampling techniques. Another future research direction is use of semantics in search operators [19,21] to improve GENETICS convergence.

**Acknowledgment.** T. Pawlak was supported by the statutory activity of Poznan University of Technology, grant no. 09/91/DSMK/0606. K. Krawiec was supported by the National Science Centre, Poland, grant no. 2014/15/B/ST6/05205.

# References

1. Aswal, A., Prasanna, G.N.S.: Estimating correlated constraint boundaries from timeseries data: the multi-dimensional German tank problem. In: 24th European Conference on Operational Research (2010). http://slideplayer.com/slide/7976536/. Accessed 09 May 2016
2. Beldiceanu, N., Simonis, H.: A model seeker: extracting global constraint models from positive examples. In: Milano, M. (ed.) 18th International Conference on Principles and Practice of Constraint Programming (CP 2012. LNCS, vol. 7514, pp. 141–157. Springer, Quebec City (2012)
3. Bellman, R.: Dynamic Programming. Dover Books on Computer Science. Dover Publications, New York (2013)

4. Bessiere, C., Coletta, R., Hebrard, E., Katsirelos, G., Lazaar, N., Narodytska, N., Quimper, C.G., Walsh, T.: Constraint acquisition via partial queries. In: International Joint Conference on Artificial Intelligence (2013)
5. Bessiere, C., Coletta, R., Koriche, F., O'Sullivan, B.: A SAT-based version space algorithm for acquiring constraint satisfaction problems. In: Gama, J., Camacho, R., Brazdil, P.B., Jorge, A.M., Torgo, L. (eds.) ECML 2005. LNCS (LNAI), vol. 3720, pp. 23–34. Springer, Heidelberg (2005). doi:10.1007/11564096_8
6. Flach, P.: Machine Learning: The Art and Science of Algorithms That Make Sense of Data. Cambridge University Press, New York (2012)
7. Gomory, R.E.: An algorithm for the mixed integer problem. Technical report, RM-2597-PR, 30 August 1960
8. Helmuth, T., Spector, L., Matheson, J.: Solving uncompromising problems with lexicase selection. IEEE Trans. Evol. Comput. **19**(5), 630–643 (2015)
9. Hernández-Lobato, J.M., Gelbart, M.A., Adams, R.P., Hoffman, M.W., Ghahramani, Z.: A General Framework for Constrained Bayesian Optimization using Information-based Search. ArXiv e-prints, November 2015
10. Hothorn, T., Hornik, K., van de Wiel, M.A., Zeileis, A.: Package 'coin': conditional inference procedures in a permutation test framework (2015). http://cran.r-project.org/web/packages/coin/coin.pdf
11. Jaccard, P.: The distribution of the flora in the alpine zone. New Phytol. **11**(2), 37–50 (1912)
12. Kanji, G.: 100 Statistical Tests. SAGE Publications, New York (1999)
13. Kolb, S.: Learning constraints and optimization criteria. In: AAAI Workshops (2016)
14. Koza, J.R.: Genetic Programming: On the Programming of Computers by Means of Natural Selection. MIT Press, Cambridge (1992). http://mitpress.mit.edu/books/genetic-programming
15. Langdon, W.B., Soule, T., Poli, R., Foster, J.A.: The evolution of size and shape. In: Spector, L., Langdon, W.B., O'Reilly, U.M., Angeline, P.J. (eds.) Advances in Genetic Programming, vol. 3, pp. 163–190. MIT Press, Cambridge (1999). (Chap. 8) http://www.cs.bham.ac.uk/ wbl/aigp3/ch08.pdf
16. Montana, D.J.: Strongly typed genetic programming. Evol. Comput. **3**(2), 199–230 (1995). http://vishnu.bbn.com/papers/stgp.pdf
17. Nordin, P., Banzhaf, W.: Complexity compression and evolution. In: Genetic Algorithms: Proceedings of the Sixth International Conference (ICGA 1995), pp. 310–317. Morgan Kaufmann, Pittsburgh, 15–19 July 1995
18. Pawlak, T.P., Krawiec, K.: Automatic synthesis of constraints from examples using mixed integer linear programming. Eur. J. Oper. Res. (2017). (in 2nd review)
19. Pawlak, T.P., Wieloch, B., Krawiec, K.: Review and comparative analysis of geometric semantic crossovers. Genet. Program. Evolvable Mach. **16**(3), 351–386 (2015)
20. Rüegg, C.: Math.NET Symbolics. http://symbolics.mathdotnet.com/
21. Vanneschi, L., Castelli, M., Silva, S.: A survey of semantic methods in genetic programming. Genet. Program. Evolvable Mach. **15**(2), 195–214 (2014). http://link.springer.com/article/10.1007/s10710-013-9210-0
22. Williams, H.: Model Building in Mathematical Programming. Wiley, New York (2013)

# Grammatical Evolution of Robust Controller Structures Using Wilson Scoring and Criticality Ranking

Elias Reichensdörfer[1,3](✉), Dirk Odenthal[2], and Dirk Wollherr[3]

[1] BMW Group, Knorrstr. 147, 80807 Munich, Germany
elias.reichensdoerfer@bmw.de
[2] BMW M GmbH, Daimlerstr. 19, 85748 Garching, Germany
dirk.odenthal@bmw-m.com
[3] Chair of Automatic Control Engineering, Technical University of Munich,
Theresienstr. 90, 80333 Munich, Germany
dw@tum.de

**Abstract.** In process control it is essential that disturbances and parameter uncertainties do not affect the process in a negative way. Simultaneously optimizing an objective function for different scenarios can be solved in theory by evaluating candidate solutions on all scenarios. This is not feasible in real-world applications, where the scenario space often forms a continuum. A traditional approach is to approximate this evaluation using Monte Carlo sampling. To overcome the difficulty of choosing an appropriate sampling count and to reduce evaluations of low-quality solutions, a novel approach using Wilson scoring and criticality ranking within a grammatical evolution framework is presented. A nonlinear spring mass system is considered as benchmark example from robust control. The method is tested against Monte Carlo sampling and the results are compared to a backstepping controller. It is shown that the method is capable of outperforming state of the art methods.

**Keywords:** Robust control · Grammatical evolution · Wilson score · Nonlinear systems · Genetic algorithms · Controller synthesis · Criticality ranking

## 1 Introduction

In industrial applications it is often necessary to enforce a certain behavior of technical systems. For example a thermostat should heat a room such that a desired room temperature is reached; a robot manipulator should follow a predefined path in an assembly line or a loading bridge should move transportation charges fast and without oscillation. Such tasks are usually achieved by a controller, that uses sensor information about the current state of the system to compute appropriate actions. However, in practical applications there is uncertainty about parameters that determine the behavior of the process. For

© Springer International Publishing AG 2017
J. McDermott et al. (Eds.): EuroGP 2017, LNCS 10196, pp. 194–209, 2017.
DOI: 10.1007/978-3-319-55696-3_13

example when steering an autonomous vehicle to follow a trajectory, the vehicle dynamics depend on parameters like payload, rolling resistance or tyre friction which are not known exactly. Additionally, external disturbance forces induced by sidewind gust or by braking on unilateral icy road act on the system. In order to achieve the desired behavior, a controller is required to be robust with respect to parameter uncertainties and disturbances. The problem of uncertainty when designing a controller for a given physical system can be difficult to deal with. Parameter variations and disturbances might have a negative effect on the performance of a control system and even result in instability. There are various ways to deal with this problem. On the one hand there are techniques from *robust control theory* (some of them outlined in Sect. 2.2). These techniques typically make restrictive assumptions on either the physical model of the control process or on the structure of the controller. On the other hand there exist *evolutionary optimization techniques* that not only allow to search for parameters but also for the structure of a solution. These methods typically make none, or a small number of relatively minor assumptions on model or solution structures.

The goal of this work is to combine ideas of both fields in order to find robust controllers for nonlinear systems with non-obvious system structure. While most publications on evolutionary controller design focus on optimizing general performance measures using multi-objective [10] or weighted sum approaches [4], robustness is rarely considered. However, in most industrial applications, robustness of a given control structure with respect to uncertainties is essential.

Evolutionary algorithms (EA) have already been used for structure optimization of robust controllers. In [16] a controller for a non-minimum phase plant and a three-lag plant was synthesized. Moreover in [17] the same system with a 5 s delay was used as a test case. However, the investigation was limited to linear, stable SISO (single-input, single-output) systems and few discrete parameter evaluations. In [21] a saturated, rate limited (and thus nonlinear) robust controller was automatically designed but again only for a linear plant and evaluation of a few uncertainties. The authors of [3] overcome the robustness problem by incorporating a stability analysis, based on Kharitonovs theorem by [13], into the optimization process. As a result, their approach is limited to linear plants and controllers. In [20] a nonlinear controller for the nonlinear unstable system of a rolling inverted pendulum is evolved. However robustness of the generated controllers is only investigated a posteriori the optimization process, not as an optimization criterion.

To the best of our knowledge there is no methodology in EAs available, to automatically synthesize structure of arbitrary controllers for robust control, with the exception of Monte Carlo simulations. This work presents a novel technique to incorporate robustness into the evolutionary process, independent of the used control structure. In particular, the method is not limited to robust control and might be applied to a large variety of optimization problems.

This paper is organized in 5 sections. Section 2 gives background information about grammatical evolution and robust control theory. Section 3 describes the proposed method in detail and outlines its key ideas. In Sect. 4, an experimental validation of the proposed approach is presented and the results are compared to other methods of robust control. In Sect. 5, a conclusion is drawn.

## 2   Background

### 2.1   Grammatical Evolution

Evolutionary algorithms [11] are tools for optimization of "difficult" target functions and have been widely used for solving real-world problems [5,12]. There are several types of EAs. In this work we will focus on genetic algorithms (GA). These algorithms are randomized search heuristics based on the simulation of natural selection. A candidate solution $c$ is called an *individual*, the function $\phi : X^n \to Y$ is called a *fitness function* of a $n$-dimensional optimization problem (in this work we use $X \subseteq \mathbb{N}_0$ and $Y \subseteq \mathbb{R}_0^+$). The fitness function is used to evaluate an individual and to assign a *fitness value* to it. Based on this value, individuals are assigned a higher or lower chance of survival and reproduction. A detailed introduction to GAs is given in [18].

While at the beginning EAs were only applicable for parameter estimation, the introduction of genetic programming (GP) [15] also accounted for structural optimization. Another method is grammatical evolution (GE) [19], which is similar to GP. In GE however, handling the structure of solutions is decoupled from the EA itself by the usage of a formal grammar in Backus-Naur Form (BNF). Here a solution is represented as a variable-dimensional vector of integers which is then transformed into a program in the fitness function itself.

Given an example context-free grammar as the 4-tuple $\mathcal{G} = (\mathcal{N}, \Sigma, \mathcal{P}, \mathcal{S})$ with $\mathcal{N}$ being the finite set of non-terminals and $\Sigma$ the finite set of terminals such that $\mathcal{N} \cap \Sigma = \emptyset$. Furthermore $\mathcal{P} \subset (V^* \setminus \Sigma^*) \times V^*$ is the finite set of production rules, where each rule follows $\mathcal{N} \to V^*$ and $\mathcal{S} \in \mathcal{N}$ is the start symbol. Here $V = \mathcal{N} \cup \Sigma$ denotes the vocabulary of $\mathcal{G}$ and $(\cdot)^* := \bigcup_{i \in \mathbb{N}_0} (\cdot)^i$ the Kleene closure of $(\cdot)$. We define the first two sets as $\mathcal{N} = \{\texttt{<e>}, \texttt{<op>}, \texttt{<var>}, \texttt{<const>}\}, \Sigma = \{\texttt{+}, \texttt{-}, \texttt{x}, \texttt{1}\}, \mathcal{S} = \texttt{<e>}$ and the set of production rules $\mathcal{P}$ as shown in Table 1a. The dispatching of the current rule $r_i$ when parsing the integer stream is defined using the modulo operator as

$$r_i = c \% p \tag{1}$$

where $c$ is the integer value of the current codon and $p$ is the number of available production rules of the current nonterminal. Given an example individual as $c = \begin{bmatrix} 9 & 5 & 8 & 17 & 4 & 1 \end{bmatrix}^T$ we can convert it by parsing, using Eq. (1) and the grammar $\mathcal{G}$ as shown in Table 1b. Performing this step shows that the example individual represents the mathematical function $f(x) = 1 - x$.

In case there remain unused codons (so-called introns), these can either be ignored or removed from the genome, using the so-called pruning operator. In the opposite case, when the parsing process is not yet finished but already run out of codons, the wrapping technique is used. It restarts reading codons from the beginning of the genome. To avoid infinite wrapping due to loops in an

**Table 1.** Mapping from integer vector to expression

(a) Example production rules

| Non-terminal | Production | Rule |
|---|---|---|
| <e> | ::= <op>(<e>,<e>) \| | $r_1$ |
|  | <var> \| | $r_2$ |
|  | <const> | $r_3$ |
| <op> | ::= + \| | $r_4$ |
|  | − | $r_5$ |
| <var> | ::= x | $r_6$ |
| <const> | ::= 1 | $r_7$ |

(b) Example parsing process

| $c$ | $p$ | $c\%p$ | Rule | Expression |
|---|---|---|---|---|
| - | - | - | $S$ | <e> |
| 9 | 3 | 0 | $r_1$ | <op>(<e>,<e>) |
| 5 | 2 | 1 | $r_5$ | −(<e>,<e>) |
| 8 | 3 | 2 | $r_3$ | −(<const>,<e>) |
| 17 | 1 | 0 | $r_7$ | −(1,<e>) |
| 4 | 3 | 1 | $r_2$ | −(1,<var>) |
| 1 | 1 | 0 | $r_6$ | −(1,x) |

individual, a maximum wrapping count is defined. If the number of wrappings exceeds that count, all recursive production rules are removed from the grammar and the parsing process is continued until finished. In the example above, only rule $r_1$ would be removed since it maps recursively back to the nonterminal <e>. More details on this mapping process can be found for example in [6].

## 2.2 Robust Control

Consider a nonlinear closed loop control system with static state feedback and disturbance like shown in Fig. 1. Single line borders indicate linear, double line borders (potentially) nonlinear blocks.

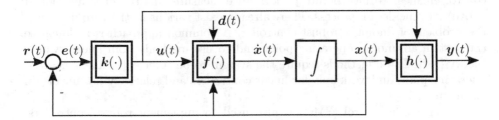

**Fig. 1.** Closed loop control system

The general dynamics of this system are given by

$$\dot{x}(t) = f(x(t), u(t), d(t), q), \quad x(0) = x_0$$
$$y(t) = h(x(t), u(t), q). \tag{2}$$

Here $t \in [0, T]$ denotes the time and $x \in X \subseteq \mathbb{R}^n$ the *state* of the system. This term refers to the minimum set of variables that describe the status the system is in, at a given time instance. For a moving object this might be the objects position and its velocity. The initial state at time $t = 0$ is $x_0 = x(0) \in X_0 \subseteq \mathbb{R}^n$. The equations of motion can be described using differential equations and the physical parameters $q \in Q \subseteq \mathbb{R}^k$ with the nonlinear mapping $f$. This mapping defines how the state variables change with time with $\dot{x}$ being the time derivative of $x$. The inputs of the system $u \in U \subseteq \mathbb{R}^m$ describe the external forces that a controller $k$ can use to interact with the system. The output $y \in Y \subseteq \mathbb{R}^h$ of the system is mapped from the state and the input using a function $h$. The controller measures the current state of the system and compares it to a reference value $r \in X \subseteq \mathbb{R}^n$. The difference from the desired value $r$ and the actual value $x$ gives an error for each state $e \in \mathbb{R}^n$ (notice that the reference signal might be set to a constant value of $0$ as well). Additionally there are disturbances $d \in D \subseteq \mathbb{R}^m$ acting on the system.

The robust control problem is to choose a controller $k$, such that the system is stable and yields "good" performance for all parameter sets $q$ and disturbances $d$. The term performance here refers to the system behavior and depends on the actual problem. Possible soft requirements of performance apart from stability can include the following:

- Fast command input responses
- Small oscillations of the control trajectories
- Rejection of disturbances
- Bounded control signals
- Robustness with respect to parameter uncertainties
- Real time capability
- ... and more.

So the task is not only to find a stabilizer of the system but also to allow the control engineer to incorporate performance measures (as the ones mentioned above) into the design process. There already exist various methods to deal with the problem of finding a robust controller. A common approach is to linearize the system around an operating point and to use methods from linear system theory [1,7]. These methods require the system to be sufficiently linear near its operating point and work only for linear controllers and relatively few uncertain parameters.

Sliding mode control (SMC) is also used to implement robust controllers. The main advantages of this method are its simplicity and its insensitivity with respect to model uncertainties. Disadvantages are its highly discontinous control signals and the so-called chattering effect (high-frequency switching of the control signal may stress the system) [8]. Another method is $\mathcal{H}_\infty$ optimal control [9,27] which has the advantages to be applicable to MIMO (multi-input, multi-output) systems and is capable of synthesizing controllers that are both robust and have good performance. Nevertheless it has the disadvantages of being unable in handling constraints [25] and generating high order controllers.

To overcome these disadvantages, grammatical evolution is used to synthesize nonlinear robust controllers, using a novel fitness function to directly include robustness into the evolutionary process.

# 3   Methodology

## 3.1   General Process

The difficulty in grammatical evolution of robust controllers is to assign a numerical value to an individual that represents its robustness on a given control problem. First we define a *control scenario*:

**Definition 1.** *A control scenario (short: scenario) $S$ is the 5-tuple $S = (q^*, x_0^*,$ $r^*(t), d^*(t), T) \in \mathbb{S} = Q \times X_0 \times R \times [0, T] \times D \times [0, T] \times \mathbb{R}^+$. $\mathbb{S}$ is the set of control scenarios.*

The goal of the evolutionary process is to find a controller that has satisfying performance on all possible control scenarios (i.e. for all combinations of possible initial conditions $x_0$, uncertain model parameters $q$, reference trajectories $r$ and disturbances $d$). The evaluation is performed in three steps:

1. A new control scenario $S$ is chosen randomly from $\mathbb{S}$ at the beginning of each generation.
2. Each individual is transformed into a feedback law (the controller) using a grammar in BNF (compare Sect. 2.1) and inserted into the system.
3. The systems are then simulated for a given time horizon on the scenario $S$ using an ODE solver. Following the fitness values of the individuals are calculated based on their performance on the control task.

After all individuals have been evaluated and selection has been performed, the robustness of each individual is evaluated separately, using the method from Sect. 3.2. Then a fraction of the most robust individuals is reinserted into the population (which is called the Wilson elite). At the end of the optimization the individual with the best robustness score is returned.

## 3.2   Wilson Scoring

To evaluate the robust performance of an individual, the equation

$$s_r = \int_{S \in \mathbb{S}} \phi(c, S) \, dS \tag{3}$$

would have to be solved. Since its usually not possible to evaluate this integral analytically, an estimate has to be used by sampling from the scenario space:

$$s_r \approx \hat{s}_i - \frac{1}{N} \sum_{S \in \mathbb{S}}^{N} \phi(c, S). \tag{4}$$

A common approach is therefore to approximate $s_r$ using Monte Carlo sampling, by drawing random samples from $\mathbb{S}$ [22,23]. The disadvantage of this method is that it still requires a large amount of simulations, potentially wasting computation time on low quality individuals. Also it can be difficult to choose a sampling count $N$. Therefore in this work a different approach is introduced which evaluates each individual only once per generation. Each individual has a current fitness value, an age and an average fitness value. The current fitness value just represents its performance on the current control scenario $S$. The age is a counter that stores how many generations the individual already survived. The average fitness is the mean value of all fitness evaluations over an individuals lifetime and can be calculated numerically stable by [14]. The "older" an individual is, the more different control scenarios it has already been tested on. So what we want to find is an individual with a small mean error on all its evaluated control scenarios and a high age, because such an individual is assumed to be a robust stabilizer. This assumption is supported by the fact that for large sampling sizes, the approximated fitness approaches the real fitness. The problem with combining the two conditions of a low error and a high age is a classical ranking problem and may be illustrated by the following example.

- Individual 1: mean error = 0.02, age = 100
- Individual 2: mean error = 0.0001, age = 1
- Individual 3: mean error = 100, age = 1000

In terms of robust stability, individual 1 seems to be best since it has a relatively low mean error on all its evaluated scenarios. Individual 2 has a lower error, but was evaluated only on 1 scenario, so little information about its robustness is available. Individual 3 has a high error (potentially due to instability on some of its scenarios). So the desired ranking here, would be 1, 2, 3. Ranking those individuals now by mean error would place individual 2 before individual 1, so robustness would not be incorporated into the ranking. Ranking by age would place individual 3 on top, which is clearly wrong. Even ranking by mean error divided by age would rank individual 2 before individual 1.

This problem is equivalent to the ranking problems on websites which rank content based on user ratings (so called "thumbs-up/thumbs-down ratings"). A common way to tackle this problem is the Wilson score [26] which is defined as the lower bound of an approximation of the binomial proportion confidence interval. The score is given by

$$s_W = \frac{\hat{p} + \frac{z^2}{2n} - z\sqrt{\frac{\hat{p}(1-\hat{p}) + \frac{z^2}{4n}}{n}}}{1 + \frac{z^2}{n}} \in (0,1). \tag{5}$$

Here $n \in \mathbb{N}$ is the number of total ratings (so in our case the age of the individual), $\hat{p}$ is the proportion of positive ratings and $z$ is the $1 - \frac{\alpha}{2}$ quantile of the standard normal distribution. In this work we use the common value of $z = 1.96$ which equals approximately a confidence level of 95%. However we do not have a thumbs-up/thumbs-down rating system but an error $\epsilon \in \mathbb{R}_0^+$. So first we need to convert this error into an equivalent *pseudo thumbs-up/thumbs-down rating*. This is done by the following mapping function

$$\hat{p} : \mathbb{R}_0^+ \to (0,1), \hat{p}(\epsilon) = \frac{a}{a + \epsilon}, \tag{6}$$

with $a \in \mathbb{R}^+$ (in this work we choose $a = 1$). This function provides a smooth mapping for the error: if $\epsilon = 0$ it follows that $\hat{p}(0) = 1$ which equals 100% "positive ratings". For errors approaching infinity, we get $\lim_{\epsilon \to \infty} \hat{p}(\epsilon) = 0$ which equals 0% "positive ratings". Using this mapping we can convert the mean control error into a percentaged pseudo rating which we can use in Eq. (5) to calculate the Wilson score for an individual. Since a Wilson score of 1 means "best" and we do minimization, we turn it around to calculate the final score:

**Definition 2.** *The robustness score $s_W^*$ of an individual is given by*

$$s_W^* = 1 - s_W \in (0,1). \tag{7}$$

Ranking by Eq. (7) we get the robustness scores (approximately) 0.069, 0.793, 0.995 for the individuals $1, 2, 3$, which gives the desired ranking.

## 3.3    Criticality Ranking

Due to the stochasticity of the optimization process, it is possible that individuals perform well on a large amount of scenarios (leading to a good robustness score) but then fail on some particular hard scenarios which occur rarely during the optimization. Since a new scenario is picked in every generation, a new individual performing equally might remain in the population for a long time until a critical scenario is hit again. To avoid this problem, a simple yet efficient heuristic is used. First we start with a definition:

**Definition 3.** *The criticality of a given control scenario $S$ at generation $t$ is given by*

$$\chi_{t,S} = \max\left(0, \frac{s_{W,t}^* - s_{W,t-1}^*}{s_{W,t-1}^*}\right) \tag{8}$$

*and defines the relative degradation of an individual's robustness score with respect to $S$.*

Using Eq. (8) we can calculate how critical a given control scenario is in terms of robustness. If an individual has a low score at generation $t-1$ (which means it has performed well on many scenarios) and a high score at generation $t$ (meaning it performed bad on this current scenario), the control scenario of generation $t$ is called critical in terms of robustness. The better the individual was before, the higher the criticality of the scenario.

Now the $C \in \mathbb{N}$ most critical and unique scenarios are collected in a global queue of maximum length $C$. The queue is updated during the optimization every time a scenario has a higher criticality than the least critical scenario in the queue. If the queue is full, the least critical scenario in the queue is removed. Each new individual is evaluated first on these critical scenarios, such that an individual of age $C$ was evaluated on all critical scenarios in the queue. Using this technique we are not only avoiding the above mentioned problem but also get a list of control scenarios that are particularly hard in terms of robustness as a side effect. Individuals that fail on those scenarios will not achieve high ages which implies poor robustness scores and therefore low criticality. This ensures that during the optimization process only individuals with a high age (and therefore a high confidence on their actual fitness) will be able to contribute to the criticality queue.

## 4    Experiments

### 4.1    Benchmark Problem

We consider a standard benchmark problem from robust control, the spring-mass system [24] in a modified version using a nonlinear cubic spring according to [22] shown in Fig. 2. Variations of this system are used in proof-of-principle investigations of different applications, like coupled mechanical systems.

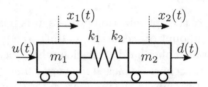

**Fig. 2.** Spring-mass system with coupled linear and nonlinear springs

The dynamics of the system are given by

$$
\begin{bmatrix} \dot{x}_1 \\ \dot{x}_2 \\ \dot{x}_3 \\ \dot{x}_4 \end{bmatrix} = \begin{bmatrix} x_3 \\ x_4 \\ -\frac{k_1}{m_1}(x_1 - x_2) - \frac{k_2}{m_1}(x_1 - x_2)^3 \\ \frac{k_1}{m_2}(x_1 - x_2) + \frac{k_2}{m_2}(x_1 - x_2)^3 \end{bmatrix} + \begin{bmatrix} 0 \\ 0 \\ \frac{1}{m_1} \\ 0 \end{bmatrix} u + \begin{bmatrix} 0 \\ 0 \\ 0 \\ \frac{1}{m_2} \end{bmatrix} d \qquad (9)
$$

with $y = x_2$. Here $x_1, x_2$ denote the positions of the two masses and $x_3, x_4$ their velocities, collected in the state vector $\boldsymbol{x}$. The uncertain parameters of the system are the masses $m_1, m_2$ and the spring constants $k_1, k_2$ which are collected in the vector $\boldsymbol{q} = \begin{bmatrix} k_1 & k_2 & m_1 & m_2 \end{bmatrix}^T$. These parameters might vary in the ranges

$$k_1 \in [0.5, 2], \ k_2 \in [-0.5, 0.2], \ m_1 \in [0.5, 1.5], \ m_2 \in [0.5, 1.5], \qquad (10)$$

with $\boldsymbol{q}^0 = \begin{bmatrix} k_1^0 & k_2^0 & m_1^0 & m_2^0 \end{bmatrix}^T = \begin{bmatrix} 1 & -0.1 & 1 & 1 \end{bmatrix}^T$ being the nominal plant parameters. The value $u$ denotes the input and $d$ the disturbance acting on the system while $y = x_2$ denotes the output of the system. In [22] a nonlinear backstepping controller is designed and parametrized for the described system using a GA and Monte Carlo sampling to optimize robustness. The task is to control the system after a unit impulse disturbance at time $t = 0$ robustly for the given parameter variations and sufficiently fast without excessive control usage. Since the backstepping controller from [22] was not able to stabilize the system on all evaluated scenarios, we define two additional benchmarks (2 and 3). These are interesting for further comparisons with the backstepping controllers, since these cannot stabilize the whole scenario space of benchmark 1.

1. System parameters are varied, unit impulse disturbance at $t_0 = 0$ (by [22]).
2. System parameters are varied, impulse disturbance with weight 0.1 at $t_0 = 0$.
3. Like 2, initial state is varied with $x_0 = \begin{bmatrix} 0 & x_{0,2} & 0 & 0 \end{bmatrix}^T$ with $x_{0,2} \in [-0.5, 0.5]$.

The proposed method as well as the standard Monte Carlo optimization method are evaluated on the 3 benchmarks. Each method is given a total amount of $10^6$ simulations and the question is, which method manages to use those evaluations most efficiently.

## 4.2   Metrics and Setup

The setup for the GE algorithm is shown in Table 2. Some remarks on this setup:

- The number of generations is set relatively high in comparison to the number of individuals. Since every individual is evaluated once per generation, this is to ensure that individuals have opportunity to reach a high age.
- If either the maximum wrapping count or the maximum depth of the parse tree are exceeded, all recursive elements are removed from the grammar in order to avoid overly complex solution structures.
- Rank selection was chosen because the fitness of individuals varies in a large range (approaching infinity for unstable controllers) and therefore fitness proportional selection schemes seem inappropriate.
- Ranking is done with a linear ranking mechanism that calculates the selection probability of an individual: $\frac{1}{N}\left(2 - SP + 2(SP - 1)\frac{N-R}{N-1}\right)$. Here $R$ is the rank of the individual, $N$ is the population size and $SP \in [1, 2]$ is the selection pressure parameter.

**Table 2.** Grammatical evolution setup

| Parameter | Value | Parameter | Value |
|---|---|---|---|
| Population size | 100 | Selection | Rank selection |
| Number of generations | 10000 | Crossover | Cut and splice |
| Recombination rate | 0.5 | Mutation | Uniform multi-point |
| Mutation rate | 0.05 | Pruning | Turned on |
| Initial genome length | $\sim \mathcal{U}\{20, 30\}$ | Elitism rate | 0.3 |
| Initial genome values | $\sim \mathcal{U}\{0, 10\}$ | Wilson elitism rate | 0.1 |
| Max. wrapping count | 5 | Selection pressure | 1.5 |
| Max. tree depth | 10 | Criticality queue size | 10 |

Numerical simulations are done in C++14 using odeint [2] from the boost libraries[1]. The numerical solver to simulate the system was the Euler forward method with a step size of $\Delta t = 0.01$. The fitness of an individual is defined by the *integral of the square of the error* (ISE) criterion plus a penalty term

$$\phi = \int_0^T x_2(t)^2 + u(t)^2 dt + p_\infty(T). \tag{11}$$

The penalty term is used to ensure that the system has no steady state error and is defined as follows

$$p_\infty(T) = \begin{cases} 0 & |x_2(T)| \leq x_{2,\max} \\ 1000 & |x_2(T)| > x_{2,\max}. \end{cases} \tag{12}$$

The simulation time $T$ is set to 100 s and the threshold for the penalty term $x_{2,\max} = 0.1$ for all benchmarks. The grammar used is

$$\begin{aligned}
\mathcal{N} &= \{\text{<e>}, \text{<op>}, \text{<func>}, \text{<var>}, \text{<const>}\} \\
\Sigma &= \{\text{+, -, *, /, pow, sin, cos, exp, log, tanh}, x_1, x_2, x_3, x_4, \\
&\quad\quad \text{f}_3(q^0, x_1, x_2), \text{f}_4(q^0, x_1, x_2), 0.1, 1, 2, 3, 4, 5, 6, 7, \text{pi}\} \\
\mathcal{S} &= \text{<e>}
\end{aligned} \tag{13}$$

with the production rules $\mathcal{P}$ listed in Table 3.

**Table 3.** Production rules $\mathcal{P}$ used for the optimization

| | |
|---|---|
| <e> | ::= <op>(<e>,<e>) \| <func>(<e>) \| <var> \| <const> |
| <op> | ::= + \| - \| * \| / \| pow |
| <func> | ::= sin \| cos \| exp \| log \| tanh |
| <var> | ::= $x_1$ \| $x_2$ \| $x_3$ \| $x_4$ \| $\text{f}_3(q^0, x_1, x_2)$ \| $\text{f}_4(q^0, x_1, x_2)$ |
| <const> | ::= 0.1 \| 1 \| 2 \| 3 \| 4 \| 5 \| 6 \| 7 \| pi |

Here $\text{f}_3(q^0, x_1, x_2)$ and $\text{f}_4(q^0, x_1, x_2)$ are the right hand sides of the system Eq. (9) as a function of $x_1$, $x_2$ and the (constant) nominal plant parameters $q^0$. The <const> values are chosen like this as different setups did not show significant differences for this specific benchmark.

We evaluate the proposed method against the standard technique of Monte Carlo sampling during the optimization on the 3 benchmarks described in Sect. 4.1. Each optimization is repeated 100 times with random initial populations according to Table 2. The number of Monte Carlo samples per fitness evaluation of a single individual was set by the authors of [22] to 500. Since we have a slightly different setup though (doing structural instead of parameter optimization), we take into account different sample numbers to avoid that result

---

[1] See http://www.boost.org.

quality is affected by an inappropriate choice of this parameter. We abbreviate this as $MC_{150}$, $MC_{500}$ and $MC_{1500}$ meaning Monte Carlo sampling with 150, 500 and 1500 samples respectively (lower samples lead to higher uncertainty about an individuals fitness but more generations can be performed before the maximum of $10^6$ fitness function evaluations is exceeded). Additionally, after every optimization run the controllers are tested on 25000 random, unseen scenarios. The backstepping controllers are abbreviated as $BS_1$ and $BS_2$ for the two parameter sets from [22].

For comparison with the backstepping controllers, we take the metric from [22]. The values $P_i, P_{ts}$ and $P_u$ denote probabilities of instability, exceeding settling time and excessive control effort respectively. Probabilities are estimated with binary indicators per sample, being 1 if $|x_2(t)| > 0.1$, $|u(t)| > 1$ or $|x_2(t)| \geq 1$ for $t = T$ respectively (zero otherwise). The probabilities are then estimated by Monte Carlo sampling the given controller with a sample count of 25000. The authors of [22] minimize the weighted sum $J = P_i^2 + 0.01 P_{ts}^2 + 0.01 P_u^2$ on benchmark 1. This can be interpreted such that stability is considered more important than speed or control effort by a factor of 100.

### 4.3   Results

First the focus is set on benchmark 1. One example controller synthesized on this benchmark with the proposed GE method is given by

$$u_{GE}(\boldsymbol{x}) = -x_2 - 3x_3 - 2f_4(\boldsymbol{q}^0, x_1, x_2) + f_3(\boldsymbol{q}^0, x_1, x_2)(1 + x_2 + 2x_4^2). \quad (14)$$

We further examine the difference between the Monte Carlo method and the proposed method. Figure 3a shows this comparison while Fig. 3b shows a typical robustness score development of one optimization run. The circles around the spikes in Fig. 3b show some example critical scenarios.

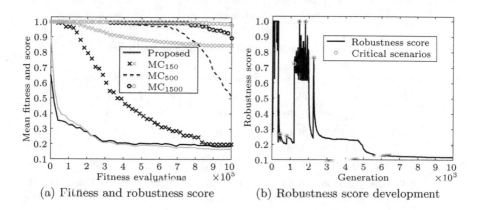

(a) Fitness and robustness score     (b) Robustness score development

**Fig. 3.** GE and backstepping controller on benchmark 1 (100 trials)

In Fig. 3a the black curves display the mean optimization error over 100 independent trials (for better visualization, all errors greater than 1 are set to 1). The gray curves display the corresponding robustness scores. One can observe that for the proposed method (solid lines) both the average error and the robustness score reach a relatively low value. So the error is minimized with a relatively high certainty. The Monte Carlo method shows a different behavior. The $MC_{150}$ method reaches a low error as well (black crosses), but its robustness score is high in comparison (gray crosses), indicating a low certainty on the result. Similar results apply to the $MC_{500}$ and $MC_{1500}$ test runs while the certainty increases with more samples. However the optimization error increases as well. The proposed method achieves a low error at relatively high confidence.

Table 4 shows the numerical results for 100 independent trials. "Opt." stands for the optimization and "Test" for the test error. The means and variances are calculated only for stable controllers (since unstable controllers have errors approaching infinity). Any controller with a mean error larger than 10 was considered unstable. Additionally we compare the probabilities of the controllers which indicate that design goals are not met. The results (mean and 95% confidence intervals) are shown in Table 5.

**Table 4.** Results for benchmarks 1–3 on 100 independent trials

| Method | Metric | Benchmark 1 | | | Benchmark 2 | | | Benchmark 3 | | |
|---|---|---|---|---|---|---|---|---|---|---|
| | | Opt. | Test | Stable | Opt. | Test | Stable | Opt. | Test | Stable |
| Proposed | Mean | 1.8e−1 | 1.7e−1 | 45% | 4.2e−4 | 2.4e−3 | 100% | 2.6e−3 | 2.6e−3 | 99% |
| | Variance | 2.2e−3 | 3.0e−3 | | 1.2e−8 | 4.0e−4 | | 1.9e−7 | 1.9e−7 | |
| $MC_{150}$ | Mean | 1.7e−1 | 2.1e−1 | 2% | 5.1e−3 | 2.8e−2 | 100% | 5.0e−3 | 5.1e−3 | 100% |
| | Variance | 5.3e−1 | 1.8e−2 | | 4.3e−8 | 2.8e−2 | | 4.2e−8 | 4.3e−8 | |
| $MC_{500}$ | Mean | 2.2e−1 | 7.6e−2 | 8% | 5.6e−3 | 6.8e−3 | 100% | 5.5e−3 | 6.1e−3 | 100% |
| | Variance | 7.4e−2 | 1.0e0 | | 2.4e−7 | 7.8e−5 | | 2.1e−7 | 2.1e−5 | |
| $MC_{1500}$ | Mean | 1.8e0 | 4.2e0 | 5% | 6.5e−3 | 1.4e−2 | 98% | 6.6e−3 | 2.2e−2 | 99% |
| | Variance | 9.2e0 | 2.0e1 | | 1.4e−6 | 2.9e−3 | | 1.5e−6 | 2.3e−2 | |

**Table 5.** Comparsion with backstepping controller

| Controller | Benchmark 1 | | | Benchmark 2 | | | Benchmark 3 | | |
|---|---|---|---|---|---|---|---|---|---|
| | $P_i$ | $P_{ts}$ | $P_u$ | $P_i$ | $P_{ts}$ | $P_u$ | $P_i$ | $P_{ts}$ | $P_u$ |
| GE | 0 | 0.1276 | 0.4389 | 0 | 0 | 0 | 0 | 0 | 0.0767 |
| | [0,0] | [0.1235, 0.1318] | [0.4328, 0.4451] | [0,0] | [0,0] | [0,0] | [0,0] | [0,0] | [0.0734, 0.0800] |
| $BS_1$ | 0.0032 | 0.5210 | 0.4513 | 0 | 0.0188 | 0 | 0 | 0.0565 | 0.2988 |
| | [0.0025, 0.0039] | [0.5148, 0.5272] | [0.4452, 0.4575] | [0,0] | [0.0171, 0.0205] | [0,0] | [0,0] | [0.0536, 0.0593] | [0.2931, 0.3045] |
| $BS_2$ | 0.0007 | 0.9543 | 0.2682 | 0 | 0.1988 | 0 | 0 | 0.2876 | 0.0798 |
| | [0.0004, 0.0010] | [0.9517, 0.9569] | [0.2627, 0.2737] | [0,0] | [0.1939, 0.2037] | [0,0] | [0,0] | [0.2820, 0.2933] | [0.0764, 0.0832] |

Looking at the data from Table 4 the GE method shows better results in all three benchmarks. The performance on benchmarks 2 and 3 are very similar for both the proposed and the Monte Carlo method. On benchmark 1, the proposed method shows a significantly higher success rate (45%) compared to the Monte Carlo method with highest success rate, $MC_{500}$ (8%). Especially the test error is lower on every benchmark for the proposed method. The data from Table 5 shows that the GE controller never exhibited unstable behavior ($P_i = 0$) on all three benchmarks. Using the metric $J = P_i^2 + 0.01P_{ts}^2 + 0.01P_u^2$ from Sect. 4.2, the GE controller outperforms the $BS_1$ and $BS_2$ controllers on all benchmarks. An analytical stability analysis of the GE controller is presented in the appendix.

# 5   Conclusion

We proposed a novel methodology using grammatical evolution for robust controller synthesis on uncertain systems. The method showed results comparable or better than state of the art methods for stochastic robust controller design. While the focus was set on robust control, the proposed method is not limited to robust control applications but can be seen as a general alternative to Monte Carlo evaluations for scenario optimization problems.

Optimization often requires sampling of test cases, because it is desirable to find solutions that perform well in the presence of uncertainty. In economic problems it might be general conditions like the market situation or political requirements; for technical problems, physical parameters might vary within some ranges; computer vision systems might require optimization for different illumination settings, just to give some examples. Whenever there are multiple test cases required to evaluate the fitness of a given solution in an optimization problem, the designer has to choose how to handle these test cases. The straightforward way to this problem is to take samples from the test cases and use the average as an approximate value. However, this method has the problems of choosing an appropriate sample count and potentially wasting evaluations on low quality solutions.

Our proposed new method can be seen as an alternative approach to draw samples in a more systematic way. The certainty about the quality of a solution is updated during the optimization process instead of being calculated at once. This was found to be beneficial in the considered benchmark.

## Appendix: Stability Analysis

We calculate the Jacobian of the system from Eq. (9), substituting $u$ with the GE controller from Eq. (14). The Jacobian for this system is given by $A(x) = \left[\frac{\partial f}{\partial x}\right]$. The characteristic polynomial can be calculated by

$$p(\lambda) = \det(\lambda I - A) = \lambda^4 + \frac{3}{m_1}\lambda^3 + \frac{k_1(m_1 + m_2) + 3m_2}{m_1 m_2}\lambda^2 + \frac{3k_1}{m_1 m_2}\lambda + \frac{k_1}{m_1 m_2}.$$

$$(15)$$

Looking at the intervals from Eq. (10) it is trivial to see that all coefficients of this polynomial are strictly positive. We can thus continue the analysis by looking at the Hurwitz matrix of the polynomial:

$$
\boldsymbol{H} = \begin{bmatrix} 1 & H_{1,2} & \frac{k_1}{m_1 m_2} & 0 \\ 0 & \frac{3}{m_1} & \frac{4k_1}{m_1 m_2} & 0 \\ 0 & 1 & H_{3,3} & \frac{k_1}{m_1 m_2} \\ 0 & 1 & \frac{3}{m_1} & \frac{3k_1}{m_1 m_2} \end{bmatrix}, \; H_{1,2} = H_{3,3} = \frac{k_1 m_1 + k_1 m_2 + 3 m_2}{m_1 m_2}. \tag{16}
$$

From the 4 principal minors of $\boldsymbol{H}$ we get the Hurwitz conditions for stability

$$
\det(\boldsymbol{H}_1) = 1 \overset{!}{>} 0, \quad \det(\boldsymbol{H}_2) = \frac{3}{m_1} \overset{!}{>} 0
$$

$$
\det(\boldsymbol{H}_3) = \frac{3(k_1 + 9)}{m_1^2} \overset{!}{>} 0, \quad \det(\boldsymbol{H}_4) = \frac{9k_1^2 + 18k_1}{m_1^3 m_2} \overset{!}{>} 0.
$$

Again since these conditions only depend on $k_1, m_1$ and $m_2$ which are strictly positive, the stability conditions hold for any parameter combination $q \in Q$. The controller $u_{\mathrm{GE}}(\boldsymbol{x})$ is thus a (local) robust stabilizer of the system.

# References

1. Ackermann, J., Blue, P., Bünte, T., Güvenc, L., Kaesbauer, D., Kordt, M., Muhler, M., Odenthal, D.: Robust Control: The Parameter Space Approach. Springer Science+Business Media, Heidelberg (2002)
2. Ahnert, K., Mulansky, M.: Odeint-solving ordinary differential equations in C++ (2011). arXiv preprint arXiv:1110.3397
3. Chen, P., Lu, Y.Z.: Automatic design of robust optimal controller for interval plants using genetic programming and Kharitonov theorem. Int. J. Comput. Intell. Syst. 4(5), 826–836 (2011)
4. Cupertino, F., Naso, D., Salvatore, L., Turchiano, B.: Design of cascaded controllers for DC drives using evolutionary algorithms. In: Proceedings of the 2002 Congress on Evolutionary Computation, 2002, CEC 2002, vol. 2, pp. 1255–1260. IEEE (2002)
5. Dasgupta, D., Michalewicz, Z.: Evolutionary Algorithms in Engineering Applications. Springer Science+Business Media, Heidelberg (2013)
6. Dempsey, I., O'Neill, M., Brabazon, A.: Foundations in Grammatical Evolution for Dynamic Environments, vol. 194. Springer, Heidelberg (2009)
7. Doyle, J.C., Francis, B.A., Tannenbaum, A.R.: Feedback Control Theory. Courier Corporation, New York (2013)
8. Edwards, C., Spurgeon, S.: Sliding Mode Control: Theory and Applications. CRC Press, Boca Raton (1998)
9. Francis, B.A.: A Course in $\mathcal{H}_\infty$ Control Theory. Springer, New York (1987)
10. Gholaminezhad, I., Jamali, A., Assimi, H.: Automated synthesis of optimal controller using multi-objective genetic programming for two-mass-spring system. In: 2014 Second RSI/ISM International Conference on Robotics and Mechatronics (ICRoM), pp. 041–046. IEEE (2014)
11. Holland, J.H.: Adaptation in Natural and Artificial Systems: An Introductory Analysis with Applications to Biology, Control, and Artificial Intelligence. University of Michigan Press, Ann Arbor (1975)

12. Hornby, G.S., Globus, A., Linden, D.S., Lohn, J.D.: Automated antenna design with evolutionary algorithms. In: AIAA Space, pp. 19–21 (2006)
13. Kharitonov, V.: Asympotic stability of an equilibrium position of a family of systems of linear differntial equations. Differntia Uravn. **14**(11), 1483–1485 (1978)
14. Knuth, D.E.: The Art of Computer Programming: Sorting and Searching, vol. 3. Pearson Education, Upper Saddle River (1998)
15. Koza, J.R.: Genetic Programming: on The Programming of Computers by Means of Natural Selection, vol. 1. MIT Press, Cambridge (1992)
16. Koza, J.R., Keane, M.A., Yu, J., Bennett, F.H., Mydlowec, W., Stiffelman, O.: Automatic synthesis of both the topology and parameters for a robust controller for a non-minimal phase plant and a three-lag plant by means of genetic programming. In: Proceedings of the 38th IEEE Conference on Decision and Control, 1999, vol. 5, pp. 5292–5300. IEEE (1999)
17. Koza, J.R., Keane, M.A., Yu, J., Mydlowec, W., Bennett, F.H.: Automatic synthesis of both the topology and parameters for a controller for a three-lag plant with a five-second delay using genetic programming. In: Cagnoni, S. (ed.) EvoWorkshops 2000. LNCS, vol. 1803, pp. 168–177. Springer, Heidelberg (2000). doi:10. 1007/3-540-45561-2_17
18. Mitchell, M.: An Introduction to Genetic Algorithms. MIT Press, Cambridge (1998)
19. Ryan, C., Collins, J.J., Neill, M.O.: Grammatical evolution: evolving programs for an arbitrary language. In: Banzhaf, W., Poli, R., Schoenauer, M., Fogarty, T.C. (eds.) EuroGP 1998. LNCS, vol. 1391, pp. 83–96. Springer, Heidelberg (1998). doi:10.1007/BFb0055930
20. Shimooka, H., Fujimoto, Y.: Generating robust control equations with genetic programming for control of a rolling inverted pendulum. In: Proceedings of the 2nd Annual Conference on Genetic and Evolutionary Computation, pp. 491–495. Morgan Kaufmann Publishers Inc. (2000)
21. Soltoggio, A.: A comparison of genetic programming and genetic algorithms in the design of a robust, saturated control system. In: Deb, K. (ed.) GECCO 2004. LNCS, vol. 3103, pp. 174–185. Springer, Heidelberg (2004). doi:10.1007/ 978-3-540-24855-2_16
22. Wang, Q., Stengel, R.F.: Robust control of nonlinear systems with parametric uncertainty. Automatica **38**(9), 1591–1599 (2002)
23. Wang, Q., Stengel, R.F.: Robust nonlinear flight control of a high-performance aircraft. IEEE Trans. Control Syst. Technol. **13**(1), 15–26 (2005)
24. Wie, B., Bernstein, D.S.: Benchmark problems for robust control design. J. Guid. Control Dyn. **15**(5), 1057–1059 (1992)
25. Wills, A.G., Bates, D., Fleming, A.J., Ninness, B., Moheimani, S.R.: Model predictive control applied to constraint handling in active noise and vibration control. IEEE Trans. Control Syst. Technol. **16**(1), 3–12 (2008)
26. Wilson, E.B.: Probable inference, the law of succession, and statistical inference. J. Am. Statist. Assoc. **22**(158), 209–212 (1927)
27. Zames, G.: Feedback and optimal sensitivity. model reference transformations, multiplicative seminorms, and approximate inverses. IEEE Trans. Autom. Control **26**(2), 301–320 (1981)

# Using Feature Clustering for GP-Based Feature Construction on High-Dimensional Data

Binh Tran[✉], Bing Xue[✉], and Mengjie Zhang

School of Engineering and Computer Science, Victoria University of Wellington,
PO Box 600, Wellington 6140, New Zealand
{binh.tran,bing.xue,mengjie.zhang}@ecs.vuw.ac.nz

**Abstract.** Feature construction is a pre-processing technique to create new features with better discriminating ability from the original features. Genetic programming (GP) has been shown to be a prominent technique for this task. However, applying GP to high-dimensional data is still challenging due to the large search space. Feature clustering groups similar features into clusters, which can be used for dimensionality reduction by choosing representative features from each cluster to form the feature subset. Feature clustering has been shown promising in feature selection; but has not been investigated in feature construction for classification. This paper presents the first work of utilising feature clustering in this area. We propose a cluster-based GP feature construction method called CGPFC which uses feature clustering to improve the performance of GP for feature construction on high-dimensional data. Results on eight high-dimensional datasets with varying difficulties show that the CGPFC constructed features perform better than the original full feature set and features constructed by the standard GP constructor based on the whole feature set.

**Keywords:** Genetic programming · Feature construction · Feature clustering · Classification · High-dimensional data

## 1 Introduction

In machine learning, there has been an immense increase in high-dimensional data such as microarray gene expression, proteomics, images, text, and web mining data [1]. These datasets usually have thousands to tens of thousands of features. This enormity leads to the curse of dimensionality that tends to limit the scalability and learning performance of many machine learning algorithms, including classification methods. Furthermore, they usually contain a significant number of irrelevant and redundant features. The existence of these features not only enlarges the search space but also obscures the effect of relevant features on showing the hidden patterns of the data. As a result, they may significantly degrade the performance of many learning algorithms [2].

Feature construction (FC) has been used as an effective pre-processing technique to enhance the discriminating ability of the feature set by creating new

© Springer International Publishing AG 2017
J. McDermott et al. (Eds.): EuroGP 2017, LNCS 10196, pp. 210–226, 2017.
DOI: 10.1007/978-3-319-55696-3_14

high-level features from the original features [3,4]. In order to create new features with better discriminating power, a FC method needs to select informative features and appropriate operators to combine the selected features. Therefore, the search space of the FC problem is very large, which requires a powerful search technique for a FC method.

With a flexible representation and a global search technique, Genetic programming (GP) [5] has been widely used in FC methods [3,4,6] in general and in high-dimensional data [7–9] as well.

GP has been proposed as a FC method with filter, wrapper, or embedded approaches [10]. While wrapper methods evaluate constructed features based on the performance of a learning algorithm, filter methods rely on the intrinsic characteristics of the training data. As a result, wrapper methods usually achieve better classification accuracy than filters. On the other hand, the later are usually faster than the former. Since GP can also be used as a classification algorithm, GP embedded FC methods have also been proposed where the constructed features are evaluated by applying GP itself as a classifier to the training data.

Although GP has been shown promising in generating better discriminating features than original features [4], its application to high-dimensional data is still challenging. In a recent study [8], Tran et al. proposed a GP-based feature construction method (or GPFC) that produced a combination of one constructed feature and a set of features selected to construct it. Experiment results demonstrated that GPFC is a promising approach on high-dimensional data. However, since these datasets may have a large number of redundant features, there may exist a high chance that GP selects redundant features to construct a new feature from the whole feature set. Therefore, the performance of GP in constructing new features may be degraded accordingly. Such a problem can be addressed if GP can avoid choosing redundant features when constructing a new feature.

Feature clustering groups similar *features* into one cluster, which is different from the common data mining task of clustering that groups similar *instances* into clusters [11]. Based on the resulting clusters, one or several features from each group can be chosen as representatives of each group. Using this approach, feature clustering has been proposed and shown promising performance in many feature selection methods [12–15]. However, applying feature clustering to FC is still limited.

**Goals.** This study proposes a cluster-based feature construction method called CGPFC for high-dimensional data. CGPFC uses feature clustering to improve the performance of GP for FC on high-dimensional data. Its performance will be compared with the standard GP for feature construction (GPFC) proposed in [8]. Specifically, we ask the following questions:

– How to automatically group features into clusters,
– Whether the proposed method can construct features with better discriminating ability (better classification accuracy) than the original full feature set and the one constructed by GPFC,

- Whether CGPFC can select a smaller number of features than GPFC to construct a better new feature, and
- Whether the CGPFC combination sets perform better than those created by GPFC.

## 2 Background

### 2.1 Genetic Programming for Feature Construction

When applying GP to FC, a constructed feature is usually represented as a *tree* in which terminal nodes are features selected from the original feature set and internal nodes are operators [3,4]. Although GP has a built-in capability to select good features based on the guide of fitness function, its performance is still affected when applied to high-dimensional data due to the large search space [8]. Therefore, it is critical to narrow this search space for GP performance improvement.

Attempt to reduce the number of features in the GP terminal set has been proposed for different applications. In [16], GP is proposed as a feature selection method for high-dimensional data. The GP terminal set is a combination of 50 top ranked features selected by Information Gain and Relief methods. Results show that GP achieves better feature subsets than baseline methods. However, it is necessary to choose a good number of top ranked features. Furthermore, many redundant features may still exist when combining selected features from the two feature selection methods. Readers are referred to [17] for more examples. In [18], GP is used to build $c$ non-dominated sets of classifiers for a $c$-class problem. In this method, GP uses a relevance measure combined with some threshold as the probability of choosing features to form the trees. During the evolutionary process, GP also eliminates features that do not appear in the population. Results show that the proposed method achieves better classification accuracy than the compared ones.

Results from the above studies have shown that helping GP select appropriate features is critical for enhancing its performance. However, this approach has not been investigated in GP for FC. Therefore, in this study, we propose to use clustering to eliminate redundant features in order to improve GP performance for FC.

### 2.2 Feature Clustering

Clustering or cluster analysis is one of the main tasks in exploratory data mining. It aims to group similar objects into the same group or cluster. Literature has proposed different clustering algorithms using different measures to evaluate the similarity between objects as well as different ways of grouping objects [19].

Clustering has been used for decades as an unsupervised task where instances are grouped into clusters based on the similarity between them [20]. In machine learning and data mining, the "clustering" terminology is usually meant *instance clustering* or instance grouping. Recently, clustering techniques have been proposed to group similar features, thus called *feature clustering*, to achieve feature selection [11]. Similar features are grouped into the same cluster. Then one or more representative features from each cluster were used to form final subsets. Clustering techniques and different strategies of using the resulting feature clusters have been proposed in feature selection methods [21–23].

Correlation coefficient (CC) is a popular measure to evaluate feature dependency or redundancy. To group redundant features, CC is used to replace the proximity measure in the k-means clustering algorithm [24]. The final feature subset is then formed by gathering the most relevant feature from each cluster. A feature is considered as the most relevant of the cluster if its CC value with the class label is the highest. Experiments on two datasets with hundreds of features show a better result than a compared method and worse than the other. However, when k-means is used, the performance of the proposed method depends on other methods in estimating the number of clusters. Feature clustering is used as an approach to dimensionality reduction not only in classification but also in regression. In [25], CC is used to calculate the input coherence $I_c$ and output incoherence $O_c$ of each feature $f$ to each feature cluster. Feature $f$ is put into the cluster with the highest $I_c$ if its $I_c$ and $O_c$ values satisfy two predefined thresholds. Then k new features are constructed from k clusters based on a weight matrix. Results on four regression problems with 13 to 363 features showed that the proposed method has better performance than the compared methods.

While CC measures the level of correlation between features, mutual information indicates how much knowledge one can get for a variable by knowing the value of another variable. Symmetric uncertainty (SU) [26], which is a normalised version of information gain (IG) [27], is also used to identify irrelevant and redundant features. A feature is considered irrelevant if it does not give any information about the target concept or its SU with the class label is very small. Two features are redundant if their mutual information or their SU is high [28]. For example, in [23], SU is combined with minimum spanning tree (MST) to group features. Firstly, features are considered as irrelevant and removed if their SU with the class label is lower than a predefined threshold. Then a MST is built on the remaining features where SU between two features are used as the cost of each edge. The MST is then partitioned into disconnected trees, each of which contains features that have SU between them higher than their SU to the class. The most relevant feature in each tree is chosen to form the final subset. Results on 35 high-dimensional data including microarray, text and images showed that the proposed method achieve better performance on micro-array data than the state of the art feature selection methods.

Different from previous approaches, statistical clustering [29] takes into account interactions between features in evaluating feature redundancy. It is used for the purpose of dimensionality reduction in [21,22]. Particle swarm optimisation is used to select features from each cluster to form the final subset. Performances on UCI datasets of the proposed methods show promising results.

However, these statistical clustering algorithms are computationally expensive, thus not suitable for high-dimensional data.

In summary, by reducing the number of redundant features in feature set, feature clustering is a promising approach to feature selection. However, it has not been investigated in GP for FC especially on high-dimensional data. In this study, we propose to use feature clustering to automatically group features into clusters and the best feature of each cluster will be chosen for FC.

# 3    The Proposed Approach

## 3.1    The Redundancy Based Feature Clustering Method: RFC

K-means is a well-known clustering algorithm. It is a simple and effective method. However, it is essential to predefine a suitable number of clusters, which is not easy especially in high-dimensional data with thousands to tens of thousands of features. An inappropriate value may lead to clusters with uncorrelated or non-redundant features. In addition, the number of clusters in feature clustering is not as meaningful as the number of clusters in instance clustering, which represents the number of different types of objects/instances. Therefore, instead of grouping features based on a predefined number of clusters, in this study, we propose a new algorithm to group features based on the redundancy levels between them (called RFC). Different from the number of clusters, the redundancy or correlation level of two features is a value in the range of 0 and 1, representing no and full correlation between them, respectively. RFC uses a simple approach to ensure that all features in the same cluster are redundant features with their correlation level higher than a predefined threshold.

The main principle of RFC is to group features that have their redundancy level higher than a given threshold. If two features $X$ and $Y$ have their $CC(X,Y)$ larger than this threshold, they will be grouped into the same cluster. In this way, the number of clusters will be automatically determined. If a dataset has a large number of redundant features, the number of clusters will be much smaller than the number of features, and vice versa. Furthermore, using this strategy enables us to ensure that the generated clusters include only features having their correlation levels higher than or equal to the predefined redundancy threshold.

Algorithm 1 shows the pseudo code of RFC for a given training set and a redundancy threshold $\theta$. First of all, we analyse features to remove irrelevant ones (lines 4–9). In this study, a feature is considered irrelevant if it does not give any information about the class label. Since the class label is a discrete variable, SU is a suitable measure for feature relevancy. Therefore, in this step, all features whose SU with the class label are equal to zero will be removed.

SU between a feature $X$ and the class $C$ is calculated based on Eq. (1) which gives a value between 0 and 1 representing no to full correlation, respectively. As SU is an entropy-based measure, it can only be applied on category or nominal data. Therefore, we discretise data before calculating SU using MDL [30], which is a popular discretisation method.

---

**Algorithm 1.** The pseudo code of RFC

---

```
   Input  : Training data, redundancy_threshold θ
   Output: Clusters of features
 1 begin
 2 │    F ← ∅ ;
 3 │    clusters ← ∅ ;
 4 │    for each feature f_i in Training data do
 5 │    │    su ← SU(f_i, class) (based on Eq.(1)) ;
 6 │    │    if su > 0 then
 7 │    │    │    F ← F ∪ {f_i};
 8 │    │    end
 9 │    end
10 │    while (F ≠ ∅) do
11 │    │    f_i ← next feature in F;
12 │    │    F ← F \ {f_i};
13 │    │    new_cluster ← {f_i};
14 │    │    while (F ≠ ∅) do
15 │    │    │    f_j ← next feature in F;
16 │    │    │    F ← F \ {f_j};
17 │    │    │    cc ← CC(f_i, f_j) (based on Eq.(3)) ;
18 │    │    │    if (cc > θ) then
19 │    │    │    │    new_cluster ← new_cluster ∪ {f_j};
20 │    │    │    end
21 │    │    │    clusters ← clusters ∪ new_cluster;
22 │    │    end
23 │    end
24 │    Return clusters;
25 end
```

---

$$SU(X, C) = 2 \left[ \frac{IG(X|C)}{H(X) + H(C)} \right] \tag{1}$$

where

$$IG(X|C) = H(X) - H(X|C) \tag{2}$$

and $H(X)$ is the entropy of $X$ and $H(X|C)$ is the conditional entropy of $X$ given $C$.

All remaining features are then grouped into exclusive or non-overlapped clusters (lines 10–23). In this step, CC is used to measure redundancy level between features because it can be directly applied to numerical data. Although CC can only measure the linear relationship between variables, it has been shown effective in many feature selection methods [24, 25]. The CC measure gives a value between $-1$ and 1 whose absolute value represents the correlation level between two features. Given that $n$ is the number of instances in the dataset, CC between feature $X$ and $Y$ is calculated based on Eq. (3) which gives a value in $[0,1]$.

$$CC(X, Y) = \left| \frac{\sum_{i=1}^{n} X_i Y_i - n\bar{X}\bar{Y}}{\sqrt{\sum_{i=1}^{n} X_i^2 - n\bar{X}^2} \sqrt{\sum_{i=1}^{n} Y_i^2 - n\bar{Y}^2}} \right| \tag{3}$$

## 3.2  The Proposed Method: CGPFC

After grouping features into clusters, collection of the best features of each cluster is used to construct one new feature. This section introduces the proposed cluster-based feature construction method, called CGPFC. Note that while the feature clustering algorithm (RFC) uses a filter measure to group features, the feature construction algorithm follows the wrapper approach.

**Representation.** CGPFC aims at constructing a single feature using a tree-based representation. Each GP individual has one tree which represents a constructed feature. We follow [8] to create a combination of the constructed and selected features from the GP tree. Figure 1 provides an example of creating this combination from the best GP tree.

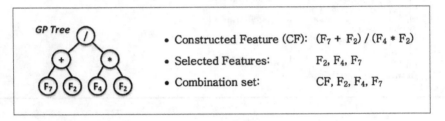

**Fig. 1.** Constructed feature and selected features

**Fitness Function.** The cluster-based feature construction method developed in this work follows the wrapper approach. All classification algorithms can be used to evaluate the performance of the constructed feature. To evaluate a GP individual, the training set is transformed based on the feature constructed by the GP tree. Then the transformed training set classification performance is tested using L-fold cross validation (CV). The average accuracy is used as the fitness of the individual.

Since many of the high-dimensional datasets are unbalanced data, the balanced classification accuracy [31] is used in this fitness function. Given $c$ as the number of classes, $TP_i$ as the number of correctly identified instances in class $i$, and $|S_i|$ as the total number of instances in class $i$, the balanced accuracy is calculated based on Eq. (4). Since no bias is given to any specific class, the weight here is set equally to $1/c$.

$$fitness = \frac{1}{c} \sum_{i=1}^{c} \frac{TP_i}{|S_i|} \tag{4}$$

**Overall Algorithm.** Algorithm 2 describes the pseudo code of CGPFC. Given a training set and a redundancy threshold, the algorithm will return the combination of one constructed feature and the selected features constructing it.

---

**Algorithm 2.** The pseudo code of CGPFC

---

    **Input**   : Training_data, redundancy_threshold $\theta$
    **Output**: Constructed feature and selected features
1  **begin**
2       $clusters \leftarrow RFC(Training\_data, \theta)$;
3       Initialise population using the best feature in each cluster of *clusters*;
4       **while** *Maximum iterations or the best solution is not found* **do**
5            **for** $i = 1$ *to Population Size* **do**
6               $transf\_train \leftarrow$ Calculate constructed feature of individual $i$ on Training data ($transf\_train$ has only one new feature) ;
7               Evaluate $transf\_train$ using the learning algorithm with L-fold CV;
8               $fitness \leftarrow$ average test accuracy based on Eq.(4);
9            **end**
10        Select parent individuals using tournament;
11        Create offspring individuals by applying crossover or mutation on the selected parents;
12        Place new individuals into the population of the next generation;
13       **end**
14       Return the constructed feature and selected features in the best individual;
15  **end**

---

First of all, the feature clustering procedure RFC is called to create a set of clusters from the training data. The most relevant feature (based on SU measure) in each cluster is employed to initialise GP individuals. Lines 5–9 are used to evaluate GP individuals. The loop of evaluation-selection-evolution (lines 4–13) is executed until the stopping criteria are met.

# 4 Experiment Design

**Datasets.** Eight binary-class gene expression datasets with thousands of features are used to examine the performance of the proposed method on high-dimensional data. These datasets are publicly available at http://www. gems-system.org, and http://csse.szu.edu.cn/staff/zhuzx/Datasets.html.

**Table 1.** Datasets

| Dataset | #Features | #Instances | Class distribution |
|---------|-----------|------------|--------------------|
| Colon | 2,000 | 62 | 35%–65% |
| DLBCL | 5,469 | 77 | 25%–75% |
| Leukemia | 7,129 | 72 | 35%–65% |
| CNS | 7,129 | 60 | 35%–65% |
| Prostate | 10,509 | 102 | 50%–50% |
| Ovarian | 15,154 | 253 | 36%–64% |
| Alizadeh | 1,095 | 42 | 50%–50% |
| Yeoh | 2,526 | 248 | 83%–17% |

Table 1 describes these datasets in detail. It can be seen that these datasets have a small number of instances compared to the number of features.

These datasets are challenging tasks in machine learning due to the curse of dimensionality issue. On top of this challenge, these are also unbalanced data with very different percentage of instances in each class as shown in the last column of the table.

Furthermore, gene expression data usually contains substantial noise generated during the data collection in laboratories. Therefore, discretisation is applied to reduce noise as suggested in [32]. Each feature is first standardised to have zero mean and unit variance. Then its values are discretised into $-1$, 0 and 1 representing three states which are the under-expression, the baseline and the over-expression of gene. Values that fall in the interval $[\mu - \sigma/2, \mu + \sigma/2]$, where $\mu$ and $\sigma$ are mean and standard deviation of the feature values, are transformed to state 0. Values that are in the left or in the right of this interval will be transformed to state $-1$ or 1, respectively.

**Parameter Settings.** Table 2 shows the parameter settings for GP. The function set includes four basic arithmetic operators and three functions that used to construct a numeric feature from the selected features and random constants. Since the numbers of features in these datasets are quite different, ranging from about two thousand to fifteen thousand, the search spaces of these problems are very different. Therefore, we set the population size proportional to the terminal set size or the number of clusters (#clusters $\cdot \alpha$). $\alpha$ is set to 20 in this experiment. The mutation rate is set to 0.2, but after generation 10 it gradually increases with a step of 0.02 in every generation to avoid stagnation in local optima. The crossover rate is also updated accordingly to ensure the sum of these two rates always equal to 1. The redundancy threshold is empirically set to 0.9. The stopping criterion is either GP reaches the maximum generation or the best solution is found.

**Table 2.** GP parameter settings

| | |
|---|---|
| Function set | $+, -, \times, \%, min, max, if$ |
| Terminal set | Features and random constant values |
| Population size | #clusters $\cdot \alpha$ |
| Maximum generations | 50 |
| Initial population | Ramped half-and half |
| Initial maximum tree depth | 2 |
| Maximum tree depth | 17 |
| Selection method | Tournament method |
| Tournament size | 7 |
| Crossover rate | 0.8 |
| Mutation rate | 0.2 |
| Elitism | True |
| Redundancy threshold | 0.9 |

**Experiment Configuration.** To test the performance of CGPFC, we compared the discriminating ability of the constructed feature versus the original features and the one constructed by the standard GP [8] based on the classification accuracy of K-nearest neighbour (KNN), Naive Bayes (NB) and Decision Tree (DT). This comparison is also conducted on the combination feature set of the constructed and the selected features.

Due to the small number of instances in each dataset, two loops of CV are used to avoid feature construction bias as described in [8]. The outer loop uses a stratified 10-fold CV on the whole dataset. One fold is kept as the test set to evaluate the performance of each method, and the remaining 9 folds are used to form the training set for feature construction and classifier training. In the fitness function, an inner loop of 3-fold CV within the training set is run to evaluate the evolved feature (see Sect. 3.2). The learning algorithm used for fitness evaluation is Logistic Regression (LR) - a statistical learning method. We choose LR because it can help in scaling the constructed feature to determine the probabilities of each class through the use of a logistic function. Therefore, it can effectively test the ability of GP-constructed feature in discriminating classes.

Since GP is a stochastic algorithm, we run CGPFC on each dataset 30 times independently with different random seeds and report the average results to avoid statistical variations.

Therefore, a totally 300 runs (30 runs combined with 10-fold CV) are executed on each dataset. Experiments were runs on PC with Intel Core i7-4770 CPU @ 3.4 GHz, running Ubuntu 4.6 and Java 1.7 with a total memory of 8 GB. The results of 30 runs from each method were compared using Wilcoxon statistical significance test [33], with the significance level of 0.05.

## 5   Results and Discussions

### 5.1   Performance of the Constructed Feature

Table 3 shows the average test results of 30 independent runs of the proposed method (CGPFC) compared with "Full" (i.e. using the original feature set) and the GPFC [8]. The number of instances in each dataset is also displayed in parentheses under its name. Column "#F" shows the average size of each feature set. The following columns display the best (B), mean and standard deviation of the accuracy (M ± Std) obtained by KNN, NB and DT on the corresponding feature set. The highest average accuracy of each learning algorithm for each dataset is bold. Columns $S_1$, $S_2$, and $S_3$ display the Wilcoxon significance test results of the corresponding method over CGPFC with significance level of 0.05. "+" or "−" means that the result is significantly better or worse than CGPFC and "=" means that their results are similar. In other words, the more "−", the better the proposed method.

**CGPFC Versus Full.** It can be seen from Table 3 that using only a single constructed feature by CGPFC, KNN obtains significantly better results than

**Table 3.** Test accuracy of the constructed feature

| Dataset | Method | #F | B-KNN | M ± Std-KNN | $S_1$ | B-NB | M ± Std-NB | $S_2$ | B-DT | M ± Std-DT | $S_3$ |
|---|---|---|---|---|---|---|---|---|---|---|---|
| Colon (62) | Full | 2000 | 74.29 | | − | 72.62 | | − | 74.29 | | − |
| | GPFC | 1 | 79.28 | 71.40 ± 4.46 | − | 78.81 | 69.64 ± 4.17 | − | 79.28 | 72.25 ± 4.07 | − |
| | CGPFC | 1 | 88.81 | **77.56 ± 4.47** | | 88.81 | **77.96 ± 4.16** | | 88.81 | **78.08 ± 4.05** | |
| DLBCL (77) | Full | 5469 | 84.46 | | − | 81.96 | | − | 80.89 | | − |
| | GPFC | 1 | 96.07 | 86.65 ± 3.76 | − | 92.32 | 86.27 ± 4.28 | − | 94.64 | 86.51 ± 4.08 | − |
| | CGPFC | 1 | 94.64 | **88.62 ± 2.92** | | 94.64 | **88.74 ± 2.90** | | 94.64 | **88.62 ± 2.92** | |
| Leukemia (72) | Full | 7129 | 88.57 | | | 91.96 | | + | 91.61 | | = |
| | GPFC | 1 | 94.46 | 89.03 ± 2.71 | − | 93.21 | 87.26 ± 4.44 | − | 95.89 | 88.97 ± 2.96 | = |
| | CGPFC | 1 | 95.89 | **90.65 ± 3.21** | | 97.32 | 90.73 ± 3.16 | | 95.89 | 90.65 ± 3.21 | |
| CNS (60) | Full | 7129 | 56.67 | | + | 58.33 | | + | 50.00 | | − |
| | GPFC | 1 | 70.00 | **57.56 ± 5.87** | = | 70.00 | **58.44 ± 5.94** | = | 70.00 | **57.78 ± 6.05** | = |
| | CGPFC | 1 | 63.33 | 55.06 ± 3.85 | | 63.33 | 56.00 ± 2.89 | | 63.33 | 56.00 ± 3.02 | |
| Prostate (102) | Full | 10509 | 81.55 | | − | 60.55 | | − | 86.18 | | − |
| | GPFC | 1 | 90.18 | 83.72 ± 3.18 | − | 90.18 | 83.18 ± 3.68 | − | 90.18 | 83.82 ± 2.85 | − |
| | CGPFC | 1 | 92.27 | **87.40 ± 3.62** | | 92.27 | **87.31 ± 3.51** | | 92.27 | **87.40 ± 3.62** | |
| Ovarian (253) | Full | 15154 | 91.28 | | − | 90.05 | | − | 98.42 | | − |
| | GPFC | 1 | 99.62 | 97.86 ± 1.22 | − | 99.62 | 97.22 ± 1.48 | − | 99.62 | 97.89 ± 1.18 | − |
| | CGPFC | 1 | 100.00 | **99.37 ± 0.48** | | 100.00 | **99.37 ± 0.48** | | 100.00 | **99.37 ± 0.48** | |
| Alizadeh (42) | Full | 1095 | 77.00 | | = | 92.50 | | + | 78.50 | | = |
| | GPFC | 1 | 86.00 | **77.88 ± 5.53** | = | 88.50 | 76.52 ± 5.85 | = | 86.00 | 77.20 ± 5.84 | = |
| | CGPFC | 1 | 87.50 | 77.12 ± 5.85 | | 87.50 | 76.88 ± 6.06 | | 87.50 | 77.20 ± 6.11 | |
| Yeoh (248) | Full | 2526 | 89.97 | | − | 93.57 | | − | 97.57 | | = |
| | GPFC | 1 | 99.17 | 97.04 ± 1.01 | = | 97.57 | 95.11 ± 2.72 | − | 99.17 | 97.05 ± 0.99 | − |
| | CGPFC | 1 | 98.77 | **97.32 ± 1.59** | | 98.77 | **97.38 ± 0.84** | | 98.77 | **97.71 ± 0.67** | |

Full on 6 out of the 8 datasets. The highest improvement in the average accuracy is 8% on Ovarian and Yeoh, and 14% in the best result of Colon dataset. On Alizadeh, it achieves a similar performance on average and 10% higher than Full in the best case. Only on CNS, does CGPFC obtain slightly worse results than Full, however, still 7% better accuracy in the best case. With 7129 features and only 60 instances, this dataset can be considered as the most challenging dataset among the eight. With a small number of training instances, it is hard for GP to construct a feature that is generalised well to the unseen data.

Similar to KNN, the constructed feature by CGPFC also helps NB achieve 4% to 27% higher accuracy than Full on 5 datasets. Using only 1 constructed feature on Prostate, the best accuracy NB can achieve is 32% higher than Full. On Leukemia and CNS, although NB obtains 1% and 2% average result lower than Full, its best accuracy is still 6% and 5% higher, respectively. On Alizadeh, the accuracy of CGPFC constructed feature is significantly lower than Full. We also note that the accuracy of NB on the Full feature set of this dataset is much higher than KNN and DT.

DT also gets benefit from the constructed feature, shown as significantly improvement in its performance on 5 datasets and obtaining a similar result on the remaining datasets. The highest improvement is on DLBCL with 8% increase on average and 14% in the best case. Although the average accuracy is slightly worse on two datasets, namely Leukemia and Alizadeh, the best accuracy DT obtained for each dataset is always higher than Full.

In general, over 24 comparisons between CGPFC and Full on 8 datasets and 3 learning algorithms, the constructed feature by CGPFC wins 16, draws 4 and loses 4. The results indicate that the CGPFC constructed feature has much higher discriminating ability than the original feature set with thousands of features.

**CGPFC Versus GPFC.** Compared with GPFC, CGPFC helps KNN further improve its results to achieve the best results on 6 out of the 8 datasets. The result is significantly better than GPFC on 5 datasets and similar on the other three. Similarly, using the CGPFC constructed feature, NB obtains significantly better results than using GPFC constructed feature on 6 datasets with the highest improvement of 8% on Colon. Applying the CGPFC constructed feature on DT also gives similar results as KNN with significantly improvement on 5 datasets and equivalent on the remaining ones.

In summary, the CGPFC constructed feature wins 16, draws 8, loses 0 out of the 24 pairs of comparisons. The results show that by reducing the irrelevant and the redundant features in the GP terminal set, the constructed feature has a better discriminating ability than the one constructed from the full feature set.

### 5.2   Performance of the Constructed and Selected Features

In GP-based method, there is a built-in feature selection process which selects informative features from the original set to construct the new feature. Results in [8] has shown that the combination of these selected features and the constructed feature from the GP tree has better performance than other combinations of the constructed and original features. This finding is also supported in this study with an even better results.

The average size of this combination created by GPFC and CGPFC over the 30 runs is shown in Fig. 2. The average accuracy of the three learning algorithms, namely KNN, NB and DT, on this combination are shown in Figs. 3, 4, and 5, respectively. In these figures, each group of bars shows the results of GPFC and CGPFC on each dataset. On the CGPFC bars, results of the significance test comparing CGPFC against GPFC are displayed. "+" and "−" mean that CGPFC is significantly better or worse than GPFC. "=" means that they are similar.

First of all, let us examine the size of this combination of features. Since both methods construct only one feature, the difference between their sizes comes from the different numbers of distinct features they select. It can be seen from Fig. 2 that CGPFC always select a much smaller number of features than GPFC. On four datasets, namely DLDCL, Leukemia, Ovarian and Alizadeh, CGPFC selects less than half the number of features selected by GPFC. With a smaller number of selected features, if the CGPFC combination sets have better classification performance than those created by GPFC, it can be inferred that the selected features by CGPFC have better discriminating power than those selected by GPFC.

Results in Fig. 3 show that the CGPFC combination sets obtain a higher KNN accuracy than those created by GPFC on 7 out of the 8 datasets. This result is significantly better on 5 datasets and similar on the other three. Similar patterns are seen in Figs. 4 and 5 for NB and DT with significantly better results on 4 and 5 datasets, respectively. In general, CGPFC either improves or maintains the performance of GPFC on all the 8 datasets.

As can be seen in Figs. 2, 3, 4 and 5, the error bars of CGPFC are always smaller than the corresponding error bar of GPFC in all datasets. This indicates that CGPFC produces more robust results than GPFC.

Results from the combination of constructed and selected features on the 8 datasets show that CGPFC uses a smaller number of features to construct a new feature with better discriminating ability than GPFC. This indicates that by reducing the number of redundant features in the terminal set, feature clustering helps GP to improve its performance.

### 5.3 Cluster Analysis

To validate the structure of the clusters generated by our algorithm, in this section, we investigate the cohesion or compactness within each cluster as well as the separation or isolation between different clusters.

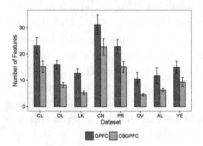

**Fig. 2.** Size of CFTer.

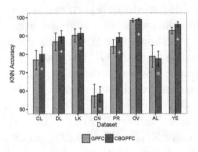

**Fig. 3.** KNN accuracy of CFTer.

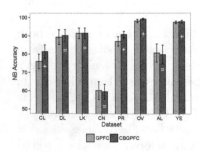

**Fig. 4.** NB accuracy of CFTer.

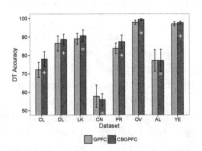

**Fig. 5.** DT accuracy of CFTer.

Silhouette analysis [34] is a popular method to study both the cohesion and the separation of clusters. Equation (5) displays the calculation of the silhouette coefficient of a feature $i$ in which $a_i$ is the average distance of feature $i$ to all other features in its cluster, and $b_i$ is the minimum average distance of feature $i$ to other clusters. Given that $c_i$ is the cluster that includes feature $i$, and $c_k$ is other clusters, $CC(f_i, f_j)$ is the CC between features $i$ and $j$, $a_i$ and $b_i$ are calculated based on Eqs. (6) and (7). Since the CC (see (Eq. 3)) measures the correlation level or similarity between 2 features and has a value between 0 and 1, we use $(1 - CC)$ as a distance or dissimilarity measure between them.

$$s_i = \frac{(b_i - a_i)}{max(a_i, b_i)} \tag{5}$$

where

$$a_i = \frac{1}{size(c_i)} \sum_{j=1}^{size(c_i)} (1 - CC(f_i, f_j)), i \neq j \tag{6}$$

$$b_i = \min_{\forall c_k \neq c_i} \left( \frac{1}{size(c_k)} \sum_{j=1}^{size(c_k)} (1 - CC(f_i, f_j)) \right) \tag{7}$$

The value of the silhouette coefficient ranges from $-1$ to 1, where $-1$ is the worst and 1 is the best case. Average silhouette coefficient (ASC) of all features is an overall measure indicating the goodness of a clustering. Since the experiments were conducted based on a 10-fold CV framework on each dataset, we calculate the ASC for each fold and report the average of 10 ASCs. Table 4 shows the original number of features, the average number of clusters generated with redundancy level of 0.9, the percentage of dimensionality reduction, and the average of ASC of clustering on each dataset.

**Table 4.** Cluster analysis

| Dataset | #Features | #Clusters | %Dimensionality reduction | Silhouette coefficient |
|---------|-----------|-----------|---------------------------|------------------------|
| Colon | 2000 | 104.10 | 0.95 | 0.80 |
| DLBCL | 5469 | 819.20 | 0.85 | 0.96 |
| Leukemia | 7129 | 901.30 | 0.87 | 0.98 |
| CNS | 7129 | 79.30 | 0.99 | 1.00 |
| Prostate | 10509 | 1634.80 | 0.84 | 0.85 |
| Ovarian | 15154 | 601.20 | 0.96 | 0.31 |
| Alizadeh | 1095 | 93.60 | 0.01 | 0.94 |
| Yeoh | 2526 | 97.60 | 0.96 | 1.00 |

As can be seen from the fourth column of Table 4, all datasets obtain at least 84% of dimensionality reduction after the proposed feature clustering algorithm is applied. The number of input features into GP is significantly reduced with the largest reduction of 99% on CNS and 96% on Ovarian and Yeoh. The third column of Table 4 also shows differences in the number of clusters generated on different datasets regardless of its original number of features. For example, CNS has much smaller number of clusters than Colon although its feature set size is more than three times larger than Colon. The silhouette coefficient of each dataset is quite good except for Ovarian, where features in different clusters are still correlated but with a smaller level than 0.9. Even though its silhouette coefficient is not good enough, the results of this dataset shown in Table 3 and Figs. 3, 4 and 5 reveal that feature clustering method enables the constructed feature perform significantly better than the feature constructed from the whole feature set.

## 6    Conclusions and Future Work

This study is the first work that aims to apply feature clustering to GP for FC in classification in order to improve its performance on high-dimensional data. The goal has been achieved by proposing a new feature clustering algorithm to cluster redundant features in one group based on a correlation or redundancy level. Then the best feature from each cluster is fed into GP to construct a single new high-level feature. Performance of the constructed and/or selected features is tested on three different classification algorithms. Results on eight gene expression datasets have shown that feature clustering helps GP construct features with better discriminating ability than those generated from the whole feature set.

The clustering technique proposed in this study has an advantage of automatically determining the number of clusters. It guarantees that features in one cluster have their correlated level higher than the given redundancy level. Although determining a redundancy level is easier than the number of clusters as in the case of K-means clustering technique, the proposed method still has some limitations, such as features in different clusters may also correlated to each other with a lower level than the given threshold. Furthermore, the proposed feature clustering method is threshold sensitive. These limitations can be solved by integrating feature clustering into feature construction process so that the performance of GP could be used to automatically adjust the feature clusters.

## References

1. Zhang, J., Wang, S., Chen, L., Gallinari, P.: Multiple Bayesian discriminant functions for high-dimensional massive data classification. Data Min. Knowl. Discov. **31**, 465–501 (2017)
2. Liu, H., Motoda, H.: Feature Extraction, Construction and Selection: A Data Mining Perspective. Kluwer Academic Publishers, Norwell (1998)
3. Krawiec, K.: Evolutionary feature selection and construction. In: Sammut, C., Webb, G.I. (eds.) Encyclopedia of Machine Learning, pp. 353–357. Springer, Heidelberg (2010)

4. Neshatian, K., Zhang, M., Andreae, P.: A filter approach to multiple feature construction for symbolic learning classifiers using genetic programming. IEEE Trans. Evol. Comput. **16**, 645–661 (2012)

5. Koza, J.R.: Genetic Programming: On the Programming of Computers by Means of Natural Selection. MIT Press, Cambridge (1992)

6. Hiroyasu, T., Shiraishi, T., Yoshida, T., Yamamoto, U.: A feature transformation method using multiobjective genetic programming for two-class classification. In: IEEE Congress on Evolutionary Computation (CEC), pp. 2989–2995 (2015)

7. Ahmed, S., Zhang, M., Peng, L., Xue, B.: Multiple feature construction for effective biomarker identification and classification using genetic programming. In: Proceedings of Genetic and Evolutionary Computation Conference, pp. 249–256. ACM (2014)

8. Tran, B., Xue, B., Zhang, M.: Genetic programming for feature construction and selection in classification on high-dimensional data. Memetic Comput. **8**, 3–15 (2015)

9. Tran, B., Xue, B., Zhang, M.: Multiple feature construction in high-dimensional data using genetic programming. In: IEEE Symposium Series on Computational Intelligence (SSCI) (2016)

10. Guyon, I., Elisseeff, A.: An introduction to variable and feature selection. J. Mach. Learn. Res. **3**, 1157–1182 (2003)

11. Butterworth, R., Piatetsky-Shapiro, G., Simovici, D.A.: On feature selection through clustering. In: ICDM, vol. 5, pp. 581–584 (2005)

12. Gupta, A., Gupta, A., Sharma, K.: Clustering based feature selection methods from fMRI data for classification of cognitive states of the human brain. In: 3rd International Conference on Computing for Sustainable Global Development (INDIACom), pp. 3581–3584. IEEE (2016)

13. Jaskowiak, P.A., Campello, R.J.: A cluster based hybrid feature selection approach. In: Brazilian Conference on Intelligent Systems (BRACIS), pp. 43–48. IEEE (2015)

14. Krier, C., François, D., Rossi, F., Verleysen, M.: Feature clustering and mutual information for the selection of variables in spectral data. In: European Symposium on Artificial Neural Networks (ESANN), Le Chesnay Cedex, France, pp. 157–162 (2007)

15. Rostami, M., Moradi, P.: A clustering based genetic algorithm for feature selection. In: Conference on Information and Knowledge Technology, pp. 112–116 (2014)

16. Ahmed, S., Zhang, M., Peng, L.: Feature selection and classification of high dimensional mass spectrometry data: a genetic programming approach. In: Vanneschi, L., Bush, W.S., Giacobini, M. (eds.) EvoBIO 2013. LNCS, vol. 7833, pp. 43–55. Springer, Heidelberg (2013). doi:10.1007/978-3-642-37189-9_5

17. Xue, B., Zhang, M., Browne, W.N., Yao, X.: A survey on evolutionary computation approaches to feature selection. IEEE Trans. Evol. Comput. **20**, 606–626 (2016)

18. Nag, K., Pal, N.: A multiobjective genetic programming-based ensemble for simultaneous feature selection and classification. IEEE Trans. Cybern. **46**, 499–510 (2016)

19. Xu, D., Tian, Y.: A comprehensive survey of clustering algorithms. Ann. Data Sci. **2**, 165–193 (2015)

20. Xu, R., Wunsch, D.: Survey of clustering algorithms. IEEE Trans. Neural Netw. **16**, 645–678 (2005)

21. Lane, M.C., Xue, B., Liu, I., Zhang, M.: Gaussian based particle swarm optimisation and statistical clustering for feature selection. In: Blum, C., Ochoa, G. (eds.) EvoCOP 2014. LNCS, vol. 8600, pp. 133–144. Springer, Heidelberg (2014). doi:10.1007/978-3-662-44320-0_12

22. Nguyen, H.B., Xue, B., Liu, I., Zhang, M.: PSO and statistical clustering for feature selection: a new representation. In: Dick, G., et al. (eds.) SEAL 2014. LNCS, vol. 8886, pp. 569–581. Springer, Heidelberg (2014). doi:10.1007/978-3-319-13563-2_48

23. Song, Q., Ni, J., Wang, G.: A fast clustering-based feature subset selection algorithm for high-dimensional data. IEEE Trans. Knowl. Data Eng. **25**, 1–14 (2013)

24. Hsu, H.H., Hsieh, C.W.: Feature selection via correlation coefficient clustering. J. Softw. **5**, 1371–1377 (2010)

25. Xu, R.F., Lee, S.J.: Dimensionality reduction by feature clustering for regression problems. Inf. Sci. **299**, 42–57 (2015)

26. Press, W.H., Teukolsky, S., Vetterling, W., Flannery, B.: Numerical Recipes in C, vol. 1, p. 3. Cambridge University Press, Cambridge (1988)

27. Quinlan, J.R.: C4.5: Programs for Machine Learning. Morgan Kaufmann Publishers, Inc., Burlington (1993)

28. Liu, H., Motoda, H.: Computational Methods of Feature Selection. CRC Press, Boca Raton (2007)

29. Pledger, S., Arnold, R.: Multivariate methods using mixtures: correspondence analysis, scaling and pattern-detection. Comput. Stat. Data Anal. **71**, 241–261 (2014)

30. Fayyad, U.M., Irani, K.B.: Multi-interval discretization of continuous-valued attributes for classification learning. In: Thirteenth International Joint Conference on Artificial Intelligence, vol. 2, pp. 1022–1027. Morgan Kaufmann Publishers (1993)

31. Patterson, G., Zhang, M.: Fitness functions in genetic programming for classification with unbalanced data. In: Orgun, M.A., Thornton, J. (eds.) AI 2007. LNCS (LNAI), vol. 4830, pp. 769–775. Springer, Heidelberg (2007). doi:10.1007/978-3-540-76928-6_90

32. Ding, C., Peng, H.: Minimum redundancy feature selection from microarray gene expression data. J. Bioinform. Comput. Biol. **3**, 185–205 (2005)

33. Wilcoxon, F.: Individual comparisons by ranking methods. Biom. Bull. **1**, 80–83 (1945)

34. Rousseeuw, P.J.: Silhouettes: a graphical aid to the interpretation and validation of cluster analysis. J. Comput. Appl. Math. **20**, 53–65 (1987)

# Posters

# Geometric Semantic Crossover
# with an Angle-Aware Mating Scheme in Genetic
# Programming for Symbolic Regression

Qi Chen[(✉)], Bing Xue, Yi Mei, and Mengjie Zhang

School of Engineering and Computer Science, Victoria University of Wellington,
Wellington, New Zealand
{Qi.Chen,Bing.Xue,Yi.Mei,Mengjie.Zhang}@ecs.vuw.ac.nz

**Abstract.** Recent research shows that incorporating semantic knowledge into the genetic programming (GP) evolutionary process can improve its performance. This work proposes an angle-aware mating scheme for geometric semantic crossover in GP for symbolic regression. The angle-awareness guides the crossover operating on parents which have a large angle between their relative semantics to the target semantics. The proposed idea of angle-awareness has been incorporated into one state-of-the-art geometric crossover, the locally geometric semantic crossover. The experimental results show that, compared with locally geometric semantic crossover and the regular GP crossover, the locally geometric crossover with angle-awareness not only has a significantly better learning performance but also has a notable generalisation gain on unseen test data. Further analysis has been conducted to see the difference between the angle distribution of crossovers with and without angle-awareness, which confirms that the angle-awareness changes the original distribution of angles by decreasing the number of parents with zero degree while increasing their counterparts with large angles, leading to better performance.

**Keywords:** Geometric semantic crossover · Angle-awareness

## 1 Introduction

In recent years, semantic genetic programming (GP) [11,18], which incorporates the semantic knowledge in the evolutionary process to improve the efficacy of search, attracts increasing attention and becomes a hot research topic in GP [6]. One popular form of semantic methods, geometric semantic GP (GSGP), has been proposed recently [12]. GSGP searches directly in the semantic space of GP individuals. The geometric crossover and mutation operators generate offspring that lies within the bounds defined by the semantics of the parent(s) in the semantic space. The fitness landscape that these geometric operators explore has a conic shape, which contains no local optimal and is easier to search. In previous research,

© Springer International Publishing AG 2017
J. McDermott et al. (Eds.): EuroGP 2017, LNCS 10196, pp. 229–245, 2017.
DOI: 10.1007/978-3-319-55696-3_15

GSGP presents a notable learning gain over standard GP [17,19]. For the generalisation improvement, GSGP shows some positive effect. However, while the geometric mutation is remarked to be critical in bringing the generalisation benefit, the geometric crossover is criticised to have a weak effect on promoting generalisation for some regression tasks [5]. One possible reason is that of the target output on the test set is beyond the scope of the convex combination of the parents for crossover [13] in the test semantic space. Another possible reason is that crossover might operate on similar parents standing in a compact volume of the semantic space, which leads to generating offspring having duplicate semantics with their parents. In this case, the population has difficulty to converge to the target output, no matter the target semantic is in or out of the covered range. Thus, the offspring produced by the geometric crossover is difficult to generalise well. Therefore, in this work, we are interested in improving the geometric crossover by addressing this issue.

The overall goal of this work is to propose a new angle-aware mating scheme to select for geometric semantic crossover to improve the *generalisation* of GP for symbolic regression. An important property of the geometric semantic crossover operator is that it generates offspring that stands in the segment defined by the two parent points in the semantic space. Therefore, the quality of the offspring is highly dependent on the positions of the two parents in the semantic space. However, such impact of the parents on the effectiveness of geometric semantic crossover has been overlooked. In this paper, we propose a new mating scheme to geometric crossover to make it operats on parents that are not only good at fitness but also have large *angle* in terms of their relative positions to the target point in the semantic space. Our goal is to study the effect of the newly proposed mating scheme to geometric crossover operator. Specific research objectives are as follows:

- to investigate whether the geometric crossover with angle-awareness can improve the learning performance of GSGP,
- to study whether the geometric crossover with angle-awareness can improve the generalisation ability of GSGP, and
- to investigate how the geometric crossover with angle-awareness influences the computational cost and the program size of the models evolved by GSGP.

## 2    Background

This section introduces geometric semantic GP in brief and reviews some state-of-the-art related work on geometric crossovers.

### 2.1    Geometric Semantic GP

Before introducing geometric semantic GP, a formal concept of individual semantics in GP needs to be given. A widely used definition of semantics in regression domain is as follows: the semantics of a GP individual is a vector, the elements of which are the outputs produced by the individual corresponding to the given instances. Accordingly, the semantics of an individual can be interpreted as a point in a $n$ dimension semantic space, where $n$ is the number of elements in the vector [9,11].

Geometric semantic GP is a relatively new branch in semantic GP. It searches directly in the semantic space, which is a notable difference from the other non-direct semantic methods, such as [2,16]. Searching in the semantic space is accomplished by its exact geometric semantic crossover and mutation. The definition of the geometric semantic crossover (GSX) is given below [12]:

**Definition 1.** *Geometric Semantic Crossover: Given two parent individuals $p_1$ and $p_2$, a geometric semantic crossover is an operator that generates offspring $p'_i (i \in (1,2))$ having semantics $s(p'_i)$ in the segment between the semantics of their parents, i.e., $\|s(p_1), s(p_2)\| = \|s(p_1), s(p'_i)\| + \|s(p'_i), s(p_2)\|$.*

Another important concept related to geometric crossover is the convex hull. It is a concept from geometry, which is the set of all convex combinations of a set of points. In geometric semantic GP, the convex hull can be viewed as the widest volume that the offspring generated by geometric crossover can cover.

Various geometric crossover operators [9,12] have been developed to satisfy the semantic constraint in Definition 1 in different ways. Locally geometric semantic crossover [9] (LGX) is a typical one with good performance.

## 2.2 Locally Geometric Semantic Crossover

Krawiec and Pawlak [8] develop the locally geometric semantic crossover (LGX), which attempts to produce offspring that satisfies the semantic constraint in Definition 1 at the subtree level. A library $L$ consisting of a set of small size trees needs to be generated before applying the LGX. Trees in the library $L$ have a maximum depth limitation $M$, and generally, each tree has unique semantics. Then given two parents $p_1$ and $p_2$, LGX tries to find their homologous subtree, which is the largest structurally common subtree of the two parents. Two corresponding crossover points are selected within the homologous subtree. Then the two subtrees $p_{c1}$ and $p_{c2}$ that root in these two crossover points are replaced by a tree $p_r$ selected from $L$. $p_r$ is randomly selected from a number of $K$ programs which are the closest neighbour to the semantics of midpoint of $p_{c1}$ and $p_{c2}$, i.e., $S(p_r) \approx \frac{(S(p_{c1}) + S(p_{c2}))}{2}$, where $S(p)$ represents the semantics of $p$. The advantage of LGX is that it can satisfy the semantic constraint by retrieving small subtrees in the library but without bringing exponential growth in the size of the offspring. The application shows that LGX brings notable improvement to the performance of GP [0].

## 2.3 Related Work

GSX performs a convex combination of the two parents. It generates offspring that lies in the segment defined by the parent points in the semantic space. Consequently, under Euclidean metric, the offspring can not be worse than the worse parent.

Moraglio et al. [12] develop the exact geometric crossover which is a transformations on the solution space that can satisfy the semantic constraint at the level of whole tree, i.e., $P_{xo} = P_1 \cdot F_r + P_2 \cdot (1 - F_r)$ where $P_i$ are parents for crossover, $F_r$ is a random real functions that outputs values in the range of $[0, 1]$. Despite the potential success of exact geometric crossover, it is criticised by leading the size of offspring to an exponential growth. Vanneschi et al. [17] propose an implementation of geometric operators to overcome the drawback of the unmanageable size of offspring. They aim to obtain the semantic of the offspring without generating the new generation in structure. The new implementation makes GSX can be applied to real-world high-dimensional problems. However, the evolved models are still hard to show and interpret.

During the evolutionary process if the target output is outside the convex hull, then surely GSX is impossible to find the optimal solution. Oliveira et al. [13] proposed a dispersion operator for GSX to address this issue. They proposed a geometric dispersion operator to move individuals to less dense areas around the target output in the semantic space. By spreading the population, the new operator increases the probability that the convex hull of the population will cover the target. Significant improvement is achieved on the learning and generalisation performance on most of the examined datasets.

However, even if the convex hull of the population covers the target, GSX may still fail and the population may still converge to a small volume far away from the target if the parents of GSX are not properly selected. It is known that due to the convexity of the squared Euclidean distance metric, the offspring cannot be worse than both parents. However, at the level of the whole population, there is still a high probability that this progress does not have much effect on guiding the population toward the target output in the semantic space, especially when a large proportion of crossovers perform on very similar parents in the semantic space. In this work, we propose a new mating scheme to geometric semantic crossover to prevent this trend and promote the exploration ability of the GSX.

## 3   Angle-Aware Geometric Semantic Crossover (AGSX)

In this work, tree based GP is employed, and we propose a new angle-aware mating scheme for Geometric Semantic Crossover (AGSX). This section describes the main idea, the detailed process, the characteristics of AGSX, and the fitness function of the GP algorithm.

### 3.1   Main Idea

How the crossover points spread in the semantic space is critical to the performance of GSGP. A better convergence to the target point can be achieved if the convex combinations cover a larger volume when the convex hull is given. AGSX should be applied to the parents that the *target output* is around the intermediate region of their semantics. Given that the semantics of the generated offspring tend to lie in the segment of the semantics of the parents as well, AGSX is expected to generate offspring that is close to the target output. To promote the convex combinations to cover a larger volume, the two parents should have a larger distance in the semantics space.

The semantic distance between the parents can be used here, but it often leads to a quick loss of semantic diversity in the population and then results in a premature solution. Therefore, we utilise the angle between the relative semantics of the parents to the target output to measure their distance in the semantic space. Specifically, suppose the target output is $T$, and the semantics of the two parents are $S_1$ and $S_2$, the angle $\alpha$ between the relative semantics of the two parents to the target output is defined as follows:

$$\alpha = \arccos \left( \frac{(S_1 - T) \cdot (S_2 - T)}{\|S_1 - T\| \cdot \|S_2 - T\|} \right) \tag{1}$$

where $(S_1 - T) \cdot (S_2 - T) = \sum_{i=1}^{n}(s_{1i} - t_i) \cdot (s_{2i} - t_i)$ and $\|S - T\| = \sqrt{\sum_{i=1}^{n}(s_i - t_i)^2}$. $i$ stands for the $i$th dimension in the $n-$dimensional semantic space. $s_{1i}, s_{2i}$, and $t_i$ are the values of $S_1, S_2$ and $T$ in the $i$th dimension, respectively.

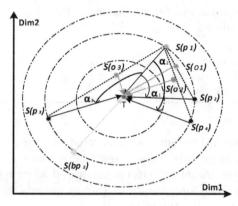

**Fig. 1.** AGSX in two dimension Euclidean semantic space. (Color figure online)

Figure 1 illustrates the mechanism of AGSX in a two-dimensional Euclidean space, which can be scaled to any $n$-dimensional space. Each point represents the semantics of one individual in GP. As shown in the figure, there are four individuals $p_1, p_2, p_3$ and $p_4$, which can be selected as the parents of AGSX. Assume $p_1$ (in blue colour) has been selected as one parent and the mate, i.e. the other parent, needs to be selected from $p_2, p_3$ and $p_4$ to perform AGSX. $\alpha_1, \alpha_2$, and $\alpha_3$ show the angles in the three pairs of parents, i.e. $\langle p_1, p_2 \rangle, \langle p_1, p_4 \rangle$ and $\langle p_1, p_3 \rangle$, respectively. The three green points, $S(o_1), S(o_2)$, and $S(o_3)$, show the three corresponding offspring of the three pairs of parents, and the green lines indicates their distances to the target point. It can be seen from the figure that the pair of parents $\langle p_1, p_3 \rangle$ has a larger angle, i.e. $\alpha_3$, and the generated offspring $S(o_3)$ is closer to the target output. In the ideal case where the yellow

---

**Algorithm 1.** Pseudo-code of AGSX

---

**Input**   : $WaitingSet[i_1, i_2, ..., i_m]$ consists of $m$ individuals on which will perform crossover. $T$ is the target semantics point.

**Output**: The generated offspring

1  **while** *WaitingSet is not empty* **do**
2      $p_1 =$ is the first individual in $WaitingSet$;
3      remove $p_1 =$ from $WaitingSet$;
4      $maxangle = 0$; /* *i.e. the maximum angle that has been found* */
5      $top$ is an empty list;
6      **for** *each individual p in WaitingSet* **do**
7          calculate the *angle* between the relative semantics of $S(p_1)$, $S(p)$ to $T$ according to Equation (1);
8          **if** *angle is equal to 180, i.e. p is the optimal mate for $p_1$* **then**
9              $top=p$;
10         **else**
11             **if** *angle is larger than maxangle* **then**
12                 $maxangle = angle$;
13                 $top=p$;
14             **else**
15                 **if** *angle is equal to the maxangle* **then**
16                     add $p$ to $top$;

17     randomly select an individual, $p_2$, from $top$;
18     perform geometric crossover on $p_1$ and $p_2$;
19     remove $p_1$ and $p_2$ from $WaitingSet$.

---

point $S(bp_2)$ is the second parent, the generated offspring is very likely to be the target point. In other words, if the parents have a larger angle between their relative semantics to the target output, the generated offspring tends to be closer to the target output. Therefore, we need to select parents with a large angle in their relative semantics to the target output.

To achieve this, we develop a new mating scheme to select parents with a large angle in their relative semantics to the target output. First, a list of candidate parents called the $WaitingSet$ is generated by repetitively applying a selection operator (e.g. tournament selection) to the current population. The size of $WaitingSet$ is determined by the population size $N$ and the crossover rate $R_X$, i.e. $|waitingset| = N \cdot R_X$. Then, the parents for each AGSX operation are selected from $WaitingSet$ without replacement so that the angles between the relative semantics of the selected parents can be maximised. The detailed process of AGSX is given in Sect. 3.2.

## 3.2   The AGSX Process

The pseudo-code of AGSX is shown in Algorithm 1. The procedure of finding the mate having the largest relative angle for a given parent $p_1$ is shown in Lines 3–18. The angles are calculated according to Eq. (1), as shown in Line 6.

## 3.3   Main Characteristics of AGSX

Compared with GSX, AGSX has three major advantages. Firstly, AGSX employs an angle-aware scheme, which is flexible and independent of the crossover process itself and can be applied to any form of the geometric semantic operator. Secondly, AGSX operates on distinct individuals in the semantic space. This way, the generated offspring are less likely to be identical with their parents in the semantic space. That is, AGSX can reduce semantic duplicates. Thirdly, by operating on parents with large angles between their relative semantics to the target output, AGSX is more likely to generate offspring that are closer to the target output.

## 3.4   Fitness Function of the Algorithm

The Minkowski metric $L_k(X,Y) = \sqrt[k]{\sum_{i=1}^{n} |x_i - y_i|^k}$, which calculates the distance between two points, is used to evaluate the performance of individuals. Typically, two kinds of Minkowski distance between the individual and the target could be used. They are Manhattan distance ($L_1$ by setting $k = 1$ in $L_k(X,Y)$) and Euclidean distance ($L_2$). According to previous research [1], Euclidean distance is a good choice and is used in this work. The definition is as follows:

$$D(X,T) = \sqrt{\sum_{i=1}^{n} |x_i - t_i|^2} \tag{2}$$

where $X$ is the semantics of the individual and $T$ is the target semantics.

## 4   Experiments Setup

To investigate the effect of AGSX in improving the performance of GP, a GP method implements the angle-awareness into one recent approximate geometric crossover, the locally geometric semantic crossover has been proposed and named GPALGX. A comparison between GPALGX and GP with locally geometric semantic crossover (GPLGX) has been conducted. We have a brief introduction of LGX in Sect. 2.2. For more details of the GPLGX, readers are referred to [9]. Standard GP is used as a baseline for comparison as well. All the compared methods are implemented under the GP framework provided by Distributed Evolutionary Algorithms in Python (DEAP) [4].

### 4.1   Benchmark Problems

Six commonly used symbolic regression problems are used to examine the performance of the three GP methods. The details of the target functions and the sampling strategy of the training data and the test data are shown in Table 1. The first two problems are the recommended benchmarks in [10]. The middle three are used in [14]. The last one is from [3] which is a modified version of the commonly used Quartic function. These six datasets are used since they have been widely used in recent research on geometric semantic GP [14,15]. The notation $rnd[a,b]$ denotes that the variable is randomly sampled from the interval $[a, b]$, while the notation $mesh([start:step:stop])$ defines the set is sampled using regular intervals. Since we are more interested in the generalisation ability of the proposed crossover operator, the test points are drawn from ranges which are slightly wider than that of the training points.

**Table 1.** Target functions and sampling strategies.

| Benchmark | Target function | Training | Test |
|---|---|---|---|
| Keijzer1 | $0.3xsin(2\pi x)$ | 20 points $x = $ mesh$((-1:0.1:1])$ | 1000 points $x = $ Rnd$[-1.1,1.1]$ |
| Koza2 | $(x^5 - 2x^3 + X)$ | | |
| Nonic | $\sum_{i=1}^{9} x^i$ | 20 points $x = $ mesh$((-2:0.2:2])$ | 1000 points $x = $ Rnd$[-2.2,2.2]$ |
| R1 | $(x+1)^3/(x^2 - x + 1)$ | | |
| R2 | $(x^5 - 3x^3 + 1)/(x^2 + 1)$ | | |
| Mod_quartic | $4x^4 + 3x^3 + 2x^2 + x$ | | |

**Table 2.** Parameter settings

| Parameter | Values | Parameter | Values |
|---|---|---|---|
| Population size | 512 | Generations | 100 |
| Crossover rate | 0.9 | Reproduction rate | 0.1 |
| #Elitism | 10 | Maximum tree depth | 17 |
| Initialisation | Ramped-half & half | Initial depth | Range (2,6) |
| Maximum tree depth in library-$M$ | 3 | Neighbourhood number-$K$ | 8 |
| Function set | $+, -, *$, protected %, log, sin, cos, exp | | |
| Fitness function | Root Mean Squared Error (RMSE) in standard GP | | |
| | Euclidean distance in GPLGX and GPALGX | | |

## 4.2 Parameter Settings

The parameter settings can be found in Table 2. For standard GP, the fitness function is different from that of GPLGX and GPALGX. Since the primary interest of this work is the comparison of the generalisation ability of the various crossover operators, all the three GP methods only have crossover operators. No mutation operator has taken apart. The values of the two key parameters $M$ and $K$ in implementing LGX, which represent for the maximum depth of the small size tree in the library and the number of the closest neighbouring trees respectively, are following the recommendation in [9].

Overall, the three GP methods are examined on six benchmarks. Each method has 100 independent runs performed on each benchmark problem.

# 5    Results and Discussions

The experiment results of GP, GPLGX and GPALGX are presented and discussed in this section. The results will be presented in terms of comparisons of RMSEs of the 100 best models on the training sets and their corresponding test RMSEs. The fitness values of models in GPLGX and GPALGX are calculated using Euclidean distance. However, for comparison purpose, the Root Mean Squared Error (RMSE) of models are also recorded. The major comparison is presented between GPLGX and GPALGX. Thus, we also compare the angle distribution of GPLGX and GPALGX. The computational time and program size are also discussed. The non-parametric Wilcoxon test is used to evaluate the statistical significance of the difference on the RMSEs on both the training sets and the test sets. The significance level is set to be 0.05.

## 5.1    Overall Results

The results on the six benchmarks are shown in Fig. 2, which displays the distribution of RMSEs of the 100 best-of-the-run individuals on the training sets and the test sets. As it shows, on all the six benchmarks, GPALGX has the best training performance among the three GP methods. For every benchmark, GPALGX has a better training performance than GPLGX and GP, by the smaller median value of the 100 best training errors and the much shorter boxplot. This indicates the training performance of GPALGX is superior to the other two methods in a notable and stable way. The results of statistical significance test confirm that the advantage of GPALGX over GPLGX and GP are all significant on the six training sets.

The overall pattern on the test sets is the same as the training set, which is GPALGX achieves the best generalisation performance on all the benchmarks. On each benchmark, the pattern in the distribution of the 100 test errors is also the same as that on the training set. GPALGX has the shortest boxplot which indicates the more consist generalisation error among the 100 runs. GPLGX has a larger distribution than GPALGX, which is still much shorter than standard GP. A significant difference can be found on the six benchmarks between GPALGX, GPLGX and GP, i.e. GPALGX generalises significantly better than GPLGX, while the two geometric methods are significantly superior to GP. The generalisation advance of LGX and ALGX over standard crossover is consistent with the previous research on LGX. In [9], the generalisation gain of LGX has been investigated and confirmed. This generalisation gain has been justified to own to the library generating process which helps reduce the semantic duplicates. The further generalisation gain of ALGX over LGX might lie in the fact that the angle-awareness helps extend the segment connecting each pair of parents for crossover, thus can reduce the semantic duplicates more intensively, and enhance the exploration ability of LGX to find better generalised solutions.

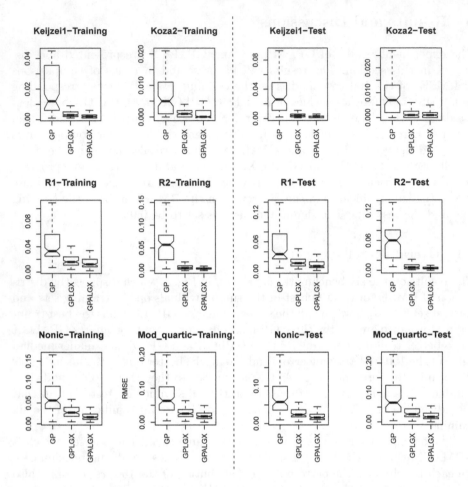

**Fig. 2.** Distribution of training RMSE and the corresponding test RMSE of the 100 best-of-run individuals.

## 5.2   Analysis on the Learning Performance

The evolutionary plots on the training sets are provided in Fig. 3. To analysis the effect of ALGX on improving the learning performance of GP. These evolutionary plots are drawn using the mean RMSEs of the best-of-generation individuals over the 100 runs.

As expected, GP with ALGX achieves the best learning performance. It is superior to the other two GP methods from the early stage of the evolutionary process, which is generally within the first ten generations. The advances of the two geometric GP methods over standard GP on the learning performance confirms that searching in the geometric space is generally much easier, since the semantic space is unimodal and has no local optimal. The comparison between the two geometric GP methods indicates ALGX is able to generate offspring

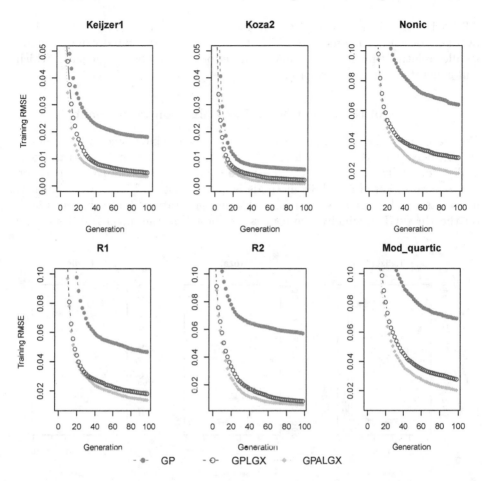

**Fig. 3.** Evolutionary plot on the test set.

which is much closer to the target point in the semantic space from the very beginning of the searching process. On all the six benchmarks, GPALGX not only has significantly smaller training RMSEs but also has higher average fitness gain from generation to generation. On $Koza_2$ and $R_2$, the two geometric GP methods can find models which are excellent approximations (the RMSE of which is smaller than 0.001), and GPALGX converges to the target semantics much faster than GPLGX. This might be because ALGX performs crossover on individuals having larger angles than GPLX, thus produces offspring closer to the target in the semantic space in an effective way. In this way, it will increase the exploitation ability of LGX and find the target more quickly. For the other four benchmarks, although none of the two geometric GP methods finds the optimal solution, on three of them, the increasingly larger difference between the two methods along with the increase of generations indicates the improvement that

ALGX brings is increasing over generations. One of the possible reasons is that, over generations, compared with LGX, ALGX will perform on individuals having smaller relative semantic distance with target output in larger angle pairs, which will generate even better offspring.

## 5.3   Analysis of the Evolution of Generalisation Performance

Compared with the training performance, we are more interested in the generalisation performance of GP with ALGX. Therefore, further analysis on the generalisation ability of GPALGX and a more comprehensive comparison between the generalisation of the three methods is carried out. In Fig. 4, the evolutionary plots on the test sets are reported along generations for each benchmark on the three GP methods. These plots are based on the mean RMSEs of the

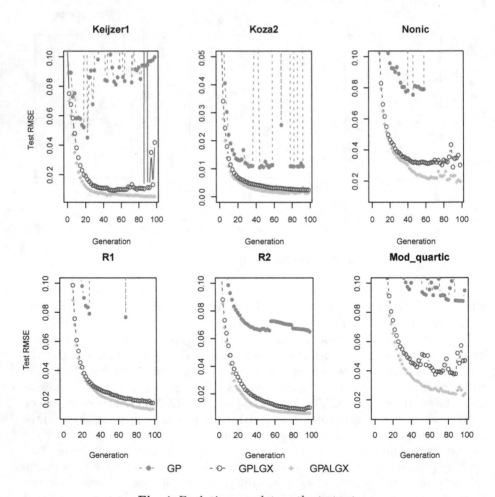

**Fig. 4.** Evolutionary plot on the test set.

corresponding test errors obtained by the best-of-generation models over 100 runs. (On each generation, the test performance of the best-of-generation model obtained from the training data has been recorded, but the test sets never take apart in the evolutionary training process).

The evolution plots confirm that GPALGX has a better generalisation gain than the other two methods on all the test sets of the considered benchmarks, which is notable. On all the six benchmarks, GPALGX can generalise quite well, while its two counterparts suffer from the overfitting problems on some datasets. On the six problems, GP overfits the training sets. The test RMSEs increase after decreasing over a small number of generations at the beginning. Also, GP generally has a very fluctuate mean RMSE on most test sets. It indicates that training the models on a small number of points (20 points), while testing the models on a larger number of points (1000 points) distributed over a slightly larger region is difficult for GP. GPLGX can generalise much better than GP but still encounters overfitting problems on three benchmarks, i.e., on Keijzer1, Nonic and Mod_quartic. On these three datasets, GPLGX has an increasing RMSEs on the last ten generations. On other three datasets, GPLGX generalises well. Overall, GPALGX generalises better than GPLGX and GP, shown as obtaining lower generalisation errors and having a smaller difference with its training errors.

The excellent generalisation ability of geometric crossover can be explained by the fact that the geometric properties of this operator are independent of the data to which the individual is exposed. Specifically, the offspring produced by LGX and ALGX lie (approximately) in the segments of parents also hold in the semantic space of the test data. Since this property holds for every set of data, no matter where the test data distributes in, the fitness of the offspring can never be worse than the worse parent. In the population level, this property can not guarantee to improve the test error on every generation for every benchmark (in fact, we can find on the last several generations, LGX has an increasing test error on three benchmarks), but during the process it surely has a high probability of generalisation gain on the test set and only a few times of getting worse generalisation over generations. That is why LGX has the ability to control overfitting and generalise better than the regular crossover.

This interpretation has a direct relationship on why ALGX is less likely to overfitting and generalises better than LGX on the test sets. In other words, ALGX puts more effect on selecting parents which consequently limits the probability of having not good enough parents to crossover, so it can lead to a large number of offspring with better generalisation at the population level. AlGX shares the same benefit with LGX, which is the geometric property leading to offspring never worse than parents on the test set. More importantly, the angle-awareness in ALGX makes the large angle between the parents also holds in the test semantic space. This leads to a higher probability to have a good process on the test data at the population level. The details of the angle distribution will be discussed in the next subsection.

## 5.4   Analysis of the Angles

To investigate and confirm the influence of ALGX to the distribution of angles of the parents, the angles between each pair of parents which performs crossover have been recorded in both GPLGX and GPALGX. In Fig. 5, the density plots show the distribution of the angles in the two GP methods. The green one is for GPLGX, and the one in orange colour is for GPALGX. The density plots are based on around $2,250,000 (\approx 225 * 100 * 100)$ values of angles in each method. While the x-axis represents the degree of angles, the y-axis is the percentage of the corresponding degree in the 2,250,000 recorded values.

From Fig. 5 we can see that the distribution of angles of parents in GPALGX is different from GPLGX in two aspects. On the one hand, it has a much smaller number of angles which are zero degrees. While in GPLGX, the peak of the distribution is at the zero degrees on all the six datasets, in GPALGX, the angle-awareness can stop the pairs of individuals with zero degrees from performing crossover. The direct consequence of this trend is the elimination of semantic duplicates, and the higher possibility of generating better offspring.

On the other hand, GPALGX has a larger number of larger angles. Most of its angles are over $90°$. The peak of the distribution is all around $120°$ on the six datasets, specifically on the last four datasets. At the first several generations, the larger angles with similar (or the same) vectors will lead to better offspring, which is represented by a shorter vector. At the last several generations, larger angles along with the shorter vectors will lead to a population of even better

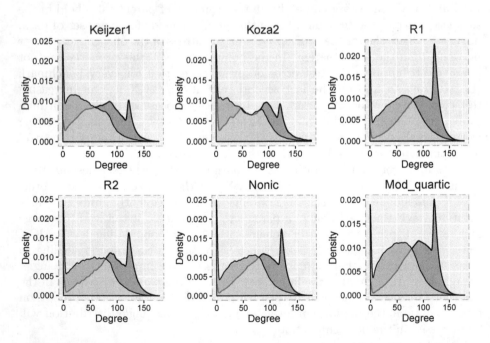

**Fig. 5.** Distribution of angles of the parents for crossover. (Color figure online)

offspring. This can explain why the distance between the training error and test error of GPLGX and GPALGX increases over generations on most of the benchmarks.

## 5.5    Comparison on Computational Time and Program Size

The comparison between the computational cost and program size of the evolved models have been performed between the two geometric methods. Table 3 shows the computational time in terms of the average training time for one GP run in each benchmark. The average program size represented by the number of nodes in the best_of_run models in each benchmark is also provided. The statistical significance results on the program size are also listed in the table. While "−"means the program size of the evolved model in GPALGX is significantly smaller than GPLGX, "+" indicates the significant larger program size of GPALGX. "= represents no significant difference can be found.

As shown in Table 3, on all the six benchmarks, the average computational time for one run in GPALGX is much higher than GPLGX, which is generally around two times as that of GPLGX. This is not surprising since GPALGX needs more effort to identify the most suitable pairs of parents during the crossover process. The longer computational time can be decreased by reducing the population size in GPALGX. Moreover, the computational time for each GP run in both methods is short, which is hundreds to two thousand second. Thus, the additional computational cost of GPALGX is affordable.

In term of the program size, on four benchmarks, i.e., Keijzer1, R1, R2, and Mod_quartic, the two methods have a similar program size and no significant difference has been found. On the other two datasets, ALGX produces offspring which are significantly larger than LGX. However, it is interesting to note that these much more complex models in term of program size still can generalise

**Table 3.** Computational time and program size.

| Benchmarks | Method | Time (in second) | Program size (node) | Significant test |
|---|---|---|---|---|
| | | Mean ± Std | Mean ± Std | (on program size) |
| Keijzer1 | GPLGX | 523 ± 83.8 | 90.52 ± 28.72 | = |
| | GPALGX | 1400 ± 317 | 87.74 ± 23.85 | |
| Koza2 | GPLGX | 560 ± 105 | 72.66 ± 29.93 | + |
| | GPALGX | 1330 ± 232 | 93.82 ± 24.18 | |
| R1 | GPLGX | 523 ± 84.5 | 88.82 ± 27.07 | = |
| | GPALGX | 1250 ± 253 | 92.18 ± 31.71 | |
| R2 | GPLGX | 524 ± 83.9 | 89.62 ± 27.41 | = |
| | GPALGX | 1250 ± 218 | 83.8 ± 28.6 | |
| Nonic | GPLGX | 571 ± 112 | 84.5 ± 32.62 | + |
| | GPALGX | 1250 ± 212 | 101.5 ± 39.33 | |
| Mod_quartic | GPLGX | 554 ± 105 | 99.98 ± 38.25 | = |
| | GPALGX | 1420 ± 369 | 105.38 ± 37.29 | |

better than its simpler counterparts on the two test sets, while the simpler model of GPLGX slightly overfits on the Nonic problem.

## 6    Conclusions and Future Work

This work proposes an angle-aware mating scheme to select parents for geometric semantic crossover, which employs the angle between the relative semantics of the parents to the target output to choose parents. The proposed ALGX performs on parents having a large angle so that the segment connecting the parents is close to the target output. Thus, ALGX can generate offspring that have better performance. To investigate and confirm the efficiency of the proposed ALGX, we run GP employed ALGX on six widely used symbolic regression benchmark problems and compare its performance with GPLGX and GP. The experimental results confirm that GPALGX has not only better training performance but also significantly better generalisation ability than GPLGX and GP on all the examined benchmarks.

Despite the improvement ALGX brings on performance, it generally is computational more expensive than GPLGX. In the future, we aim to improve the angle detecting process. Instead of using the deterministic method to calculate the angle between two individuals iteratively, we can introduce some heuristic search methods to find the best parent pairs to reduce the computational cost. We also would like to explore a further application of ALGX, for example, to introduce the angle-awareness to other forms of geometric crossover, such as the exact geometric semantic crossover [12] and Approximate geometric crossover [7], to investigate their effectiveness. In addition, this work involves solely crossover and no mutation. The effect of angle-awareness to mutation and using both crossover and mutation are also interesting topics to work on.

## References

1. Albinati, J., Pappa, G.L., Otero, F.E.B., Oliveira, L.O.V.B.: The effect of distinct geometric semantic crossover operators in regression problems. In: Machado, P., Heywood, M.I., McDermott, J., Castelli, M., García-Sánchez, P., Burelli, P., Risi, S., Sim, K. (eds.) EuroGP 2015. LNCS, vol. 9025, pp. 3–15. Springer, Heidelberg (2015). doi:10.1007/978-3-319-16501-1_1
2. Beadle, L., Johnson, C.G.: Semantically driven crossover in genetic programming. In: IEEE Congress on Evolutionary Computation, pp. 111–116 (2008)
3. Burks, A.R., Punch, W.F.: An efficient structural diversity technique for genetic programming. In: Proceedings of the 2015 Annual Conference on Genetic and Evolutionary Computation, pp. 991–998. ACM (2015)
4. Fortin, F.A., Rainville, F.M.D., Gardner, M.A., Parizeau, M., Gagné, C.: DEAP: Evolutionary algorithms made easy. J. Mach. Learn. Res. **13**, 2171–2175 (2012)
5. Gonçalves, I., Silva, S., Fonseca, C.M.: On the generalization ability of geometric semantic genetic programming. In: Machado, P., Heywood, M.I., McDermott, J., Castelli, M., García-Sánchez, P., Burelli, P., Risi, S., Sim, K. (eds.) EuroGP 2015. LNCS, vol. 9025, pp. 41–52. Springer, Heidelberg (2015). doi:10.1007/978-3-319-16501-1_4

6. Koza, J.R.: Genetic Programming: On the Programming of Computers by Means of Natural Selection, vol. 1. MIT press, Cambridge (1992)
7. Krawiec, K., Lichocki, P.: Approximating geometric crossover in semantic space. In: Proceedings of the 11th Annual conference on Genetic and evolutionary computation, pp. 987–994. ACM (2009)
8. Krawiec, K., Pawlak, T.: Locally geometric semantic crossover. In: Proceedings of the 14th Annual Conference Companion on Genetic and Evolutionary Computation, pp. 1487–1488. ACM (2012)
9. Krawiec, K., Pawlak, T.: Locally geometric semantic crossover: a study on the roles of semantics and homology in recombination operators. Genetic Program. Evol. Mach. **14**(1), 31–63 (2013)
10. McDermott, J., White, D.R., Luke, S., Manzoni, L., Castelli, M., Vanneschi, L., Jaskowski, W., Krawiec, K., Harper, R., De Jong, K., et al.: Genetic programming needs better benchmarks. In: Proceedings of the 14th Annual Conference on Genetic and Evolutionary Computation, pp. 791–798. ACM (2012)
11. McPhee, N.F., Ohs, B., Hutchison, T.: Semantic Building Blocks in Genetic Programming. In: O'Neill, M., Vanneschi, L., Gustafson, S., Esparcia Alcázar, A.I., Falco, I., Cioppa, A., Tarantino, E. (eds.) EuroGP 2008. LNCS, vol. 4971, pp. 134–145. Springer, Heidelberg (2008). doi:10.1007/978-3-540-78671-9_12
12. Moraglio, A., Krawiec, K., Johnson, C.G.: Geometric semantic genetic programming. In: Coello, C.A.C., Cutello, V., Deb, K., Forrest, S., Nicosia, G., Pavone, M. (eds.) PPSN 2012. LNCS, vol. 7491, pp. 21–31. Springer, Heidelberg (2012). doi:10.1007/978-3-642-32937-1_3
13. Oliveira, L.O.V., Otero, F.E., Pappa, G.L.: A dispersion operator for geometric semantic genetic programming. In: Proceedings of the Genetic and Evolutionary Computation Conference, pp. 773–780 (2016)
14. Pawlak, T.P., Wieloch, B., Krawiec, K.: Review and comparative analysis of geometric semantic crossovers. Genetic Program. Evol. Mach. **16**(3), 351–386 (2015)
15. Szubert, M., Kodali, A., Ganguly, S., Das, K., Bongard, J.C.: Reducing antagonism between behavioral diversity and fitness in semantic genetic programming. In: Proceedings of the 2016 on Genetic and Evolutionary Computation Conference, pp. 797–804. ACM (2016)
16. Uy, N.Q., Hien, N.T., Hoai, N.X., O'Neill, M.: Improving the generalisation ability of genetic programming with semantic similarity based crossover. In: Esparcia-Alcázar, A.I., Ekárt, A., Silva, S., Dignum, S., Uyar, A.Ş. (eds.) EuroGP 2010. LNCS, vol. 6021, pp. 184–195. Springer, Heidelberg (2010). doi:10.1007/978-3-642-12148-7_16
17. Vanneschi, L., Castelli, M., Manzoni, L., Silva, S.: A new implementation of geometric semantic GP and its application to problems in pharmacokinetics. In: Krawiec, K., Moraglio, A., Hu, T., Etaner-Uyar, A.Ş., Hu, B. (eds.) EuroGP 2013. LNCS, vol. 7831, pp. 205–216. Springer, Heidelberg (2013). doi:10.1007/978-3-642-37207-0_18
18. Vanneschi, L., Castelli, M., Silva, S.: A survey of semantic methods in genetic programming. Genetic Program. Evol. Mach. **15**(2), 195–214 (2014)
19. Vanneschi, L., Silva, S., Castelli, M., Manzoni, L.: Geometric semantic genetic programming for real life applications. In: Riolo, R., Moore, J.H., Kotanchek, M. (eds.) Genetic Programming Theory and Practice XI, pp. 191–209. Springer, New York (2014)

# RECIPE: A Grammar-Based Framework for Automatically Evolving Classification Pipelines

Alex G.C. de Sá[✉], Walter José G.S. Pinto, Luiz Otavio V.B. Oliveira, and Gisele L. Pappa

Computer Science Department, Universidade Federal de Minas Gerais, Belo Horizonte, Minas Gerais, Brazil
{alexgcsa,walterjgsp,luizvbo,glpappa}@dcc.ufmg.br

**Abstract.** Automatic Machine Learning is a growing area of machine learning that has a similar objective to the area of hyper-heuristics: to automatically recommend optimized pipelines, algorithms or appropriate parameters to specific tasks without much dependency on user knowledge. The background knowledge required to solve the task at hand is actually embedded into a search mechanism that builds personalized solutions to the task. Following this idea, this paper proposes RECIPE (REsilient ClassifIcation Pipeline Evolution), a framework based on grammar-based genetic programming that builds customized classification pipelines. The framework is flexible enough to receive different grammars and can be easily extended to other machine learning tasks. RECIPE overcomes the drawbacks of previous evolutionary-based frameworks, such as generating invalid individuals, and organizes a high number of possible suitable data pre-processing and classification methods into a grammar. Results of f-measure obtained by RECIPE are compared to those two state-of-the-art methods, and shown to be as good as or better than those previously reported in the literature. RECIPE represents a first step towards a complete framework for dealing with different machine learning tasks with the minimum required human intervention.

**Keywords:** Grammar-based genetic programming · Classification · Automatic Machine Learning

## 1 Introduction

When genetic programming was first framed within the context of machine learning, its main idea was to generate complete programs following a "learning from examples" approach [1]. Although we are still far from generating complete programs without a considerable level of human intervention, the gap between program automation and the amount of expertise required to produce these programs has been progressively reduced [2]. Many initiatives in this direction are currently denominated Automatic Machine Learning (Auto-ML).

© Springer International Publishing AG 2017
J. McDermott et al. (Eds.): EuroGP 2017, LNCS 10196, pp. 246–261, 2017.
DOI: 10.1007/978-3-319-55696-3_16

The area of Auto-ML has as its main objective to automatically recommend pipelines, algorithms or appropriate parameters to tasks without much dependency on user knowledge [3]. In reality, the background knowledge required to learn the task is embedded into a search mechanism that, considering the task at hand, builds personalized solutions to the problem. The importance of Auto-ML techniques is undeniable, given the constant growth of data generated and the need of interpreting, classifying and contextualizing this data into useful information. This area becomes even more important given the limited number of experts in the area and an increasing number of enthusiastic practitioners that follow a complete ad-hoc process to deal with their data.

Considering all the problems the area of Auto-ML deals with, we are interested in building classification pipelines. A classification pipeline is defined as a sequence of tasks that needs to be performed to classify the instances belonging to a given dataset into a set of predefined categories. The tasks included in the pipeline may represent different ways of transforming or pre-processing the dataset, as well as a classification algorithm and its associated parameters.

Among the methods previously proposed in the literature to deal with this task are Auto-WEKA [4], Auto-SKLearn [5] and Tree-based Pipeline Optimization Tool (TPOT) [6]. While the first two use a hierarchical Bayesian method to build a classification pipeline, TPOT uses a tree-based genetic programming algorithm to solve the problem. All these methods use a list of predefined components (tasks) that can be considered during the search. Auto-WEKA and Auto-SKLearn perform a local search to explore these components, and add a few constraints to be able to avoid invalid combinations. TPOT, in turn, exploits the advantages of a global search but considers an unconstrained search, where resources can be spent into generating and evaluating invalid solutions. The latter can be considered one of its main drawbacks.

When background knowledge is available to guide the search, grammar-based genetic programming (GGP) algorithms [7] appear as a better alternative than a canonical GP. The main difference between a canonical and a grammar-based GP, as the name indicates, is the definition of a grammar. The grammar is used to generate the initial population, as well as to constraint the crossover and mutation operations, which always need to be valid according to the grammar. In this direction, this paper proposes a GGP framework to automatically evolve classification pipelines, customized to the dataset of interest, called RECIPE (REsilient ClassifIcation Pipeline Evolution). The method uses a more complete set of components that can be considered during the search, and constitute the non-terminal symbols of the grammar. Although the focus of the paper is in classification task, the framework can be easily adapted to deal with a number of tasks, including regression, ranking or clustering.

RECIPE is compared with the current state-of-the-art aforementioned methods and with a random search, and tested with different grammars generating very different search spaces. We show that the results produced by the proposed method are as good as or outperform the state-of-the-art methods in terms of f-measure. RECIPE also solves the problem of generating invalid solutions but can still be further improved by adding new mechanisms to ensure diversity.

The reminder of this paper is organized as follows. Section 2 reviews related works in the Auto-ML area. Section 3 details the proposed method, while Sect. 4 presents and discusses the results obtained. Finally, Sect. 5 draws some conclusions and discusses directions of future work.

## 2    Related Work

Although the term Auto-ML was recently coined, the area itself is not new, and has received different names, including hyper-heuristics, hyper-parameter optimization and constructive meta-learning [2]. This happened because the main idea behind Auto-ML appeared in the fields of ML and optimization at different time frames, and were developed mostly independently.

In this work, we are particularly interested in Auto-ML for classification problems, and will focus on works conceived to solve this task. Most of the literature focuses in searching the components of specific classification algorithms, instead of complete pipelines with pre-processing, classification and post-processing methods. We identified six different classification models that had their components optimized using this approach: (i) artificial neural networks [8–10], (ii) rule induction algorithms [11], (iii) support vector machines [12,13], (iv) decision trees [14], (v) Bayesian network classifiers [15,16] and (vi) Bayesian neural networks [17]. However, as the main objective of this paper is to deal with complete pipelines, we go into detail in methods that deal with this task, including Auto-WEKA [4], Auto-SKLearn [5] and Tree-based Pipeline Optimization Tool (TPOT) [6].

Auto-WEKA and Auto-SKLearn are methods based on Bayesian optimization and their prime objective is to find the best combination between complete ML pipelines and their respective parameters. Both methods follow a hierarchical method to find the "best" pipeline to the dataset at hand. In this case, the Auto-ML method first chooses the classification algorithm (or the pre-processing method) and, only after this step, its parameters are optimized. The use of this hierarchical optimization approach can be advantageous in the sense that it divides the search space into two, but it may also left out algorithms that, with the right parameters, could generate better results than the selected ones.

Auto-WEKA automates the process of selecting the best ML pipeline in WEKA [18], whereas Auto-SKLearn aims to optimize the pipelines in SciKit-Learn library [19]. The choice of this library by current Auto-ML methods is motivated by the great number of methods already implemented and the popularity of this Python library. Besides, Auto-SKLearn has some improvements when compared to Auto-WEKA. For instance, Auto-SKLearn can be initialized via meta-learning [20] and can also construct an ensemble to combine the classification results in a post-processing stage. Hence, we will use Auto-SKLearn in the comparisons in Sect. 4 instead of Auto-WEKA.

TPOT, in turn, uses a genetic programing (GP) as its search method to choose the more suitable pipeline for a ML problem, performing a global search approach. It also searches for methods available in the SciKit-Learn library, but has a smaller search space than Auto-SKLearn. One difference in the search space of TPOT compared to the aforementioned methods and the one proposed here is that it allows the use of many copies of the dataset, which are processed in parallel by different pre-processing methods and later combined. For example, a pipeline can have two or more feature selection methods, and then a combination method is used to verify what are the common and distinct features found by the techniques. In addition, TPOT considers a Pareto selection (NSGA-II) [21] to perform a multi-objective search. Two separate objectives are considered: maximizing the final accuracy measure of the pipeline as well as minimizing the pipeline's overall complexity, given by the total number of pipeline operators, in order to avoid overfitting.

One of the major drawbacks of TPOT is that it can create ML pipelines that are arbitrary/invalid, i.e., it can create a ML pipeline that fails to solve a classification problem, as there are no constraints in which type of components can be combined. For example, TPOT can create a pipeline without a classification algorithm [3]. This also leads to a waste of computational resources, as these individuals are identified as invalid and given a very low fitness value during the evaluation of the pipelines.

RECIPE has three important improvements when compared to Auto-SKLearn and TPOT. First, it uses a grammar to organize the knowledge acquired from the literature on how successful ML pipelines look like. The grammar avoids the generation of invalid pipelines, and can speed up search. Second, it works with a bigger search space of ML pipelines than Auto-SKLearn and TPOT. Although this makes search more challenging, it also gives the opportunity of finding a greater variety of pipelines. Finally, the guided global search allows us to simultaneously evaluate the whole pipeline instead of looking at discrete and then at continuous search parameters as Auto-SKLearn does.

On the other hand, this first version of RECIPE does not use any sophisticated initialization scheme, such as Auto-SKLearn, and the results show the algorithm finds its way from a random initialization. It also considers a single-objective fitness function, and does not seem to be prone to overfitting, as TPOT was before adding the complexity of the pipelines to its fitness. Finally, RECIPE currently does not include post-processing steps and, although it can use different data pre-processing methods in sequence, it does not consider them in parallel, as TPOT does.

## 3   Automatically Evolving Classification Pipelines

This section introduces the GGP method proposed to automatically generate classification pipelines, illustrated in Fig. 1. RECIPE receives as input a dataset and a grammar, which is used to initialize the population. Each individual is represented by a derivation-tree built from the context-free grammar (CFG), which encompasses all the knowledge gathered from specialists on how to generate an effective classification pipeline. The individuals are mapped into pipelines implemented by the SciKit-Learn library, which are executed into a data sample from the application being solved and evaluated according to a metric of accuracy. Crossover and mutation operators are applied after a tournament selection, and guarantee that the new individuals generated also respect the production rules of the grammar. Elitism is also used, and evolution goes on until a maximum number of generations is reached or the best individual does not improve after a predefined number of generations. RECIPE was implemented using the library *Libgges* [22], and is available for download.[1] The next subsections describe in details the main components of the framework.

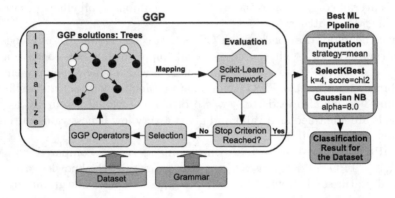

**Fig. 1.** Framework followed by RECIPE.

## 3.1  Grammar: Representing Effective Classification Pipelines

In GGP systems, a grammar is an effective way of representing background knowledge about the problem being solved. A grammar $G$ is represented by a four-tuple $<N, T, P, S>$, where $N$ represents a set of non-terminals, $T$ a set of terminals, $P$ a set of production rules and $S$ (a member of $N$) the start symbol. The production rules define the language the grammar represents by combining the grammar symbols.

In RECIPE, the grammar represents a set of pipelines that can be used to solve a classification problem. As previously mentioned, a pipeline is a sequence of tasks that should be applied to a given dataset to produce accurate solutions to a given input problem. Recall that the pipelines produced are customized to the data at hand.

Hence, the first step is to define which steps should be included in the grammar. Previously proposed systems have divided the pipelines into three main steps: data pre-processing, data processing and data post-processing. We also follow this basic framework, illustrated in Fig. 2. The data pre-processing steps include making transformations to the input data to make it more suitable for the learning task, once the great majority of data is not generated with learning in mind. These methods include data normalization, feature selection,

---

[1] https://github.com/RecipeML/Recipe.

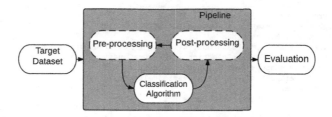

**Fig. 2.** Main components of a pipeline, where dashed lines represent optional components.

data imputation, among others. The second step is the core of the pipeline and the only compulsory component. It involves the choice of the classification algorithm and its parameters. Among the methods that can be used are those generating different types of knowledge models, such as Naive Bayes, SVM, decision trees, neural networks, among others. Finally, the post-processing stage can be applied when more than one algorithm is tested in the second step, and their results can be combined into an ensemble-like method. However, for the sake of simplicity, in this first version of our proposed grammar we do not include post-processing techniques.

There is a lot of options of tasks or building blocks that can be considered in this three-step approach, as the area of machine learning is in constant development. At the same time, there is no right or wrong, although a few methods are well-known for requiring normalized or standardized data (e.g. neural networks). Although it is always good to have choice, within a Auto-ML framework, the greater the number of building blocks, the larger the size of the search space. At the same time, a very limited number of choices may not reflect the appropriate pipelines.

Figure 3 presents a sample of the produced grammar using the Backus Naur Form (BNF), where each production rules has the form $<Start> ::= [<Pre-processing>] <Algorithm>$. Symbols wrapped in "$<>$" represent non-terminals, and the special symbols "|", "[ ]" and "( )" represent a choice, an optional symbol and a set of grouped symbols that should be used together. The proposed grammar has 147 production rules, in a total of 146 non-terminals and 239 terminals. A complete version of the grammar can be found online,[2] and details about the methods implemented are defined in the SciKit-Learn API[3].

In order to make a fair comparison with systems previously proposed in the literature, we also organized the building blocks present in Auto-SKLearn and TPOT search spaces (but that do not necessarily follow any constraints as the grammar imposes) into two other grammars. Table 1 summarizes the number of building blocks considered in each of the three main steps of the frameworks and the search space size of the grammars built over these components, highlighting their similarities and differences. Note that all components from TPOT are

---

[2] https://github.com/RecipeML/Recipe/tree/master/grammars.
[3] http://scikit-learn.org/stable/modules/classes.html.

```
<Start> ::= [<Pre-processing>] <Algorithm>
<Pre-processing> ::= [<Imputation>] <DimensionalityDefinition>
<Imputation> ::= Mean | Median | Max
<DimensionalityDefinition> ::= <FeatureSelection>   [<FeatureConstruction>]
                       [<FeatureSelection>] <FeatureConstruction>
<FeatureSelection> ::= <Supervised> | <Unsupervised>
<Supervised> ::= SelectKBest <K> <score> | VarianceThreshold | [...]
<score> ::= f-classification | chi2
<K> ::= 1 | 2 | 3 | [...] | NumberOfFeatures - 1
<perc> ::= 1 | 2 | 3 | [...] | 99
<Unsupervised> ::= PCA | FeatureAgglomeration <affinity> | [...]
<affinity> ::= Euclidian | L1 | L2 | Manhattan | Cosine
<FeatureConstruction> ::= PolynomialFeatures
<Algorithm> ::= <NaiveBayes> | <Trees> | [...]
<NaiveBayes> ::= GaussianNB | MultinomialNB | BernoulliNB
<Trees> ::= DecisionTree | RandomForest | [...]
[...]
```

**Fig. 3.** Sample of the defined grammar.

**Table 1.** Comparison of the proposed grammar with others generated from the building blocks used by previous methods, organized in a grammar.

| Building blocks | RECIPE | Auto-SKLearn | TPOT |
|---|---|---|---|
| Pre-process | 33 | 20 | 20 |
| Process | 23 | 17 | 12 |
| Post-process | 0 | 1 | 0 |
| Intersections | - | 35 | 32 |
| Differences | - | 3 | 0 |
| Search space size | $\approx 4.10 \times 10^{34}$ | $\approx 2.47 \times 10^{17}$ | $\approx 7.53 \times 10^{7}$ |

within the grammar used by RECIPE, while three of the components of Auto-SKLearn are not present in RECIPE: the post-processing step, the generalized eigenvector extraction and the one hot encoder, mainly due to computational complexity. The search space size was calculated considering the number of possible combinations of all discrete parameters of the grammar. The continuous parameters, which can generate an infinite number of combinations, were disregarded in this analysis. Note that the search space grows significantly from TPOT to Auto-SKLearn and RECIPE.

## 3.2   Individual Representation

As previously mentioned, individuals represent machine learning pipelines focused on the classification task. These individuals are generated from the grammar using a set of derivation steps. Figure 4 shows an example of an individual created by deriving the production rules of the grammar presented in Fig. 3. In this case, from the *Start* symbol of the grammar, which initializes the *Pipeline*, the optional non-terminal *Pre-processing* is selected together with the classification non-terminal *Algorithm*. In the following step, the optional *Imputation*

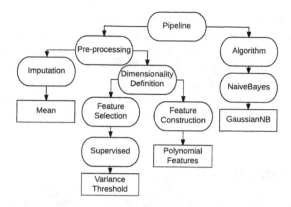

**Fig. 4.** Example of individual generated by the use of genetic programming applying the developed grammar.

is added to the tree, together with *DimensionalityDefinition*. *Imputation* is then replaced by the terminal *Mean*, and from *DimensionalityDefinition* we select *FeatureSelection* followed by *Supervised* and *VarianceThreshold*. From *DimensionalityDefinition*, the optional symbol *FeatureConstruction* is also chosen with a *PolynomialFeatures* as its terminal. With the pre-process step defined, the *NaiveBayes* algorithm is selected, followed by the *GaussianNB* terminal.

## 3.3  Individual Evaluation

Recall that each individual is a representation of the data classification pipeline, and its evaluation involves executing the pipeline in a sample of the dataset of interest. Therefore, the first step of the fitness evaluation process is to generate an executable pipeline from the individual representation. This executable pipeline is generated using the Python SciKit-Learn library [19]. This library is interesting because it offers a wide number of different methods for all phases included in the framework, and allows us to generate easy to read pipelines that can be later executed by non-experts to extract information from their data. The pipelines are generated by reading the leaf nodes of the individual, showed within squares in Fig. 4.

After this translation process from the grammar derivation tree to an executable pipeline, the pipeline is run using a 3-fold cross-validation procedure created over the training set, i.e., three new training and validation subsets are generated. During this phase, each training set is used to generate the model represented by the pipelines, and the validation set used to calculate the fitness, which will be the average f-measure over the three repetitions. This strategy was used to increase the confidence of the results obtained by the approach. Additionally, the training and validation sets are resampled every five generations in order to avoid overfitting.

The f-measure [23], used as the fitness function, is the harmonic mean between the precision and recall metrics defined under a binary class problem. The precision is calculated by dividing the number of examples correctly classified for a given class over all the examples that were assigned that class. Recall, in turn, divides the number of examples correctly classified for a given class by the number of examples that actually belong to that class. This metric was chosen because it accounts for different levels of class imbalance when evaluating the model.

## 4    Experimental Results

This section reports experimental results of RECIPE in 10 datasets, including five classical UCI (University of California Irvine) benchmarks [24] and five bioinformatics datasets introduced in [25–27]. These bioinformatics datasets bring real-world problems about longevity, DNA repair and carcinogenic gene expression. Table 2 presents the main characteristics of the datasets, including the number of instances (inst.), the number of features (feat.) and classes, the types of attributes and presence of missing values.

All experiments were run using a 10-fold cross-validation with three repetitions. The results reported in this section correspond to the average and standard deviations obtained for the 30 executions in the test set. All results were compared using a Wilcoxon Signed-Rank test [28] with 5% of significance.

RECIPE was executed using the following parameters: 100 individuals evolved for 100 generations, tournament selection of size two, elitism of five individuals and uniform crossover and mutation probabilities of 0.9 and 0.1, respectively. If the best individual remains the same for over five generations, we stop the evolutionary process and return its respective pipeline. We also consider a time budget for each ML pipeline generated by RECIPE. Given the size of the datasets used in the experiments, this timeout was set to five minutes. TPOT

**Table 2.** Datasets used in the experiments.

| Dataset | # inst. | # feat. | Feat. types | # classes | Missing? |
|---|---|---|---|---|---|
| Breast-Cancer (BC) | 286 | 9 | Numeric (integer) | 2 | Yes |
| Car Evaluation (CAR) | 1,728 | 6 | Nominal | 4 | No |
| Caenorhabditis Elegans (CE)* | 478 | 765 | Binary | 2 | No |
| Chen-2002 (CHEN)* | 179 | 85 | Real | 2 | No |
| Chowdary-2006 (CHOW)* | 104 | 182 | Real | 2 | No |
| Credit-G (CRED) | 1,000 | 20 | Numeric/nominal | 2 | No |
| Drosophila Melanogaster (DM)* | 104 | 182 | Real | 2 | No |
| DNA-No-PPI-T11 (DNA)* | 135 | 104 | Numeric/nominal | 2 | Yes |
| Glass (GLS) | 214 | 9 | Real | 7 | No |
| Wine Quality-Red (WQR) | 1,599 | 11 | Real | 10 | No |

The symbol * indicates bioinformatics datasets.

was also run with these same parameters, except for the number of generations that was set to 40, which corresponds to the highest value that RECIPE reached in the experiments. Auto-SKLearn requires a single parameter – a time budget – which was set to one hour for each of the 30 executions, resulting in 30 h per dataset. This number was defined after measuring the average time of the other two algorithms in each dataset.

## 4.1 Comparison with Other State-of-the-Art Methods

Before going into detail in the results of f-measure obtained by all methods considered, we first show the numbers that motivated the development of RECIPE: the generation of invalid solutions by TPOT. Figure 5 shows the number of invalid solutions generated by TPOT at each generation, considering a population of 100 for three datasets: *car*, *glass* and *breast cancer*. Note that, in the first generations, at least 10% of the solutions are invalid for all datasets, with *glass* reaching 30%. As evolution progresses, these numbers vary substantially, but invalid solutions never disappear. RECIPE, on the other hand, always guarantees all solutions generated are valid according to the grammar provided. The only way it can generate an invalid solution is if a pipeline reaches a user-defined timeout, and is assigned fitness 0.

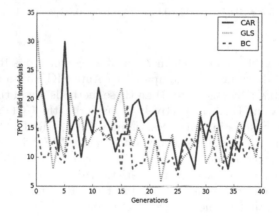

**Fig. 5.** Invalid pipelines produced by TPOT over the GP generations.

Following, the results of f-measure obtained by RECIPE were compared with those of TPOT and Auto-SKLearn, and also with a completely random strategy (RAND) using the proposed grammar, where the number of pipelines generated was equivalent to the maximum number of individuals evaluated during RECIPE's search process. This last comparison was motivated by the studies of Mantovani *et al.* [13] and Bergstra *et al.* [29,30], which showed that a random search is sufficient to optimize ML algorithm components in several domains.

Table 3 presents the results of average and standard deviation of f-measure when comparing the proposed method with the three aforementioned baselines under the parameters previously described. For the comparisons, the symbol ▲ denotes a statistically significant positive variation for the method in that column relative of RECIPE and ▼ a statistically significant negative variation relative to RECIPE according to the Wilcoxon Signed Rank test.

**Table 3.** Comparison of f-measures obtained by RECIPE (with the proposed grammar) and the baselines in the test set.

| Dataset | RECIPE | Auto-SKLearn | TPOT | RAND |
|---------|--------|--------------|------|------|
| BC | 0.939 (0.04) | 0.940 (0.04) | 0.942 (0.04) | 0.936 (0.05) |
| CAR | 0.991 (0.01) | 0.990 (0.01) | 0.999 (0.00)▲ | 0.994 (0.01)▲ |
| CE | 0.486 (0.08) | 0.444 (0.12) | 0.475 (0.09) | 0.484 (0.08) |
| CHEN | 0.959 (0.04) | 0.933 (0.06)▼ | 0.951 (0.05) | 0.942 (0.06)▼ |
| CHOW | 0.993 (0.03) | 0.977 (0.05) | 1.000 (0.00) | 0.986 (0.03) |
| CRED | 0.793 (0.03) | 0.806 (0.04)▲ | 0.802 (0.03)▲ | 0.789 (0.03) |
| DM | 0.758 (0.12) | 0.756 (0.09) | 0.759 (0.10) | 0.754 (0.11) |
| DNA | 0.827 (0.07) | 0.824 (0.04) | 0.788 (0.07)▼ | 0.786 (0.18) |
| GLS | 0.741 (0.09) | 0.734 (0.08) | 0.789 (0.09)▲ | 0.745 (0.08) |
| WQR | 0.642 (0.05) | 0.642 (0.03) | 0.679 (0.03)▲ | 0.636 (0.04) |

The results in Table 3 show that in 7 out of 20 comparisons RECIPE results present statistical different when compared to Auto-SKLearn and TPOT. The two cases where RECIPE was better than these methods are in the bioinformatics datasets CHEN and DNA, respectively. Considering the UCI datasets, TPOT presented the best results in four cases, namely CAR, CRED, GLS and WQR. Auto-SKLearn, in turn, was statistically better in the CRED dataset. When comparing RECIPE to the random approach, both with an identical search space, in 8 out of 10 comparisons we see no statistical difference, while the random search had a better result in one UCI dataset (CAR), and RECIPE improved the classification output in one real-world dataset (CHEN).

These preliminary results, although interesting, might be biased. This is because the search spaces of the methods (apart from random and RECIPE ) are different, and this might represent an advantage for TPOT, which has the smallest search space. Hence, in a second experiment, we compared the effectiveness of the search mechanism of RECIPE when using grammars that encompass the set of valid solutions that can be generated by the state-of-the-art methods (see Table 1), making the comparison among the methods more natural.

Table 4 shows the results of RECIPE with TPOT (RECIPE-TP) and Auto-SKLearn (RECIPE-AS) search spaces, as well as the results of randomly generated solutions with these same search spaces. We can observe that the results of RECIPE improved when compared to the ones reported in Table 3. When

**Table 4.** Results of f-measure obtained by RECIPE and a random method when using grammars that encompass the search spaces of TPOT and Auto-SKLearn.

| Dataset | TPOT search space | | | Auto-SKLearn search space | | |
|---|---|---|---|---|---|---|
| | RECIPE | TPOT | RAND | RECIPE | Auto-SKLearn | RAND |
| BC | 0.944 (0.04) | 0.942 (0.04) | 0.906 (0.09)▼ | 0.949 (0.04) | 0.940 (0.04) | 0.935 (0.04)▼ |
| CAR | 1.000 (0.00) | 0.999 (0.00) | 1.000 (0.00) | 0.998 (0.00) | 0.990 (0.01)▼ | 0.999 (0.00) |
| CE | 0.463 (0.08) | 0.475 (0.09) | 0.451 (0.11) | 0.482 (0.07) | 0.444 (0.12)▼ | 0.495 (0.08) |
| CHEN | 0.949 (0.05) | 0.951 (0.05) | 0.964 (0.05) | 0.937 (0.05) | 0.933 (0.06) | 0.939 (0.06) |
| CHOW | 1.000 (0.00) | 1.000 (0.00) | 0.988 (0.04) | 0.997 (0.01) | 0.977 (0.05)▼ | 0.997 (0.01) |
| CRED | 0.807 (0.04) | 0.802 (0.03) | 0.803 (0.03) | 0.801 (0.03) | 0.806 (0.04) | 0.799 (0.03) |
| DM | 0.734 (0.09) | 0.759 (0.10) | 0.709 (0.09) | 0.732 (0.11) | 0.756 (0.09) | 0.743 (0.12) |
| DNA | 0.818 (0.08) | 0.788 (0.07)▼ | 0.798 (0.09) | 0.823 (0.08) | 0.824 (0.04) | 0.806 (0.08) |
| GLS | 0.779 (0.10) | 0.789 (0.09) | 0.778 (0.09) | 0.734 (0.09) | 0.734 (0.08) | 0.738 (0.12) |
| WQR | 0.685 (0.03) | 0.679 (0.03) | 0.675 (0.03)▼ | 0.664 (0.04) | 0.642 (0.03)▼ | 0.660 (0.07) |

compared to Auto-SKLearn, RECIPE presents statistically superior results in 4 out of 10 datasets. In the case of TPOT, it is better in the DNA dataset. The most surprising results, however, are again those obtained by the random search. In only two cases RECIPE results were statistically significantly better with the TPOT grammar, and only one case in the case of the Auto-SKLearn grammar. One way to look at it is that the components of the grammar are always suitable for the task, hence the variations in f-measure should be within a predefined range that is not too large. In this case, the use of different species within RECIPE in the early generations may favor diversity.

In order to summarize the results, Table 5 shows a comparison between the best absolute classification result for RECIPE and all the others methods. In this table, the symbol ♦ indicates there is no statistical significance between the

**Table 5.** Comparison between the best result of RECIPE (using different versions of the grammar) and the baselines.

| Dataset | RECIPE | | Auto-SKLearn | TPOT | Random | | |
|---|---|---|---|---|---|---|---|
| | Grammar | F-measure | | | Auto-SKLearn | TPOT | New |
| BC | Auto-SKLearn | 0.949 (0.04) | ♦ | ♦ | ▼ | ▼ | ▼ |
| CAR | TPOT | 1.000 (0.00) | ▼ | ♦ | ♦ | ♦ | ▼ |
| CE | New | 0.486 (0.08) | ♦ | ♦ | ♦ | ▼ | ♦ |
| CHEN | New | 0.959 (0.04) | ▼ | ♦ | ▼ | ♦ | ▼ |
| CHOW | TPOT | 1.000 (0.00) | ▼ | ♦ | ♦ | ♦ | ▼ |
| CRED | TPOT | 0.807 (0.04) | ♦ | ♦ | ♦ | ♦ | ▼ |
| DM | New | 0.758 (0.12) | ♦ | ♦ | ♦ | ▼ | ♦ |
| DNA | New | 0.827 (0.07) | ♦ | ▼ | ♦ | ♦ | ♦ |
| GLS | TPOT | 0.779 (0.10) | ▼ | ♦ | ▼ | ♦ | ▼ |
| WQR | TPOT | 0.685 (0.03) | ▼ | ♦ | ▼ | ▼ | ▼ |

compared methods. RECIPE is statistically better in 6 out of 20 cases when compared to the state-of-the-art methods, and in 15 cases out of 30 when compared to random search. Also notice that different versions of the grammar presented different results. The proposed grammar, which has the larger search space, was better in 4 out of 10 dataset, while the TPOT grammar was better in 5 and the Auto-SKLearn in a single dataset. This might indicate that the method can be further improved to work better with a larger search space, as previously suggested. Another interesting thing to observe is that, as the search space increases, so does the difference in performance from the random approach to the other methods. Notice that for all grammars but the proposed the methods obtained better results than random in 4 out of 10 dataset. As we increase the search space size with the complete grammar, this number increases to 7 out of 10.

## 4.2   Analysis of the Evolutionary Process of RECIPE

The results in the previous section showed the values of f-measure obtained by RECIPE can be as good as or better than those obtained by the state-of-the-art methods, with the guarantees of always generating valid solutions. This section looks into more detail at the evolutionary process and the solutions generated by RECIPE. As previously discussed, the results of fitness of different pipelines do not vary significantly and, with few exceptions, are within a small interval.

Figure 6 shows the f-measure of the best and worst individuals of the population in the training set, as well as the average fitness over all individuals. Notice that the values of averages can be brought down by a few individuals with fitness 0, which are those that exceed the time limit for the classification model to be built. The fluctuations in the values of the best individual are due to the data resampling strategy, implemented to avoid overfitting.

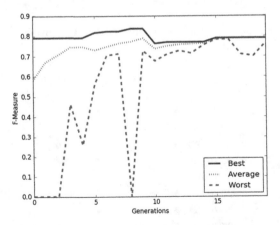

**Fig. 6.** RECIPE convergence curves for DNA dataset over the generations.

Additionally, Fig. 7 shows the distribution of algorithms and pre-processing methods along the generations. Notice that very quickly one classification method ends up overtaking others, while the pre-processing methods have a greater variation. Although the point in evolution where this happens varies from one dataset to the other, we observed that the first combination of pre-processing, algorithm and parameters that works slightly better than the others is likely to be present in the final solution. Again, this opens up opportunities to include in the method a varied range of diversity mechanisms and speciation, which are left as our first direction of future work. This loss of diversity is one of the reasons the methods present results sometimes too similar to those obtained by a random search.

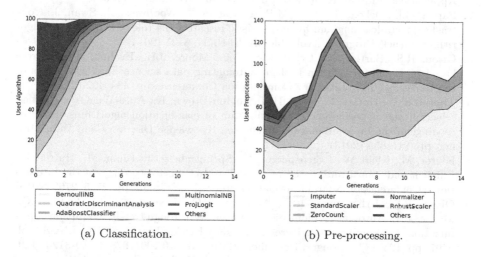

(a) Classification.                                (b) Pre-processing.

**Fig. 7.** RECIPE coverage in classification and pre-processing stages over the generations for DNA dataset.

## 5    Conclusions and Future Work

This paper introduced RECIPE, a grammar-based genetic programming framework conceived to evolve complete machine learning pipelines to solve classification tasks. The method improves over previous ones by introducing a grammar to the search process, which organizes previous knowledge in the literature and avoids the generation and evaluation of invalid solutions. The framework is flexible enough to deal with different sets of grammars, and can be easily extended to solve other machine learning tasks different from classification.

The results show that the method does solve the problem of generating invalid solutions, while being as good as or better than previously proposed approaches in the literature. However, they also indicate that the substantial growth of the search space introduced by the proposed grammar in the new method might require the implementation of more sophisticated mechanisms within RECIPE to increase diversity and ensure a minimum coverage of the search space. A more in-depth study of the methods parameters and a larger study in a wider range of datasets are also steps of future work.

**Acknowledgments.** This work was partially supported by the following Brazilian Research Support Agencies: CNPq, CAPES and FAPEMIG.

# References

1. Banzhaf, W., Francone, F.D., Keller, R.E., Nordin, P.: Genetic Programming - An Introduction: On the Automatic Evolution of Computer Programs and Its Applications. Morgan Kaufmann Publishers Inc., Burlington (1998)
2. Pappa, G.L., Ochoa, G., Hyde, M.R., Freitas, A.A., Woodward, J., Swan, J.: Contrasting meta-learning and hyper-heuristic research: the role of evolutionary algorithms. Genet. Program. Evolvable Mach. **15**(1), 3–35 (2014)
3. Olson, R.S., Bartley, N., Urbanowicz, R.J., Moore, J.H.: Evaluation of a tree-based pipeline optimization tool for automating data science. In: Proceedings of the Genetic and Evolutionary Computation Conference, pp. 485–492 (2016)
4. Thornton, C., Hutter, F., Hoos, H.H., Leyton-Brown, K.: Auto-WEKA: combined selection and hyperparameter optimization of classification algorithms. In: Proceedings of the International Conference on Knowledge Discovery and Data Mining, pp. 847–855 (2013)
5. Feurer, M., Klein, A., Eggensperger, K., Springenberg, J., Blum, M., Hutter, F.: Efficient and robust automated machine learning. In: Proceedings of the International Conference on Neural Information Processing Systems, pp. 2755–2763 (2015)
6. Olson, R.S., Urbanowicz, R.J., Andrews, P.C., Lavender, N.A., Kidd, L.C., Moore, J.H.: Automating biomedical data science through tree-based pipeline optimization. In: Squillero, G., Burelli, P. (eds.) EvoApplications 2016. LNCS, vol. 9597, pp. 123–137. Springer, Heidelberg (2016). doi:10.1007/978-3-319-31204-0_9
7. McKay, R., Hoai, N., Whigham, P., Shan, Y., O'Neill, M.: Grammar-based genetic programming: a survey. Genet. Program. Evolvable Mach. **11**(3), 365–396 (2010)
8. Mendoza, H., Klein, A., Feurer, M., Springenberg, J., Hutter, F.: Towards automatically-tuned neural networks. In: Proceedings of the ICML AutoML Workshop (2016)
9. Stanley, K.O., Miikkulainen, R.: Evolving neural networks through augmenting topologies. Evol. Comput. **10**(2), 99–127 (2002)
10. Yao, X.: Evolving artificial neural networks. Proc. IEEE **87**(9), 1423–1447 (1999)
11. Pappa, G.L., Freitas, A.A.: Automating the Design of Data Mining Algorithms: An Evolutionary Computation Approach. Springer, Heidelberg (2009)
12. Dioşan, L., Rogozan, A., Pecuchet, J.P.: Improving classification performance of support vector machine by genetically optimising kernel shape and hyper-parameters. Appl. Intell. **36**(2), 280–294 (2012)
13. Mantovani, R.G., Rossi, A.L.D., Vanschoren, J., Bischl, B., de Carvalho, A.: Effectiveness of random search in SVM hyper-parameter tuning. In: Proceedings of the International Joint Conference on Neural Networks, pp. 1–8 (2015)
14. Barros, R.C., Basgalupp, M.P., de Carvalho, A.C.P.L.F., Freitas, A.A.: Automatic design of decision-tree algorithms with evolutionary algorithms. Evol. Comput. **21**(4), 659–684 (2013)

15. Sá, A.G.C., Pappa, G.L.: Towards a method for automatically evolving bayesian network classifiers. In: Proceedings of the Conference Companion on Genetic and Evolutionary Computation, pp. 1505–1512 (2013)
16. Sá, A.G.C., Pappa, G.L.: A hyper-heuristic evolutionary algorithm for learning bayesian network classifiers. In: Bazzan, A.L.C., Pichara, K. (eds.) IBERAMIA 2014. LNCS (LNAI), vol. 8864, pp. 430–442. Springer, Heidelberg (2014). doi:10.1007/978-3-319-12027-0_35
17. Springenberg, J.T., Klein, A., Falkner, S., Hutter, F.: Bayesian optimization with robust bayesian neural networks. In: Proceedings of the Conference on Neural Information Processing Systems (2016)
18. Hall, M., Frank, E., Holmes, G., Pfahringer, B., Reutemann, P., Witten, I.H.: The WEKA data mining software: an update. ACM SIGKDD Explor. Newsl. **11**, 10–18 (2009)
19. Pedregosa, F., Varoquaux, G., Gramfort, A., Michel, V., Thirion, B., Grisel, O., Blondel, M., Prettenhofer, P., Weiss, R., Dubourg, V., Vanderplas, J., Passos, A., Cournapeau, D., Brucher, M., Perrot, M., Duchesnay, E.: SciKit-learn: machine learning in Python. J. Mach. Learn. Res. **12**, 2825–2830 (2011)
20. Feurer, M., Springenberg, J.T., Hutter, F.: Initializing bayesian hyperparameter optimization via meta-learning. In: Proceedings of the AAAI Conference on Artificial Intelligence, pp. 1128–1135 (2015)
21. Deb, K., Pratap, A., Agarwal, S., Meyarivan, T.: A fast and elitist multiobjective genetic algorithm: NSGA-II. IEEE Trans. Evol. Comput. **6**(2), 182–197 (2002)
22. Whigham, P.A., Dick, G., Maclaurin, J., Owen, C.A.: Examining the "best of both worlds" of grammatical evolution. In: Proceedings of the Conference on Genetic and Evolutionary Computation, pp. 1111–1118 (2015)
23. Witten, I.H., Frank, E., Hall, M.A.: Data Mining: Practical Machine Learning Tools and Techniques, 3rd edn. Morgan Kaufmann Publishers Inc., Burlington (2011)
24. Asuncion, A., Newman, D.: UCI machine learning repository (2007)
25. Freitas, A.A., Vasieva, O., de Magalhães, J.P.: A data mining approach for classifying DNA repair genes into ageing-related or non-ageing-related. BMC Genomics **12**(1) (2011)
26. Souto, M., Costa, I., Araujo, D., Ludermir, T., Schliep, A.: Clustering cancer gene expression data: a comparative study. BMC Bioinf. **9**(1), 497 (2008)
27. Wan, C., Freitas, A.A., De Magalhães, J.P.: Predicting the pro-longevity or anti-longevity effect of model organism genes with new hierarchical feature selection methods. IEEE/ACM Trans. Comput. Biol. Bioinform. **12**(2), 262–275 (2015)
28. Wilcoxon, F., Katti, S., Wilcox, R.A.: Critical values and probability levels for the Wilcoxon rank sum test and the Wilcoxon signed rank test. Sel. Tables Math. Stat. **1**, 171–259 (1970)
29. Bergstra, J., Bardenet, R., Bengio, Y., Kégl, B.: Algorithms for hyper-parameter optimization. In: Proceedings of the International Conference on Neural Information Processing Systems, pp. 2546–2554 (2011)
30. Bergstra, J., Bengio, Y.: Random search for hyper-parameter optimization. J. Mach. Learn. Res. **13**(1), 281–305 (2012)

# A Grammar Design Pattern for Arbitrary Program Synthesis Problems in Genetic Programming

Stefan Forstenlechner[✉], David Fagan, Miguel Nicolau, and Michael O'Neill

Natural Computing Research and Applications Group, School of Business,
University College Dublin, Dublin, Ireland
stefan.forstenlechner@ucdconnect.ie,
{david.fagan,miguel.nicolau,m.oneill}@ucd.ie

**Abstract.** Grammar Guided Genetic Programming has been applied to many problem domains. It is well suited to tackle program synthesis, as it has the capability to evolve code in arbitrary languages. Nevertheless, grammars designed to evolve code have always been tailored to specific problems resulting in bespoke grammars, which makes them difficult to reuse. In this study a more general approach to grammar design in the program synthesis domain is presented. The approach undertaken is to create a grammar for each data type of a language and combine these grammars for the problem at hand, without having to tailor a grammar for every single problem. The approach can be applied to arbitrary problem instances of program synthesis and can be used with any programming language. The approach is also extensible to use libraries available in a given language. The grammars presented can be applied to any grammar-based Genetic Programming approach and make it easy for researches to rerun experiments or test new problems. The approach is tested on a suite of benchmark problems and compared to PushGP, as it is the only GP system that has presented results on a wide range of benchmark problems. The object of this study is to match or outperform PushGP on these problems without tuning grammars to solve each specific problem.

**Keywords:** Grammar design · Program synthesis · Genetic Programming

## 1 Introduction

Genetic Programming (GP) has been used in many different areas and on many problems as it is a flexible approach to evolve solutions. GP does not require a specific representation that has to be used, like tree or linear representation. Many GP systems exist with different representations that fit certain types of problems. Grammar Guided Genetic Programming (G3P) systems are fairly independent of a problem representation, as the solutions produced only require

© Springer International Publishing AG 2017
J. McDermott et al. (Eds.): EuroGP 2017, LNCS 10196, pp. 262–277, 2017.
DOI: 10.1007/978-3-319-55696-3_17

their syntax to be specified by a grammar. Grammars can be used to mimic tree-based GP and GAs [28]. Terminal and non-terminal symbols used in the grammar are of no importance to G3P itself. Only the evaluation, which can be done externally by an interpreter or virtual machine, needs to be able to handle the individual's syntax.

Program synthesis is a problem domain to which G3P seems especially well suited for the following reasons. (1) Programming languages are already defined in grammars as grammars are used to check syntactical correctness. (2) The same or similar problems have to be solved in many programming languages. As many programming language have similar syntax, it should be easy to adapt grammars from one language to another. (3) Grammars can be exchanged between researchers and used in all kinds of GP systems without reimplementing code. Therefore, experiments can be reproduced in an easy way.

The major drawback of previously used grammars to tackle program synthesis so far is that they have been written for a specific problem instance [20,21]. Therefore they cannot be reused for another problem instance without heavy adaptation, as they only use certain data types or data structures and only use a small subset of the available functionality of a programming language. These grammar approaches limit the search space, but are not applicable to a wide variety of problems. In this study, a design pattern for the creation of grammars for programming languages is presented so that arbitrary program synthesis problems can be tackled, that the grammars can be extended to use any libraries already existing in a language, and that invalid individuals, due to runtime errors etc., are kept to a minimum to avoid wasting resources. The grammars can be used in any grammar-based GP system with minimal implementation overhead. A similar idea of a general-purpose method to solve problems was proposed by Koza in the form of a Genetic Programming Problem Solver [11]. It should be noted that other approaches to program synthesis than using grammars or GP exist and for more information the reader is referred to [6].

In the remainder of the paper, a design pattern to tackle program synthesis problems is proposed in Sect. 2. Other GP systems that have been created to tackle program synthesis are reviewed in Sect. 3. An experimental setup to test the proposed design pattern is described in Sect. 4 and discussed in Sect. 5. Conclusion and future work is outlined in Sect. 6.

## 2    System Description

The concept proposed to tackle program synthesis is using a standard tree-based grammar guided GP system [18], which will be referred to as G3P, with a few additions to make it flexible to evolve code for any programming language and extensible to use libraries of those languages. The approach presented in this study consists of two parts that have to be provided, a Grammar and a Skeleton.

The evolved code is not abstract code or pseudocode, but code for a specific programming language. Therefore, syntactical differences make it necessary to adapt the grammar and skeleton for other languages. In this study, the focus is on evolving Python code for arbitrary problems, but the general concept applies to many languages. Specific differences for Python will be discussed in Sect. 2.3.

## 2.1    Grammar

A grammar defines a search space and how valid individuals will look. The design pattern for grammars presented in this paper enables the creation of grammars that can tackle arbitrary program synthesis problems. The advantages of using grammars are that every individual is automatically syntactically correct and can be executed and that grammars can be exchanged easily between systems. Therefore, the grammars created can be used in any other G3P system.

Poli et al. [23] explained that closure is an important property for most representations for GP to work effectively. To achieve closure, type consistency and evaluation safety are required. Evaluation safety will be explained in the next Section. Grammars are suited to address type consistency, which is important because "type awareness reduces the search space and makes genetic operators more effective." [7]. Programming languages have multiple data types and different operations are defined for these types. For example, an integer might not have an append function. Languages which have to be compiled should not compile if append is called on an integer, while languages which are interpreted may raise an exception. This problem should be avoided in grammars as individuals which cannot be compiled or raises an exception has to be given a penalty. This problem should be avoided via the design of grammars such that individuals that fail to compile or raise an exception due to type errors are simply not produced.

Not every problem requires the use of all available data types. Therefore, a grammar for every data type was created as well as a general grammar for the structure of the program. For example, with a data structure like a list, a grammar was created for each basic data type with the data structure. This has the advantage that lists are typed in the grammar and there is no ambiguity of the type of the contents in the lists. Depending on the input and output defined for a problem and the knowledge about the problem, these grammars can be combined to cover the data types required to solve a certain problem. Types of grammars created for this study are shown in Fig. 1a. An example of one of the grammars is shown in Fig. 1b. The grammars are defined in the context-free Backus-Naur Form (BNF). As the rules in the grammars can create program expressions which return data of types defined in different grammars (e.g. integer.bnf defines the comparison of two integer values, which results in a boolean value), these options have to be added to the correct rule, when combining multiple grammars. Although lists containing multiple types are available in some programming languages, they are not used in this study to provide type safety on a grammar level. All grammars are available online [5].

Another advantage of using grammars is that they can easily be adapted. If a code library already exists that should be used for a problem, it can be added directly to the grammar without major changes. Therefore, any library which is available for a programming language can be used and more complex programs

structure.bnf

bool.bnf          list_bool.bnf

integer.bnf       list_integer.bnf

float.bnf         list_float.bnf

string.bnf        list_string.bnf

(a) Grammars per data type.

```
<bool_assign> ::= <bool_var>' = '<bool>
<bool_var> ::= 'b0' | 'b1' | 'b2'
<bool> ::= <bool_var> | <bool_const>
         | <bool_pre> <bool>
         | '(' <bool> <bool_op> <bool> ')'
<bool_const> ::= 'True' | 'False'
<bool_pre> ::= 'not'
<bool_op> ::= 'and' | 'or'
```

(b) Boolean grammar (bool.bnf)

**Fig. 1.** Grammars created in this study and an example of one of the grammars.

can be evolved, instead of evolving everything from scratch. Even code snippets can already be integrated in the grammar, if desired.

One disadvantage of using grammars is that individuals only consist of parts defined in the grammar. Therefore, variable names have to be defined beforehand. As input and output for a problem instance is known, the number of input and output variables is known. But all other variables that shall be used in the code to store data temporarily have to be defined beforehand as well, although the number of variables required is most likely not known. As shown in Fig. 1b, the rule <bool_var> contains three choices, therefore three boolean variables are available. If this grammar is used to evolve a program, as many variables as necessary can be added to that rule. If an input or output variable is also of type bool, it is required to add it to the rule <bool_var> as well. Input and output variables of other types have to be added to the according rules of their type.

At the moment it is not possible to adjust the number of variables during the run. All available variables can be used by an individual, but do not have to. Adjusting the variables during runtime is a topic that is left for future work, see Sect. 6. In this study, three additional variables of each type were added to the grammar for every problem.

## 2.2   Skeleton

The skeleton is the part of the design pattern that is executed in the end to evaluate the individual. It defines the method header of the code that is going to be evolved, the fitness function, and the protected methods (evaluation safety, as mentioned in Sect. 2.1). An example of a shortened version of a skeleton is shown in Fig. 2. The method header defines the return type as well as the input variables of the method that is evolved. The body of this method is empty at this stage, except for the return statement. During the evolutionary run the body is replaced with the code from an individual and executed to determine the fitness. The fitness function is already part of the skeleton rather than the GP system, so that it has to be in the same programming language as the grammar and therefore it is easier to exchange the skeleton and the grammar with other researchers and run experiments. The protected methods are the most important part of the skeleton. As in regression problems where normally instead of division and logarithm protected operations of these operators are used, the same is done in this system for program synthesis to provide evaluation safety. The protected method calls have to be used in the grammar instead of the unprotected versions.

```python
import maths
import Levenshtein
# Protected Methods
def div(nom, denom):
    return num if denom <= 0.00001 else num / denom
    ...
#  evolved function
def evolve(in0, in1):
    <insert_code_here>
    return res

def fitness(in_val, out_val):
    fit = []
    for (i, o) in zip(in_val, out_val):
        fit.extend(abs(evolve(i[0], i[1]) - o[0]))
    return fit
```

**Fig. 2.** Example skeleton in Python

## 2.3   Python Specific Differences

One of the main differences to other programming languages on the syntax level is that python uses indentation instead of brackets to separate code blocks. Unfortunately, indentation for nested loops and if-statements cannot be added in a context free grammar, because context information of the level of indentation of the previous block would be required, also known as off-side rule [12]. A Context Sensitive Grammar (CSG) would be needed. To avoid CSG's the characters "{:" and ":}" are used as brackets to identify lines that have to be indented. The indentation is done, before the code is executed. As already mentioned, protected methods are used to provide execution safety. This has to be done only once for a programming language as they are part of the skeleton and not part of the grammar. The skeleton used for this paper has protected methods written in Python. For other languages, adaptations have to be made depending on the default behaviour of the language to certain exceptions. Other changes that have to be made to run similar experiments as in this paper, are syntactical changes in the grammar due to syntax differences in programming languages.

## 2.4   Implementation Details

Although every individual is syntactically correct and many methods can be replaced with protected versions if desired, runtime exceptions and long evaluations can still occur. The general approach chosen to avoid these problems is to run the evolved code in a separate process, other options like running the code in virtual machines are also possible. If an individual takes too long to finish, the process is killed and the individual gets assigned the worst possible fitness. A time out parameter to specify the amount of time an individual is allowed to run is used. The time out should not be seen as a way to improve code performance, but as a last resort to stop evaluation. Another way to stop the evaluation is already included in the grammar in the bodies of loops. Every loop contains a code block which checks the number of iterations executed and breaks after a certain number. All loops share the same variable for counting the iterations, because nested loops increase the runtime exponentially and if every loop would have its own counter, even a small number of nested loops might increase runtime dramatically and the time out will be reached. If any other runtime exceptions happen, like memory overflow etc., the individual will also simply be assigned the worst possible fitness, as it cannot be evaluated.

# 3  Previous Approaches to Program Synthesis

The following subsections describe related genetic programming systems which have tackled program synthesis problems, either by creating code from scratch or via using existing code to improve a program, as it is done in Genetic Improvement [14]. A description of the systems as well as advantages and disadvantages are explained. A comparison with the approach, presented in this study, is shown at the end to summarize this Section.

## 3.1  PushGP

PushGP [25] is a GP system that uses the Push programming language. Push was solely invented for the purpose of using it in evolutionary computation systems. It uses stacks to save data instead of using variables. One stack per data type exists. A program consists of instructions that are executed by an interpreter. Depending on the instruction, elements are taken from the stacks of the correct data type and the result is pushed back on top of a stack. If no values can be taken from a stack, the instruction is not executed. Implementations of Push are available in many programming languages. The reference implementation is available in Clojure. In 2004, results achieved with PushGP were awarded a gold medal in the Human Competitive Results competition at GECCO [24].

   On the one hand, PushGP has great advantages as it does not need to declare variables and the number of variables is not limited in any way, because all data is saved on stacks. Another benefit of PushGP is that it does not produce runtime exceptions as instructions are already implemented in a protected way and nothing happens if certain data types are not available. On the other hand, as the language has been invented for evolutionary computation systems, it is not used by programmers or in real world applications. Therefore, to use existing libraries to evolve more complex programs, functions available in a Push implementation's host language have to be wrapped in Push instructions. Additionally, evolved Push code cannot be directly integrated in already existing programs, as the programs are written in another programming language. Integration of the Push interpreter would also be needed, adding additional overhead.

## 3.2 Strongly Formed Genetic Programming

Strongly Formed Genetic Programming (SFGP) [3] is an extension of Strongly Typed Genetic Programming (STGP) [19]. Initialization, crossover and mutation are implemented in a way that they check the type of nodes and what type the subtrees require. The operators are only allowed to connect nodes of the appropriate type. Hence, type consistency is implicitly given when evaluating an individual like in grammars. SFGP extends STGP with several nodes that allow the generation of generic code. Every node has to be implemented for the system as it is not based on any programming language. Code blocks are limited to a fixed number of child nodes, which has to be predefined. Variables are part of the terminals, hence the number of variables is predefined.

Due to the generic code that SFGP evolves, the code has to be translated to another programming language, if it shall be used in a real world program. This can be achieved for many languages, as the provided nodes follow common structures of other programming languages.

## 3.3 Grammar Guided Genetic Programming

G3P systems, like Grammatical Evolution [22], have already been able to use grammars to evolve programs for arbitrary languages. Due to the flexibility of grammars and that many problems can be defined with grammars (any GP and GA problem can be transformed to a grammar [28]), G3P has been applied to a wide range of problems, evolving music [16], creating truss design [4], optimising pylon structures [2], evolving aircraft models [1], controlling femtocell network coverage [10] and also program synthesis [20].

Even though G3P can be applied to program synthesis, no general grammar or concept has been established to tackle arbitrary program synthesis problems. A new grammar has been designed for every new program synthesis problem tackled. This disadvantage will be solved with the concept presented in Sect. 2.

## 3.4 Program Synthesis via Code Reusage

Another way to approach program synthesis in GP is to use existing code and improve a program, e.g. by improving non-functional requirements or by fixing bugs. The system GISMOE [13] automatically creates a BNF grammar from existing code. Terminal symbols in the grammar are complete lines from that code. The block structure of the code cannot be changed in GISMOE, e.g. opening and closing brackets in C++, but the contents of these blocks can be. GISOME is used for *Genetic Improvement* and therefore its main goal is to improve exiting code. It does not create code from scratch, but reuses the existing code of a program and adapts it.

Abstract Syntax Trees (ASTs) are another representation that can be used for program synthesis instead of grammars. GenProg [15] or Gen-O-Fix [26] are two such systems that utilise ASTs to synthesize code by reusing existing code. As the name already suggests, ASTs represent the syntax of code in a tree structure, similar to a derivation tree of a grammar. The tree can then be modified in a manner like trees in STGP. Crossover and mutation are only allowed to exchange nodes of the same type to keep the chance of compilation errors low.

Both approaches have the disadvantage that new individuals might be invalid in the sense that they might not even compile due to syntactical errors. Nevertheless, both approaches have been used on real world applications, GISMOE to improve performance of programs [13] and GenProg to fix bugs [15]. GISMOE, GenProg and Gen-O-Fix are specialised to modify existing code, which is a different kind of problem compared to the systems presented so far.

## 3.5 Comparison of Program Synthesis Approaches

The presented design pattern in Sect. 2 makes it easy to reuse grammars for arbitrary program synthesis problems and exchange them between G3P systems instead of creating new ones for every problem instance. The concept ensures type and evaluation safety and because grammars are used the evolved individuals are syntactically correct so that there are no compilation or syntax errors. Due to the skeleton, which also contains the fitness function, functions can be implemented in a protected way that otherwise might throw errors. Grammars and skeletons have to be created only once per programming language and can be reused for arbitrary problems, except the fitness function in the skeleton might have to be exchanged. The concept is not specific for a certain programming language, which makes it possible to use it for almost any programming language. Two additional advantages of the proposed approach are firstly, that in contrary to PushGP and SFGP, which have to wrap functions of existing libraries in extra instructions, libraries can be integrated in grammars without overhead. Secondly, that evolved code of the proposed approach can be integrated in real world program without the need of an extra interpreter as it can already be evolved in the required language, whereas a Push interpreter is needed for code evolved with PushGP.

The downside of the proposed concept is that grammars and skeletons have to be created by hand. This has to be done only once for any programming language as they can be reused. Every grammar has to be made type safe to avoid compiler/runtime exceptions, even if a language is not strongly typed. Program languages which are strongly typed require type safety, whereas loosely types languages would not. Another disadvantage is that variables have to be predefined. That means that before running the search process, the number of variables needed to solve a problem has to be estimated. In contrast to bespoke grammars, the search space may be larger with the concept introduced in this study. Advantages and disadvantages of this concept are listed below.

Pros:

- All individuals are syntactically correct
- Protected methods keep exceptions to a minimum
- Real world programming languages can be used
- Using libraries of programming languages
- Grammars and skeletons are reusable for arbitrary problems

Cons:

- Grammar and skeleton have to be created by hand and for the first time for a new language
- Grammars have to be made type safe
- Variables have to be predefined
- Search space might be larger than required

Type safe grammars for Python have been used in this paper to evolve programs and are available online[1] as well as the system used to run the experiments.

## 4  Experimental Setup

This Section outlines the benchmark suite and the parameter settings used for the experiments as well as some minor differences to PushGP.

### 4.1  Benchmark Suite

The benchmark suite used in this study was first presented with PushGP [9]. The benchmark suite will be used to test the concept presented in Sect. 2 and to compare the results to PushGP, as PushGP is the only system tested on this extensive benchmark suite. The suite consists of 29 well defined problems from introductory computer science textbooks. The benchmark suite is not only well defined, but also defines how to generate training and test cases, number of cases used for training and testing as well as which data types should be used. For each problem, the general grammars described in Sect. 2.1 are combined according to the data types required without further tailoring. One exception to this is if additional terminal symbols were also provided for PushGP, in which case they have been added to the relevant grammars. For the language, Push, all used methods for the data types are listed. In order for a fair comparison to be made only functions similar to the function set used in PushGP are contained in the grammars. Therefore, not all built in functionality of Python is available. A general description of the fitness function is also given. Penalty values that have to be applied, for example when a list output is shorter than the expected outcome, have been taken from the PushGP GitHub[2] repository, where the exact fitness functions can be found.

---

[1] https://github.com/t-h-e/HeuristicLab.CFGGP.

[2] https://github.com/lspector/Clojush.

## 4.2 Experimental Parameter Settings

The parameter settings for the experiments are as close to the PushGP settings as possible to be able to do a comparison and analyse how well derivation tree based G3P does. PushGP has been tested with three different selection operators. Tournament selection, Implicit Fitness Sharing [17] and Lexicase selection [8]. For the experiments conducted in this study tournament selection and lexicase selection have been used. Tournament selection is one of the most common selection operators in GP and lexicase selection [8] has achieved the best results on the benchmark suite with PushGP [9]. All other parameters are typical settings for a GP system. The settings are shown in Table 1. Two additional parameters are required. Number of variables per type and maximum execution time. As explained in Sect. 2, variables have to be defined in the grammar beforehand. So additionally to the input and output variables, three variables per type are used to store temporary data, which is more than enough for all problems. The maximum number of nodes was set to 250, which was established during initial experiments. A depth limit was not used as the trees are already limited by the number of nodes. Maximum execution time is the amount of time the execution of a single individual on all training cases is allowed to take, before it is assigned the worst fitness. This parameter is used, because it is not possible to infer if the code of an individual will eventually finish or how long the execution might take. One second seems reasonable, as most problems can be solved within less than a tenth of a second on the machines running the experiments. PushGP used a certain number of steps the interpreter was allowed to execute before stopping, which is not possible in this case, as the default interpreter or compiler of the desired language is used, for this paper the CPython interpreter.

**Table 1.** Experimental parameter settings

| Parameter | Setting |
|---|---|
| Runs | 100 |
| Generations | 300[a] |
| Population size | 1000 |
| Tournament size | 7 |
| Crossover probability | 0.9 |
| Mutation probability | 0.05 |
| Elite size | 1 |
| Node limit | 250 |
| Variables per type | 3 |
| Max execution time | 1 s |

[a] 200 Generations for "NumberIO" as set in [9].

### 4.3   PushGP Differences

As PushGP and G3P system are different, there are two important aspects that should be noted. The result of a program in PushGP is often the printed output of a program. The code evolved with G3P does not print anything as it evolves a method which has a return value. This return value is always the output of a program. Therefore, the grammars do not contain any print statements as these are not required. Certain terminals, which have merely been added to the problem definition in [9] to format the output, are not included in the grammar as well. Due to these differences and that the grammars in this study do not allow data structures containing multiple different types, the problem "String Differences" was excluded from the problem set, as it requires a data structure as output value containing multiple types. "String Differences" has not yet been solved with PushGP.

## 5   Results

In this Section, the results of G3P are shown on the benchmark problems and compared to results achieved with PushGP. The main goal of the comparison is not to beat PushGP, but to show that the presented concept is able to do well compared to PushGP. Additionally, to find advantages and disadvantages to improve the concept. The results for PushGP have been taken from [9].

### 5.1   Comparison to PushGP on Tournament Selection

The left-hand side of Table 3 shows the number of times a problem has been successfully solved on all test cases with tournament selection. Problems which have never been solved by G3P or PushGP have been discarded. A count of how many different problems have been solved at least once, how many times a system has been better on a problem than the other with the same selection method as well as the average rank of the four settings is described in Table 2. On the one hand, PushGP was able to solve five problems where G3P was no able to find a correct solution compared to two problems that G3P solved and PushGP failed to find correct solution with tournament selection. On the other hand, in many cases where a solution was found by G3P, the number of correct

**Table 2.** Statistics of how many problems have been solved, how often G3P or PushGP did better than the other with the same selection method and the average rank of each method.

|  | G3P | | PushGP | |
|---|---|---|---|---|
|  | Tour | Lex | Tour | Lex |
| # of solved problems | 10 | 16 | 13 | 22 |
| Better than same selection method | 8 | 11 | 7 | 12 |
| Average rank | 2.61 | 1.83 | 2.91 | 1.74 |

solutions in 100 runs was considerably higher than with PushGP, like "Median", "Negative To Zero" or "Super Anagrams". Whereas, this is only the case for the problem "Mirror Image" for PushGP. The difference of the number of found solutions is rather small on problems where PushGP is doing better than G3P, except "Mirror Image" and "Vector Average".

## 5.2  Comparison to PushGP on Lexicase Selection

Table 3 shows all the results achieved with G3P and PushGP with tournament and lexicase selection. As PushGP, G3P was able to solve more problems with

**Table 3.** Number of times a correct individual was found that solves all test cases for all 29 problems with tournament and lexicase selection. PushGP results taken from [9]. The best results are marked in bold. The column "Diff" shows the difference between G3P and PushGP. Differences were G3P was more successful are in green, otherwise red was used.

| Problem | Tournament | | | Lexicase | | |
|---|---|---|---|---|---|---|
| | G3P | PushGP | Diff | G3P | PushGP | Diff |
| Compare String Lengths | 1 | 3 | -2 | 3 | **7** | -4 |
| Count Odds | 0 | 0 | 0 | **10** | 8 | +2 |
| Digits | 0 | 0 | 0 | 0 | 7 | -7 |
| Double Letters | 0 | 0 | 0 | 0 | 6 | -6 |
| Even Square | 0 | 0 | 0 | 1 | **2** | -1 |
| For Loop Index | 0 | 0 | 0 | **25** | 1 | +24 |
| Grade | 0 | 0 | 0 | **14** | 4 | +10 |
| Last Index of Zero | 3 | 8 | -5 | **24** | 21 | +3 |
| Median | 62 | 7 | +55 | **65** | 45 | +20 |
| Mirror Image | 0 | 46 | -46 | 1 | **78** | -77 |
| Negative To Zero | 94 | 10 | +84 | **98** | 45 | +53 |
| NumberIO | 95 | 68 | +27 | 96 | **98** | -2 |
| Replace Space with Newline | 0 | 8 | -8 | 0 | **51** | -51 |
| Scrabble Score | 0 | 0 | 0 | 0 | **2** | -2 |
| Small Or Large | **25** | 3 | +22 | 24 | 5 | +19 |
| Smallest | **96** | 75 | +21 | 89 | 81 | +8 |
| String Lengths Backwards | 28 | 7 | +21 | **70** | 66 | +4 |
| Sum of Squares | 0 | 2 | -2 | 3 | **6** | -3 |
| Super Anagrams | **38** | 0 | +38 | 28 | 0 | +28 |
| Syllables | 0 | 1 | -1 | 0 | **18** | -18 |
| Vector Average | 0 | 14 | -14 | 0 | **16** | -16 |
| Vectors Summed | 7 | 0 | +7 | **85** | 1 | +84 |
| XWord Lines | 0 | 0 | 0 | 0 | **8** | -8 |

lexicase selection than with tournament selection and in most cases lexicase selection finds more correct solutions, except for "Small Or Large" and "Smallest". On the one hand G3P is only able to solve one problem that PushGP is not able to solve at all, whereas PushGP is able to solve seven that G3P cannot solve. On the other hand, G3P is able to find more correct solutions on most problems that both systems are able to solve. The average rank shown in Table 2 shows that lexicase selection is a great improvement over using tournament selection as already shown in [9]. G3P and PushGP achieve quite similar ranks.

According to [27], PushGP produced many runs where solutions were found with zero training error, which did not generalize and failed the unseen test set. It is interesting that G3P shows a similar behaviour. G3P produced on seven problems more than 20 runs where a solution with zero error on training was found which failed on the test set. In case of "Compare String Lengths" and "Super Anagrams" more than 70 runs were not able to generalize. This phenomena might be caused by lexicase selection, as overfitting does not happen that regularly with tournament selection.

Additionally, these experiments confirm the superiority of lexicase selection over tournament selection in the program synthesis domain. To our knowledge, this is the first time lexicase selection has been used in a G3P system.

## 5.3   Generational Progress

The number of generations chosen to run G3P was the same PushGP. In contrary to standard GP settings, the number of generations was set rather high. When analysing at which generations correct solutions have been found, most problems show that the majority of the correct solutions have already been found after half of the generations have been finished with tournament and lexicase selection. For "Negative To Zero", "Number IO" and "Smallest" more than 50 correct solutions have been found up to generation 12 with tournament selection and 7 with lexicase selection. More than 90% of the solutions that are discovered up to generation 300 have already been found before generation 200 for most problems with both selection operators. This shows that even with a reduced number of generations G3P would yield similar results in many cases.

## 5.4   Invalids

As mentioned in Sect. 3, depending on the representation and system used new individuals might not compile, produce runtime errors or the execution has to be aborted as it takes too long to evaluate. Even with the concept described in Sect. 2, individuals may take too long to evaluate or may throw an exception, which will be called invalids. More than two thirds of the problems did not even produce one percent of invalid individuals summed over all runs on the problems tested in this study. For each of the two selection operations tested, only one problem produced more than two percent of invalids. Only a fraction of the invalids is due to runtime errors, like MemoryError and OverflowError which are more difficult to avoid. 0.1% of invalids are due to runtime errors on 90% of the problems. Most invalids are due to the execution limit of one second used for the experiments, which occur due to reaching the maximum limit of loop iterations for every single training case.

The small amount of invalids is due to the type safety of the grammars as well as the protected methods added to the skeleton discussed in Sect. 2, which makes the search more effective [7].

## 6    Conclusion and Future Work

In this paper, an approach to tackle general program synthesis problems with BNF grammars was presented that can create code from scratch instead of reusing existing programs by creating type-safe grammars that can be combined to fit the specification of a problem instance. These grammars can be used by most grammar-based GP systems, with no or only little implementation overhead. This approach allows researchers to easily reproduce experiments under similar conditions, even in their own grammar-based GP system.

Experiments were conducted on a suite of program synthesis problems to evaluate the system's performance. G3P with the grammar design introduced in Sect. 2 was able to achieve similar results as PushGP. G3P was able to find more correct solutions than PushGP on some of the problems and even solve "Super Anagrams", which has not been solved by PushGP at all. PushGP was able to find at least one solution on more problem instances than G3P. Both systems have the capabilities to solve arbitrary program synthesis problems without needing special changes for every single problem instance. As this paper only introduces a new approach and settings were chosen to be able to compare with PushGP, performance improvements are expected with new operators or more adjusted parameter settings. The experiments also confirmed the superiority of lexicase selection in the program synthesis domain with G3P.

As this is the first paper on reusing general grammars for program synthesis problems, many open questions remain, which are going to be addressed in the future. Can this approach be extended to adjust the number of variables dynamically? At the moment, parameter settings comparable to PushGP's settings were used to test this approach. How robust is this approach with other parameter settings and is it possible to improve the results? As the benchmark suite only required basic functionality to solve problems, no external libraries have been used to solve them. Experiments using external libraries to create more complex programs have to be conducted.

**Acknowledgments.** This research is based upon works supported by the Science Foundation Ireland, under Grant No. 13/IA/1850.

## References

1. Byrne, J., Cardiff, P., Brabazon, A., O'Neill, M.: Evolving parametric aircraft models for design exploration and optimisation. Neurocomputing **142**(0), 39–47 (2014). SI Computational Intelligence Techniques for New Product Development

2. Byrne, J., Fenton, M., Hemberg, E., McDermott, J., O'Neill, M.: Optimising complex pylon structures with grammatical evolution. Inf. Sci. (0), (2014). http://www.sciencedirect.com/science/article/pii/S0020025514002904

3. Castle, T., Johnson, C.G.: Evolving high-level imperative program trees with strongly formed genetic programming. In: Moraglio, A., Silva, S., Krawiec, K., Machado, P., Cotta, C. (eds.) EuroGP 2012. LNCS, vol. 7244, pp. 1–12. Springer, Heidelberg (2012). doi:10.1007/978-3-642-29139-5_1. http://www.cs.kent.ac.uk/pubs/2012/3202/content.pdf

4. Fenton, M., McNally, C., Byrne, J., Hemberg, E., McDermott, J., O'Neill, M.: Automatic innovative truss design using grammatical evolution. Autom. Constr. **39**(0), 59–69 (2014). http://www.sciencedirect.com/science/article/pii/S0926580513002124

5. Forstenlechner, S.: Github repository: HeuristicLab.CFGGP: provides context free grammar problems for HeuristicLab (2016). https://github.com/t-h-e/HeuristicLab.CFGGP. Accessed 24 Jan 2017

6. Gulwani, S.: Dimensions in program synthesis. In: Proceedings of 12th International ACM SIGPLAN Symposium on Principles and Practice of Declarative Programming, PPDP 2010, pp. 13–24. ACM, New York (2010). http://doi.acm.org/10.1145/1836089.1836091

7. Harman, M., Langdon, W.B., Jia, Y., White, D.R., Arcuri, A., Clark, J.A.: The GISMOE challenge: constructing the pareto program surface using genetic programming to find better programs (keynote paper). In: Proceedings of 27th IEEE/ACM International Conference on Automated Software Engineering (ASE), pp. 1–14, September 2012

8. Helmuth, T., Spector, L., Matheson, J.: Solving uncompromising problems with lexicase selection. IEEE Trans. Evol. Comput. **19**(5), 630–643 (2015)

9. Helmuth, T., Spector, L.: General program synthesis benchmark suite. In: GECCO 2015: Proceedings of 2015 on Genetic and Evolutionary Computation Conference, pp. 1039–1046. ACM, Madrid, 11–15 July 2015. http://doi.acm.org/10.1145/2739480.2754769

10. Hemberg, E., Ho, L., O'Neill, M., Claussen, H.: A comparison of grammatical genetic programming grammars for controlling femtocell network coverage. Genet. Program. Evol. Mach. **14**(1), 65–93 (2013). http://dx.doi.org/10.1007/s10710-012-9171-8

11. Koza, J.R., Andre, D., Bennett, F.H., Keane, M.A.: Genetic Programming III: Darwinian Invention and Problem Solving, 1st edn. Morgan Kaufmann Publishers Inc., San Francisco (1999)

12. Landin, P.J.: The next 700 programming languages. Commun. ACM **9**(3), 157–166 (1966). http://doi.acm.org/10.1145/365230.365257

13. Langdon, W.B., Harman, M.: Optimizing existing software with genetic programming. IEEE Trans. Evol. Comput. **19**(1), 118–135 (2015)

14. Langdon, W.B., Ochoa, G.: Genetic improvement: a key challenge for evolutionary computation. In: Li, Y. (ed.) Key Challenges and Future Directions of Evolutionary Computation, pp. 3068–3075. IEEE, Vancouver, 25–29 July 2016. http://www.cs.ucl.ac.uk/staff/W.Langdon/ftp/papers/langdon_2016_cec.pdf

15. Le Goues, C., Nguyen, T., Forrest, S., Weimer, W.: GenProg: a generic method for automatic software repair. IEEE Trans. Softw. Eng. **38**(1), 54–72 (2012). http://www.cs.virginia.edu/~weimer/p/weimer-tse2012-genprog.pdf

16. Loughran, R., McDermott, J., O'Neill, M.: Tonality driven piano compositions with grammatical evolution. In: IEEE Congress on Evolutionary Computation, CEC 2015, Sendai, Japan, 25–28 May 2015, pp. 2168–2175. IEEE (2015). http://dx.doi.org/10.1109/CEC.2015.7257152

17. McKay, R.I.B.: Fitness sharing in genetic programming. In: Proceedings of 2Nd Annual Conference on Genetic and Evolutionary Computation, GECCO 2000, pp. 435–442. Morgan Kaufmann Publishers Inc., San Francisco (2000). http://dl.acm.org/citation.cfm?id=2933718.2933800

18. McKay, R., Hoai, N., Whigham, P., Shan, Y., O'Neill, M.: Grammar-based genetic programming: a survey. Genet. Program. Evol. Mach. 11(3–4), 365–396 (2010). http://dx.doi.org/10.1007/s10710-010-9109-y

19. Montana, D.J.: Strongly typed genetic programming. Evol. Comput. 3(2), 199–230 (1995). http://dx.doi.org/10.1162/evco.1995.3.2.199

20. O'Neill, M., Nicolau, M., Agapitos, A.: Experiments in program synthesis with grammatical evolution: a focus on integer sorting. In: 2014 IEEE Congress on Evolutionary Computation (CEC), pp. 1504–1511, July 2014

21. O'Neill, M., Ryan, C.: Automatic generation of caching algorithms. In: Miettinen, K., Mäkelä, M.M., Neittaanmäki, P., Periaux, J. (eds.) Evolutionary Algorithms in Engineering and Computer Science, pp. 127–134. Wiley, Jyväskylä, 30 May–3 June 1999. http://www.mit.jyu.fi/eurogen99/papers/oneill.ps

22. O'Neill, M., Ryan, C.: Grammatical Evolution: Evolutionary Automatic Programming in an Arbitrary Language. Kluwer Academic Publishers, Norwell (2003)

23. Poli, R., Langdon, W.B., McPhee, N.F.: A field guide to genetic programming (2008). (With contributions by J.R. Koza). Published via http://lulu.com and freely available at http://www.gp-field-guide.org.uk

24. Spector, L.: Automatic Quantum Computer Programming: A Genetic Programming Approach, vol. 7. Kluwer Academic Publishers, Boston (2004)

25. Spector, L., Robinson, A.: Genetic programming and autoconstructive evolution with the push programming language. Genet. Program. Evol. Mach. 3(1), 7–40 (2002). http://hampshire.edu/lspector/pubs/push-gpem-final.pdf

26. Swan, J., Epitropakis, M.G., Woodward, J.R.: Gen-O-Fix: an embeddable framework for dynamic adaptive genetic improvement programming. Technical report CSM-195, Computing Science and Mathematics, University of Stirling, UK, 17 January 2014. http://www.cs.stir.ac.uk/~kjt/techreps/recent.html

27. Helmuth, T.M., L.S.: Detailed problem descriptions for general program synthesis benchmark suite. Technical report, School of Computer Science, University of Massachusetts Amherst (2015)

28. Whigham, P.A.: Grammatical bias for evolutionary learning. Ph.D. thesis, New South Wales, Australia (1996). AAI0597571

# Improving the Tartarus Problem
# as a Benchmark in Genetic Programming

Thomas D. Griffiths[✉] and Anikó Ekárt

Aston Lab for Intelligent Collectives Engineering (ALICE), Aston University,
Aston Triangle, Birmingham B4 7ET, UK
{grifftd1,a.ekart}@aston.ac.uk

**Abstract.** For empirical research on computer algorithms, it is essential
to have a set of benchmark problems on which the relative performance of
different methods and their applicability can be assessed. In the majority
of computational research fields there are established sets of benchmark
problems; however, the field of genetic programming lacks a similarly
rigorously defined set of benchmarks. There is a strong interest within
the genetic programming community to develop a suite of benchmarks.
Following recent surveys [7], the desirable characteristics of a benchmark
problem are now better defined. In this paper the Tartarus problem is
proposed as a tunably difficult benchmark problem for use in Genetic
Programming. The justification for this proposal is presented, together
with guidance on its usage as a benchmark.

**Keywords:** Genetic programming · Benchmark · Tartarus

## 1   Introduction

The Genetic Programming (GP) research community has recently recognised
that there is a lack of serious and structured benchmark test problems [3]. The
need to define and establish a suite of benchmark problems is becoming greater
as the popularity of GP increases and the field matures. The most prominent
step so far has been the use of community surveys to establish which problems
and tests are most popular and widely used [3,7]. There are certain attributes
and characteristics that all benchmark problems are expected to exhibit. In this
paper we analyse how these apply to Genetic Programming and argue that the
*Tartarus problem* satisfies all the relevant characteristics, and therefore can be
used as a reliable benchmark problem.

A benchmark is defined as a point of reference, against which given artifacts
can be compared and contrasted. In computer science, entire computer programs
or specific software components are often tested and compared using so-called
benchmark problems. Many fields in computer science have established suites of
benchmark problems on which proposed methods and approaches are tested.

As a research field matures, it is common for a suite of benchmark problems
to emerge. This is important for the field as it provides a standardised way

© Springer International Publishing AG 2017
J. McDermott et al. (Eds.): EuroGP 2017, LNCS 10196, pp. 278–293, 2017.
DOI: 10.1007/978-3-319-55696-3_18

to compare and measure the relative performance of different solution methods and approaches. When a new algorithm is developed and tested, its performance must be compared to other state-of-the-art algorithms within the field. The only practical way to do this is using a suite of benchmark problems.

For example, in evolutionary algorithms, there is a benchmark repository [4] to which newly created benchmarks are continuously added, via benchmarking competitions and special sessions at conferences. However, for GP specific problems this has not been the case, there is a lack of agreement within the GP community about which problems should be used. Often the problems on which new methods are demonstrated are outdated, with little applicability to real world situations, or they are so-called toy problems of trivial difficulty.

In this paper we propose that the Tartarus problem become adopted as a GP benchmark and we justify how it satisfies all required characteristics for a benchmark. The paper is organised as follows: Sect. 2 details the desirable characteristics of a GP benchmark problem, Sect. 3 outlines previous attempts at GP benchmarks and the issues with the usage of toy problems in GP. Section 4 justifies the usage of the Tartarus problem as a benchmark for GP and Sect. 5 draws conclusions.

## 2   Desirable GP Benchmark Characteristics

For many years there has been no agreement on what makes an effective benchmark in GP. More recently, surveys of the GP community by White et al. [7] have outlined some of the important characteristics that must be present across a suite of benchmark problems. The intention was not to create an exhaustive list of every desirable feature of any benchmark suite, but a list of the key features that must be present in order for the set of benchmark problems to be useful and effective for GP. These desirable characteristics are discussed below.

*Tunable Difficulty.* One of the most important characteristics of an effective benchmark problem is tunable difficulty. A problem is said to be tunably difficult if there are methods by which the difficulty of instances can be changed and altered relative to each other. This provides scope for the benchmark to be used across a wide range of GP methods, while maintaining comparability between the results. The creation of instances of increasing difficulty is essential in order to push the boundaries of current research [3].

*Precisely Defined.* A benchmark should be well-defined and documented, outlining the problem constraints and boundaries. It is common for a benchmark to be accompanied by a set of recommended resource constraints, such as an upper limit on the number of available evaluations or placing a time limitation on the program execution.

*Accommodating to Implementors.* In order for a benchmark problem to be accepted by the research community, it must be accommodating to the practitioners who implement it, and straight forward to use. The benchmark problem must be self-contained and all its elements must be open-source and accessible, to ensure universal access without the need for specific domain knowledge.

*Representation Independent.* An effective benchmark should attempt, as far as it is reasonably practical, to be representation independent in terms of the programming language used and the programming style. As the field of GP expands and matures, we expect that the number of programming languages and representations being used will increase. Benchmarks should be flexible enough to allow adaptation between various languages and representations, while still being effective. Attempts should be made to ensure that the benchmark does not rely on any specific attributes from a language.

*Easy to Interpret and Compare.* It is also important that the results generated by the benchmark are easy to interpret and can be compared without ambiguity. This clarity in understanding the results is vital, it allows for trends and relationships to be established between different sets of results, and reliable conclusions to be drawn on the data.

*Relevant.* A benchmark problem should contain elements which are directly relevant to the wider field in which it operates. Ideally for GP this would be the wider machine learning and optimisation communities. A well structured benchmark should not be vulnerable to paradigm shifts in the underlying application domain, rendering it irrelevant.

*Fast.* Due to the fact that GP individuals are programs which must be executed, often many times per individual, fitness evaluations can be slow. Therefore a GP benchmark should be fast enough to allow the large number of runs required to create meaningful comparisons between approaches, to be carried out in a reasonable time frame.

In order for a problem to be considered an effective benchmark candidate it must satisfy the majority of the aforementioned characteristics, combining them together to make an effective benchmark. However, many of the problems currently being used as de-facto benchmarks in GP only satisfy a small number of these characteristics. It is important to define a suite of benchmark problems, which collectively satisfy the entire range of desirable characteristics. The main benefit of having a suite of benchmark problems instead of using just a single problem is that it allows for different types of problems from various areas to be tested and the approaches compared. The range of problem domains also allows for the portability and scalability of a solution approach to be tested across the field of research.

## 3    GP Benchmarks

In order to demonstrate how GP worked, in 1992 Koza defined a set of test problems, including the *k-even parity, symbolic regression, artificial ant* and *lawnmower* problems [2]. Over the following 20 years these problems became widely used within the GP community, emerging as the de-facto problems for GP experimental research. More recently, however, the GP community has started to expand and has attempted to use a more varied set of test problems. As a consequence, there is a growing desire within the GP community to establish a suite of well-defined benchmark problems for universal use [7].

One of the main issues with Koza's problems, as outlined by Vanneschi et al. [6], is the fact that many of them lack rigorous evaluation methods, and more importantly, the majority of the problems have their own level of *fixed structural complexity* built into the problem. It is often hard to create several instances of these problems with a predictably varied level of complexity and difficulty. The inability to create a range of difficulties across instances by altering the parameters of the problem is a major drawback for many of Koza's problems. Many of these problems are referred to as 'toy problems' as they are usually simple in nature and are trivially easy to solve using modern GP methods (as well as manually, by a human).

It has been said that 'GP has a toy problem problem' and as a field of research it is often too reliant on these simple, trivially easy problems. It has been argued that the use of toy problems as de-facto benchmarks does little to advance the field, and may in fact be responsible for holding it back in some regards [3]. As mentioned previously, one of the main drawbacks of using toy problems as de-facto benchmarks is that many of the instances may be disproportionate to each other in terms of difficulty. This can lead to poor coverage of the problem space in which they operate, limiting their effectiveness as a benchmark [1]. Therefore, the results obtained from toy problems may be misleading and create the illusion of success, when the relative performance of an approach may actually be measurably worse. In order to have practical significance it is often important for a benchmark to be able to simulate some aspects of real world problems. However, this is one area in which toy problems are rather weak: they are often unreliable at predicting the success of a methodology or solution in real world situations [8].

## 3.1   The Lawnmower Problem

One of Koza's test problems which has been widely used is the *Lawnmower problem* [10]. In this simple problem a $n \times m$ toroidal grid representing a lawn of grass and a controllable agent representing a lawnmower are given. The essence of the problem is to find a program to control the movement of the lawnmower so that it traverses the entire lawn. The lawnmower has a state consisting of its current location in the $n \times m$ grid and its current orientation (North, East, South or West). Figure 1 shows examples of the Lawnmower problem of size $10 \times 10$.

The shaded area shows the traversed lawn, while the line within indicates the actual trajectory taken. The lawnmower has no sensors with which to collect information from its surroundings. This lack of sensory information has little impact on the agent, as there are no obstacles in the environment and the toroidal nature of the grid effectively removes any boundaries which the agent might

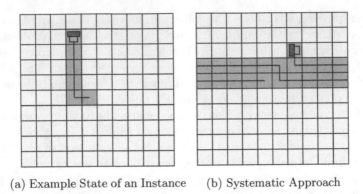

(a) Example State of an Instance    (b) Systematic Approach

**Fig. 1.** The Lawnmower problem

otherwise have encountered. There are four operations which are used to change the state of the lawnmower: *turn left, turn right, move forward one square* and *move forward k squares.*

The Lawnmower problem has been used to test and prove the effectiveness of different GP methods and representations. Current GP methods are very effective and can solve the problem easily. For this problem the only way in which the difficulty can be changed is by altering the size of the instance. Given that there are no constraints to satisfy, an increase in the size of the problem results in a proportional increase in the time taken to solve the problem. Often solutions to the Lawnmower problem can simply be scaled up along with the instance size with no increase in complexity.

The solutions that are most successful at solving the Lawnmower problem tend to be systematic in their approach, an example of which is shown in Fig. 1(b). They repeatedly execute a relatively small set of instructions until the problem's termination criteria are met, an example GP solution that leads to the simple systematic approach used in Fig. 1(b) is shown in Fig. 2. There are two reasons why systematic controllers are successful at solving the Lawnmower problem. Firstly due to the fact that the grid is toroidal in nature, a solution can

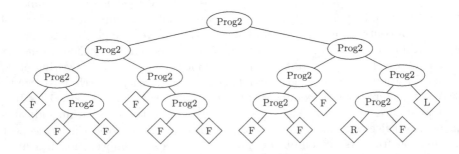

**Fig. 2.** Simple systematic solution to the Lawnmower problem

move freely in any direction without being impeded. This allows for a randomly generated solution, which would otherwise have become stuck against a wall or corner, to exploit the toroidal nature of the grid and generate adequate solutions. Secondly, there is little external information available to the lawnmower, and no information relating to the state of its surrounding environment, and in fact no need for such information.

The Lawnmower problem can be seen as part of a wider family of grid-based problems, including the *Maze problem* and the *Tartarus problem*. There are many parallels that can be drawn across these, they all operate in a grid environment and involve an agent following some path, but according to different objectives:

- *Lawnmower Problem*   The objective of the agent is to cover as much of the lawn as possible.
- *Maze Problem*   The objective of the agent is to find a trajectory from one position to another, given obstacles (walls).
- *Artificial Ant Problem*   The objective of the agent is to locate all food pellets on a trajectory, possibly with gaps.
- *Tartarus Problem*   The objective of the agent is to explore the environment in order to find the blocks. The agent must then move the found blocks to prescribed areas of the grid.

The Lawnmower problem in its default setting is clearly the simplest of the three problems. It is possible however to alter the problem definition slightly and add obstacles, increasing the complexity of the problem instance. Another method of increasing the complexity of the Lawnmower problem is to use a non-toroidal grid. Removing the toroidal nature of the grid will force the agent to take the edges of the grid into consideration when creating solutions. This new constrained Lawnmower problem becomes a special case of the Maze problem where there are multiple potential start and finish locations. The Maze problem can be seen as a generalisation of a constrained Lawnmower problem. On the other hand, the Tartarus problem involves locating blocks and moving them to the sides of the grid, which is similar to finding routes in the maze, while picking up blocks on the way. The added complexity comes from the fact that the locations of the blocks are not known to the agent. The problems do differ however in terms of their applicability to real world situations.

In unconstrained problems such as the lawnmower, with limited available external information, it is hard to create a strategic controller that is able to strategically plan ahead the actions of the agent. Therefore, systematic controllers become dominant in terms of performance. However as the problems become more constrained (i.e. via the addition of obstacles) strategic controllers become more appropriate, utilising the higher level of external information available to them to create successful and effective solutions, as shown in Fig. 3.

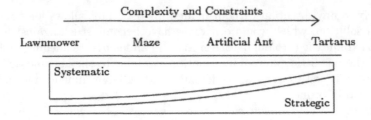

Fig. 3. Systematic and strategic controllers for grid-based problems

## 4    The Tartarus Problem

The Tartarus problem, originally introduced by Teller [5] is a grid-based optimisation problem. In the Tartarus problem a $n \times n$ non-toroidal grid representing an enclosed environment, a number of movable blocks and a controllable agent representing a bulldozer are given. The aim of the problem is to locate and move the blocks to the edge of the environment. Unlike in the Lawnmower problem, the agent in the Tartarus problem has no initial knowledge of its location in the grid or its orientation. The agent does however, have eight sensors, which are able to sense the environment in the surrounding eight gridsquares. There are three operations which can be used to change the state of the dozer: *turn left, turn right* and *move forwards one square.*

The canonical Tartarus instance, an example initial state of which is shown in Fig. 4(a), consists of a $6 \times 6$ grid, six blocks and one dozer. The blocks and the dozer must be placed in a valid start position in the central $4 \times 4$ gridsquares, shown by the shaded area in Fig. 4(b). During execution of a $6 \times 6$ instance, the agent is allowed a total of 80 moves to attempt to create a viable solution, with each of the three operations counting as an individual move. The final state of a successfuly completed Tartarus instance, where all of the blocks have been pushed to the edges of the grid, is shown in Fig. 4(c).

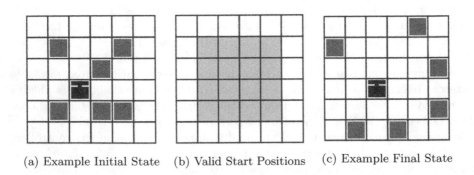

(a) Example Initial State    (b) Valid Start Positions    (c) Example Final State

Fig. 4. Example Tartarus states

## 4.1  Satisfying the Desirable Benchmark Characteristics

With the modified state evaluation proposed in Sect. 4.3, the Tartarus problem satisfies all of the characteristics outlined in Sect. 2:

*Tunable Difficulty.* The difficulty of the Tartarus problem can be altered by changing the parameters of the problem (grid size, number of blocks) and the restrictions placed upon the agent (number of allowed moves). This leads to a predictable shift in the complexity and difficulty of the generated instance. A detailed explanation of this process and guidelines on the tuning of the difficulty in the Tartarus problem is outlined in Sect. 4.6.

*Precisely Defined.* The Tartarus problem is a constrained problem with well defined boundaries for both the problem instance and the agent. The original problem definition outlined a maximum number of allowed moves, in this paper we suggest minimum and maximum constraints for allowed number of moves, number of blocks and the size of the instance, providing a comprehensive list of the operating constraints for the Tartarus problem.

*Accommodating to Implementors.* Due to the fact that the Tartarus problem is a grid based problem it is simple to implement and use. There are no specialist skills or software tools that are required to create and use an implementation of the Tartarus problem.

*Representation Independent.* As the implentation of the Tartarus problem is simple, it does not require any specific languages or software tools. The problem is representation independent and it should be possible to implement in any modern programming language and paradigm.

*Easy to Interpret and Compare.* In order to increase the ease with which Tartarus problem instances and solutions could be interpreted and compared we suggest an improved method of state evaluation in Sect. 4.3. The improved method of state evaluation allows for a more fine-grained evaluation of instances, allowing for a more accurate comparison to be made.

*Relevant.* The Tartarus problem has scope for real applications outside of GP. For example, the recently announced Emergency Robots competition[1] (building upon the success of EuRathlon[2]), inspired by the 2011 Fukushima accident focuses on realistic emergency response scenarios. In these scenarios missing workers have to be found, critical hazards have to be identified in limited time.

---

[1] http://eu-robotics.net/robotics_league.

[2] http://www.eurathlon.eu. An outdoor robotics challenge for land, sea and air.

*Fast.* Given the improved method of state evaluation for the Tartarus problem outlined in this paper, together with the simple nature of the implementation, tests can be carried out and meaningful comparisons can be made in a reasonable time frame. The fitness evaluation is fast, allowing for multiple executions per individual to be carried out.

## 4.2   Current State Evaluation

For the Tartarus problem a solution consists of a series of instructions, which control the actions of the agent in the environment. In order to test the efficacy of a solution, there needs to be a way to measure its outcome on an instance environment. This is usually done by evaluating the end position (state) after executing the complete series of instructions. In fact, the same evaluation method can be used to evaluate any state, for different purposes:

– evaluate the initial state, in order to evaluate the problem instance
– evaluate an intermediate state, after a set period of time or number of moves, in order to measure progress
– evaluate the end state in order to evaluate solution quality.

The current method of state evaluation only rewards the number of blocks, which are located at the edges of the grid. This rather binary success or fail approach works well for many benchmark problems where the absolute score achieved by a candidate solution is the only desired success measure. However, for GP, rewarding part-way solutions is essential during evolution, so that better solutions can evolve. For example, the cluster of Fig. 5(a) is very different from the dispersed blocks of Fig. 5(b). Both these states would have the same evaluation score of zero. The dispersed blocks example in Fig. 5(b) is visibly closer to an optimal end state when compared to the cluster example in Fig. 5(a) as the blocks are closer to the edges of the grid. Specifically, from the state in Fig. 5(a) 32 moves would be required to move all blocks to an edge position, while to do the same from the state in Fig. 5(b) 27 moves would be needed. Therefore

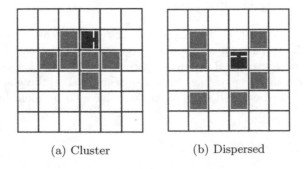

(a) Cluster                (b) Dispersed

**Fig. 5.** Example Tartarus states

a solution that arrived at the latter position had made more progress than one that arrived at the earlier position.

The current binary assignment of success or fail is too coarse. It misses differences in performance between candidate solutions in the population. Solutions that move no blocks at all would be evaluated the same as solutions that make some progress, but fall short of actually moving a block to the edge of the grid (see Fig. 5(b)). Solutions that have not actually moved blocks to the edges of the grid could still have made some progress by moving blocks *closer* to the edges and this should be recognised and rewarded in some manner. According to the same rationale, initial states with the same number of blocks are not equivalent, as some require less moves than others to solve. Therefore the difficulty of the problem instance is not only dependent on the grid size, number of blocks or allowed number of moves, but also the initial distribution of blocks on the grid.

### 4.3   Proposed Improved State Evaluation

We propose a new evaluation method that rewards states according to how close they are to the desired final states, by including how close to the edge each block in the given state is:

$$
State\_value = C_1 \cdot \left( B - \frac{2}{n} \sum_{i=1}^{B} d_i - C_2 \right) \tag{1}
$$

where $B$ is the total number of blocks, $n$ is the size of the grid and $d_i$ is the distance of block $i$ from an edge in the given problem instance. $C_1$ and $C_2$ are scaling and translation constants based on $B$ and $n$ in order to make the value range consistent with the current evaluation method. For a $6 \times 6$ instance with 6 blocks $C_1 = 1.8$ and $C_2 = \frac{8}{3}$.

We generated 1000 random states for grid size $6 \times 6$ containing 6 blocks and evaluated them with the two methods (there was no need to generate candidate solutions, as the candidate solutions would be evaluated on the basis of the end position that they lead to). In Table 1 we present 30 examples of the state evaluation. As shown, the new method of state evaluation allows for a more fine-grained evaluation.

**Table 1.** Fitness data comparing evaluation methods

| Individual | 1 | 2 | 3 | 4 | 5 | 6 | 7 | 8 | 9 | 10 | 11 | 12 | 13 | 14 | 15 |
|---|---|---|---|---|---|---|---|---|---|---|---|---|---|---|---|
| Old value | 0.0 | 0.0 | 0.0 | 1.0 | 1.0 | 1.0 | 1.0 | 1.0 | 1.0 | 1.0 | 1.0 | 1.0 | 1.0 | 1.0 | 2.0 |
| New value | 1.2 | 1.2 | 1.8 | 0.6 | 1.8 | 1.8 | 1.8 | 1.8 | 2.4 | 2.4 | 2.4 | 2.4 | 2.4 | 3.0 | 3.0 |
| Individual | 16 | 17 | 18 | 19 | 20 | 21 | 22 | 23 | 24 | 25 | 26 | 27 | 28 | 29 | 30 |
| Old value | 2.0 | 2.0 | 2.0 | 2.0 | 2.0 | 2.0 | 3.0 | 3.0 | 3.0 | 3.0 | 3.0 | 3.0 | 4.0 | 4.0 | 4.0 |
| New value | 3.0 | 3.0 | 3.0 | 3.6 | 3.6 | 3.6 | 3.0 | 3.6 | 3.6 | 4.2 | 4.2 | 4.2 | 3.6 | 4.2 | 4.8 |

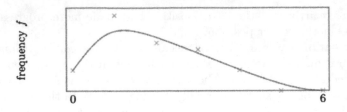

**Fig. 6.** Current evaluation distribution

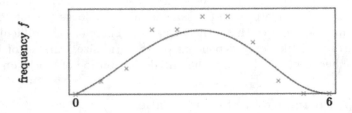

**Fig. 7.** Proposed evaluation distribution

The distributions of the results are shown in Figs. 6 and 7. The current evaluation method rates all states with no blocks at the edges as equally poor, when in reality some can lead to end states that are just a few moves away from optimal end states. The proposed evaluation method rewards these. The distribution in Fig. 7 better reflects the actual performance, at a much more fine grained level.

### 4.4   Baseline Values for Tartarus Instances

A baseline is a set of instances that are generated corresponding to a set of values for the parameters defining difficulty. This is structured to allow clean, controlled comparison of difficulty between generated instances. In the Tartarus problem the difficulty increases with size, therefore when comparing difficulty of problem instances, by fixing the number of moves and blocks, the only variable altered in the tuning of the baseline difficulty is the grid size $n$. When establishing the functions for the number of moves $m(n)$ and the number of blocks $B(n)$, we endeavour to maintain the difficulty relative to grid size.

An actual solution will need to include moving to a block, then pushing the block to an edge and repeating these steps for all blocks, taking into account obstacles (non-movable blocks). Let us consider the step of pushing a block to an edge. The expected distance of a block randomly placed on the inner grid of size $n - 2$ to an edge is:

$$\text{Dist} = \frac{1}{(n-2)^2} \sum_{i=1}^{\frac{n-2}{2}} 4(n-1-2i)i = \frac{n(n-1)}{6(n-2)}, \tag{2}$$

where $4(n-1-2i)$ is the number of positions on the inner grid of size $n-2$ that are at distance $i$ from an edge of the grid of size $n$. For grid size 6, the agent can expect to have to move 1.25 for each block, and the number of moves increasing with grid size will be 1.55 for grid size 8, and 4.35 for grid size 25. At the same time, if the agent *does not know the direction in which to move* to get to the closest edge, the expected distance to have to push a block randomly placed on the grid at location $(x, y)$ to any edge becomes:

$$\text{Dist} = \frac{x + (n-1-x) + y + (n-1-y)}{4} = \frac{n-1}{2}. \tag{3}$$

Comparatively, for grid size 6, the number of moves becomes 2.5, for grid size 8, 3.5 and for grid size 25, 12. The fraction of the grid that the agent can be expected to travel to reach the edge will be:

$$\frac{\text{Dist}}{n} = \frac{n-1}{2n}. \tag{4}$$

For grid size 6 the agent without global vision can expect to have to travel a proportion of 0.416 of the grid size $n$ in order to move one block to the edge (not accounting for obstacles and travelling from the edge after successful move of one block to the next block). As $n$ increases, this proportion approaches 0.5. A size $6 \times 6$ instance is thus solvable with less effort than a $7 \times 7$ instance (0.429) and substantially less effort than an instance of size $25 \times 25$ (0.48). This does not account for the occurrence of impossible to move blocks in an instance. We can conclude that the effort expected to move one block to the edge is increasing with the increase in grid size. So far the tunability of Tartarus is similar to the tunability of the Lawnmower problem. However, when the number of blocks and their location is considered, more fine tuning becomes possible with the Tartarus problem.

The number of blocks in the environment which must be moved, $B$, contributes to the overall difficulty of the problem instance. A baseline was determined through regression, following generation and evaluation of a set of 1000 instances each of the sizes (6,7,8,16) and with varying numbers of blocks, but fixed numbers of moves (80,109,142,569). This baseline with $C = \frac{2}{9}$ is:

$$B_{baseline} = \left\lceil (n-1)^2 \cdot C \cdot \left( \frac{3}{2} - \frac{n-1}{2n} \right) \right\rceil. \tag{5}$$

The number of allowed moves $m$ can be used to tune the difficulty of the Tartarus problem. It makes sense to link this resource limitation to the grid size, as the number of steps required for simply traversing the grid, or moving one block to an edge, depends on the grid size. In order to establish the relationships between the problem parameters and produce reliable results, we determined that the following quadratic function would be suitable:

$$m_{baseline} = \left\lceil 10\, C \cdot n^2 \right\rceil. \tag{6}$$

## 4.5    Generating Tartarus Instances

When generating Tartarus instances it is important to note that not all generated instances can be solved perfectly. As the placement of the blocks in the instance is random, it is likely that some instances may contain blocks, which are impossible to move, making the instance partly or completely impossible to solve. The two most common of these configurations are shown in Fig. 8(a) and (b). In the Wilson configuration in Fig. 8(a), although it is possible to move some of the blocks, doing so will create a cluster of four blocks which cannot be moved any further. An example of an instance containing a four-block cluster is outlined in Fig. 8(b). Although the four block square cluster is the most common situation in a 6 × 6 instance, it is also possible for all six blocks to be placed together in a non-movable cluster.

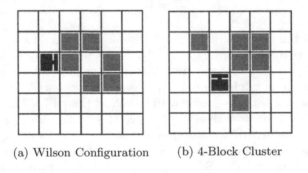

(a) Wilson Configuration     (b) 4-Block Cluster

**Fig. 8.** Partially solvable Tartarus instances

In previous work on the Tartarus problem [11–13], the authors identified the instances that contained unmovable block configurations, and removed them from the study. Their argument for removing the impossible to solve instances was that $\frac{2}{3}$ of all the blocks in the instances were impossible to move and studying these would not lead to solutions. However, with the increase in size of the problem, the impact of the four-block clusters becomes more negligible. For example, in the case of a 32 × 32 instance with over 200 blocks, the presence of 5 four-block clusters still allows for a substantial proportion of blocks to be moved. It is for this reason that we advise against discarding Tartarus instances which are partially solvable for instances above the canonical size of 6 × 6.

We next identified a reasonable working range of parameters for the Tartarus problem based on a set of 1000 generated instances. Beyond this range the results that are produced are of little to no value, due to the fact that the instances become too easy or too hard to reasonably solve. For example, a Tartarus instance with $m = 20$ allowed moves would be extremely difficult, if not impossible, to reliably find an effective solution. Therefore a set of *baseline* values were established. It is recommended that these be used when creating instances of the Tartarus problem. We located a usable working range of approximately $\pm 25\%$ around the baseline values. It is proposed that Table 2 should be used as a reference when tuning the individual difficulty of a Tartarus instance, in order to maintain a relative level of practicality and usability for any instances created.

**Table 2.** Reference guide for generating Tartarus instances

| $n$ | Moves | | | Blocks | | |
|---|---|---|---|---|---|---|
| | $-25\%$ | $m_{baseline}$ | $+25\%$ | $-25\%$ | $B_{baseline}$ | $+25\%$ |
| 6 | 60 | 80 | 100 | 4 | 6 | 8 |
| 7 | 82 | 109 | 136 | 7 | 9 | 11 |
| 8 | 107 | 142 | 177 | 9 | 12 | 15 |
| 16 | 427 | 569 | 711 | 39 | 52 | 65 |
| 32 | 1707 | 2276 | 2845 | 163 | 217 | 271 |

## 4.6  Tuning Difficulty

In addition to the established baseline for difficulty for each given grid size $n$, the difficulty can be adjusted in a more fine grained manner to achieve a more complete set of problem instances of different levels of increasing difficulty. This can be achieved by modifying the number of blocks $B$ and the number of moves $m$ away from the baseline values. It is best to systematically make changes and follow a Design of Experiments approach [9]. We propose a generic method to estimate the relative difficulty of a Tartarus instance, outlined in Eq. 7:

$$
D = \begin{cases} 0.5 \cdot \dfrac{m_{baseline}}{m} + 0.5 \cdot \dfrac{B_{baseline}}{B} + \dfrac{B_I}{B} & \text{if} \quad B_I < B \\ \textbf{impossible} & \text{if} \quad B_I = B \end{cases} \tag{7}
$$

where $m_{baseline}$ is the baseline number of moves defined in Eq. 6, $m$ is the user set number of allowed moves, $B_{baseline}$ is the baseline number of blocks defined in Eq. 5, $B$ is the user set number of blocks, $B_I$ is the number of impossible-to-move blocks.

The formula ensures a clear separation between instances that only contain impossible-to-move blocks and instances that have some movable blocks. In the case when there are movable blocks, it accounts equally for the relative changes in number of moves away from the baseline and number of blocks away from the baseline. It also factors in the difficulty added by having some impossible-to-move blocks. It has been designed to indicate difficulty of approximately 1 for baseline values. The value 1 cannot be ensured with exactness due to the small number of impossible-to-move blocks that could be included in randomly generated problem instances.

Table 3 provides a range of settings for difficulty levels from very easy to very hard for each grid size. These examples reflect how reducing the number of allowed moves makes an instance harder and also how this coupled with reducing the number of blocks could make an instance even harder (intuitively it takes longer to locate blocks which are spaced further apart).

In order to demonstrate equivalent difficulty levels across different grid sizes, we ran 100 experiments of standard linear GP for each grid size and corresponding difficulty level according to Table 3. For every difficulty level, we performed the experiments on randomly generated problem instances with the

**Table 3.** Example instances and their difficulty

|        |     | Very easy | Easy | Baseline | Hard | Very hard |
|--------|-----|-----------|------|----------|------|-----------|
| 6 × 6  | $m$ | 96        | 88   | 80       | 66   | 55        |
|        | $B$ | 7         | 7    | 6        | 6    | 6         |
|        | $D$ | 0.85      | 0.88 | 1        | 1.11 | 1.23      |
| 8 × 8  | $m$ | 170       | 156  | 142      | 116  | 104       |
|        | $B$ | 14        | 13   | 12       | 11   | 11        |
|        | $D$ | 0.85      | 0.92 | 1        | 1.11 | 1.22      |
| 16 × 16| $m$ | 704       | 662  | 569      | 486  | 422       |
|        | $B$ | 58        | 55   | 52       | 50   | 47        |
|        | $D$ | 0.85      | 0.90 | 1        | 1.11 | 1.23      |

proposed number of moves and number of blocks. The instructions *left, right, forward,* crossover rate 0.5, mutation rate 0.05, population size 100 over 50 generations were used. A snapshot of the fitness values at the end of the runs is presented in Fig. 9. The confidence intervals at 95% confidence level ranged between 0.03 and 0.06. This reinforces the hypothesis that as the difficulty and grid size increase, the quality of solutions and associated fitness obtained with equivalent resources decrease.

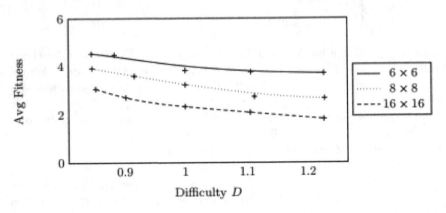

**Fig. 9.** Average fitness vs difficulty in linear GP experiments

## 5  Conclusion

In this paper we proposed using the Tartarus problem for benchmarking purposes in GP. We first proposed a more fine grained fitness evaluation measure. Then we demonstrated its tunable difficulty character, introduced a method to estimate the difficulty of instances and proposed suitable ranges of parameters to tune the difficulty. We are planning to publish methods to automatically generate suites of problem instances of desired varied difficulty levels.

A remaining aspect of an effective benchmark to be satisfied is to explore in more detail how the Tartarus problem can cross the gap between the theoretical and practical domains. This is one essential feature distinguishing between toy problems and effective benchmarks. We are confident that real life robotics scenarios (such as handling dangerous equipment) can be found where theoretical results can be directly applied. Therefore future work should focus upon ways in which results of the Tartarus problem can be extended to a real world context. Once this is achieved, the Tartarus problem will be set apart from other seemingly easy toy problems within GP.

# References

1. Korkmaz, E.E., Üçoluk, G.: Design and usage of a new benchmark problem for genetic programming. In: Yazıcı, A., Şener, C. (eds.) ISCIS 2003. LNCS, vol. 2869, pp. 561–567. Springer, Heidelberg (2003)
2. Koza, J.R.: Genetic Programming: On the Programming of Computers by Means of Natural Selection (1992)
3. McDermott, J., White, D.R., Luke, S., Manzoni, L., Castelli, M., Vanneschi, L., Jaskowski, W., Krawiec, K., Harper, R., De Jong, K., O'Reilly, U.M.: Genetic programming needs better benchmarks. In: Soule, T., et al. (eds.) Proceedings of the 14th International Conference on Genetic and Evolutionary Computation, GECCO 2012, pp. 791–798 (2012)
4. Sendhoff, B., Roberts, M., Yao, X.: Evolutionary computation benchmarking repository. IEEE Comput. Intell. Mag. 1, 50–60 (2006)
5. Teller, A.: The evolution of mental models. In: Kinnear Jr. K.E. (ed.) Advances in Genetic Programming, pp. 199–217 (1994)
6. Vanneschi, L., Castelli, M., Manzoni, L.: The K landscapes: a tunably difficult benchmark for genetic programming. In: Krasnogor, N., et al. (eds.) Proceedings of the 13th Annual Conference on Genetic and Evolutionary Computation, GECCO 2011, pp. 1467–1474 (2011)
7. White, D.R., McDermott, J., Castelli, M., Manzoni, L., Goldman, B.W., Kronberger, G., Jaśkowski, W., O'Reilly, U.M., Luke, S.: Better GP benchmarks: community survey results and proposals. Genet. Program Evolvable Mach. 14(1), 3–29 (2013)
8. Woodward, J., Martin, S., Swan, J.: Benchmarks that matter for genetic programming. In: Woodward, J., et al. (eds.) 4th Workshop on Evolutionary Computation for the Automated Design of Algorithms, GECCO 2014, pp. 1397–1404 (2014)
9. Jiju, A.: Design of Experiments for Engineers and Scientists. Elsevier, Amsterdam (2003)
10. Koza, J.R.: Scalable learning in genetic programming using automatic function definition. In: Kinnear Jr. K.E. (ed.) Advances in Genetic Programming, pp. 99–117 (1994)
11. Ashlock, D., Joenks, M.: ISAc lists: a different program induction method. In: Koza, J.R., et al. (eds.) Proceedings of the Second Annual Conference on Genetic Programming, pp. 18–26 (1998)
12. Ashlock, D., Freeman, J.: A pure finite state baseline for Tartarus. In: Proceedings of the 2000 Congress on Evolutionary Computation, pp. 1223–1230 (2000)
13. Dick, G.: An effective parse tree representation for Tartarus. In: Blum, C., et al. (eds.) Proceedings of the 15th Annual Conference on Genetic and Evolutionary Computation, GECCO 2013, pp. 1397–1404 (2013)

# A New Subgraph Crossover for Cartesian Genetic Programming

Roman Kalkreuth[(✉)], Günter Rudolph, and Andre Droschinsky

Department of Computer Science, TU Dortmund University, Dortmund, Germany
{Roman.Kalkreuth,Guenter.Rudolph,Andre.Droschinsky}@tu-dortmund.de

**Abstract.** While tree-based Genetic Programming is often used with crossover, Cartesian Genetic Programming is mostly used only with mutation as genetic operator. In this paper, a new crossover technique is introduced which recombines subgraphs of two selected graphs. Experiments on symbolic regression, boolean functions and image operator design problems indicate that the use of the subgraph crossover improves the search performance of Cartesian Genetic Programming. A preliminary comparison to a former proposed crossover technique indicates that the subgraph crossover performs better on our tested problems.

**Keywords:** Cartesian Genetic Programming · Crossover · Recombination

## 1  Introduction

Genetic Programming (GP) as popularized by Koza [4–6] uses syntax trees as program representation. Cartesian Genetic Programming (CGP) as introduced by Miller et al. [10] offers a novel graph-based representation which in addition to standard GP problem domains, makes it easy to be applied to many graph-based applications such as electronic circuits, image processing, and neural networks. CGP is often used only with mutation as genetic operator. The reason for this is that previous work on crossover in CGP showed ambivalent results which will be surveyed in this paper.

Tree-based GP was originally introduced with a crossover technique which swaps randomly chosen sub-branches of the parent trees to produce new offsprings. Koza considered crossover as the dominant genetic operator as a result of his experiments [5,6]. However, later research with more comprehensive and detailed experiments found that the beneficial effects of crossover can not be generalised in GP [7,8,17].

In contrast to comprehensive knowledge about crossover in tree-based GP, the state of knowledge in CGP appears to be ambiguous and ambivalent. Furthermore, the potential and understanding of crossover in CGP seems to be an open and remaining question. As a first step, we introduce a new method for crossover in CGP. Our new method recombines subgraphs of two parent graphs and preserves active paths by preventing the activation of passive nodes.

© Springer International Publishing AG 2017
J. McDermott et al. (Eds.): EuroGP 2017, LNCS 10196, pp. 294–310, 2017.
DOI: 10.1007/978-3-319-55696-3_19

Our crossover technique is tested on symbolic regression, boolean function, and image operator design problems. The results of our experiments show that it improves the search performance of CGP for these problems. Furthermore, in comparison to a former proposed crossover technique, the subgraph crossover performs better on our tested problems.

Section 2 of this paper describes CGP briefly and surveys previous work on crossover in CGP. In Sect. 3 we introduce our new crossover technique. Section 4 is devoted to the experimental results and the description of our experiments. In Sect. 5 we discuss the results of our experiments. Finally, Sect. 6 gives a conclusion and outlines future work.

## 2  Related Work

### 2.1  Cartesian Genetic Programming

Cartesian Genetic Programming is a form of Genetic Programming which offers a novel graph-based representation. In contrast to tree-based GP, CGP represents a genetic program via genotype-phenotype mapping as an indexed, acyclic, and directed graph. Originally the structure of the graphs was a rectangular grid of $N_r$ rows and $N_c$ columns, but later work also focused on a representation with one row. The genes in the genotype are grouped, and each group refers to a node of the graph, except the last one which represents the outputs of the phenotype. Each node is represented by two types of genes which index the function number in the GP function set and the node inputs. These nodes are called *function nodes* and execute functions on the input values. The number of input genes depends on the maximum arity $N_a$ of the function set. The last group in the genotype represents the indexes of the nodes which lead to the outputs. A backward search is used to decode the corresponding phenotype. The backward search starts from the outputs and processes the linked nodes in the genotype. In this way, only active nodes are processed during the evaluation procedure. The number of inputs $N_i$, outputs $N_o$, and the length of the genotype is fixed. Every candidate program is represented with $N_r * N_c * (N_a + 1) + N_o$ integers. Even when the length of the genotype is fixed for every candidate program, the length of the corresponding phenotype in CGP is variable which can be considered as a significant advantage of the CGP representation.

CGP is traditionally used with a $(1 + \lambda)$ evolutionary algorithm. The new population in each generation consists of the best individual of the previous population and the $\lambda$ created offspring. The breeding procedure is mostly done by a point mutation which swaps genes in the genotype of an individual in the valid range by chance.

### 2.2  Previous Work on Crossover in CGP

First attempts of crossover in CGP included four variations of crossover which were tested on the simple regression problem $x^2 + 2x + 1$. Clegg et al. [2] reported that all four techniques failed to improve the convergence of CGP.

Compared to running CGP with mutation only, the addition of these crossover techniques hindered the performance. The four techniques were tested on the standard integer-based representation of CGP. For instance, the genetic material was recombined by swapping parts of the genotypes of the parent individuals or randomly exchanging selected nodes. Clegg et al. [2] reported that merely swapping the integers (in whatever manner) in the CGP representation disrupts the performance.

This was the motivation for the introduction of a real-valued representation and new crossover technique for CGP by Clegg et al. [2]. The real-valued representation of CGP represents the directed graph as a fixed length list of real-valued numbers in the interval [0,1]. The genes are decoded to the integer-based representation by their normalisation values (e.g. number of functions or maximum input range). The recombination of two genotypes is performed by an arithmetic crossover with a random weighting factor which can also be found in the field of real-valued Genetic Algorithms. Clegg et al. showed that the new representation in combination with crossover improves the convergence behaviour of CGP. However, for the latter generations, Clegg et al. demonstrated that the use of crossover in real-valued CGP disrupts the convergence on one of the two tested problems. The improved convergence of the arithmetic crossover was evaluated in the domain of symbolic regression and has been found useful in this problem domain [2].

Slaný and Sekanina [13] analysed the fitness landscapes of functional-level CGP on image operator design problems including single and multipoint crossover operators. It was demonstrated that the mutation operator and the single-point crossover operator generate the smoothest landscapes for the tested problems.

For a multi-chromosome approach to CGP, Walker et al. [15] investigated a multi-chromosome crossover operator which joins the best chromosome parts from all individuals. This crossover technique was found useful for problems with multiple outputs and independent fitness assignment.

Another positive effect of crossover in CGP was obtained by the use of an implicit context representation for CGP in which recombination is useful for the Even Parity-3 problem [1].

CGP has been extended for the Automatic Definition and reuse of Functions by Walker and Miller [14] and Kaufmann and Platzner [3]. Kaufmann and Platzner adopted the module creation mechanisms for a cone- and age-based CGP crossover [3]. Cone-based crossover showed good results for functions with repetitive inner patterns, while age-based crossover excel for randomized inner structures.

To our best knowledge, the arithmetic crossover seems to be the only approach which aims at standard CGP and has been able to demonstrate a better convergence if two chromosomes are recombined directly. However, for the harder problem of the two tested symbolic regression problems, Clegg et al. concluded that the arithmetic crossover technique does not have a good effect on convergence. Furthermore, it was concluded that the arithmetic crossover causes occasional runs with a huge number of generations to converge. This has been the motivation for the development of a new crossover technique.

# 3  The Proposed Method

The proposed subgraph crossover for CGP is inspired by the subtree crossover found in tree-based GP. However, a directed acyclic graph enables more connections between the nodes, merely choosing one crossover point within the graph is not sufficient. Furthermore, to recombine subgraphs by just swapping parts of the genotype would be a disastrous approach according to the reportings by Clegg et al. Our approach to recombine two directed acyclic graphs is performed by respecting the CGP phenotype. The phenotype of each individual is represented by the active path of the graph and is determined through the evaluation process. Furthermore, the active path of a graph leads to the semantic value of a certain individual in CGP. As a consequence, we exclusively want to recombine the genetic material of the active paths.

For the description of the subgraph crossover procedure, let $N_{\text{inputs}}$ be the predefined number of input nodes. In CGP, the nodes are indexed from 0 to $N - 1$ and the input nodes of each graph are indexed from 0 to $N_{\text{inputs}} - 1$. The nodes which lie between the input and output nodes are denoted as function nodes. The crossover is done with two parents which are denoted as $P_1$ and $P_2$. For the crossover procedure, the node numbers of the active function nodes are necessary. The node numbers of the active nodes of $P_1$ and $P_2$ are stored in two arrays $M_1$ and $M_2$. The active nodes are determined by the backward search in the evaluation procedure.

In order to define one suitable crossover point we define two possible crossover points $C_{\text{P1}}$ and $C_{\text{P2}}$ of the two parents. With information about the active nodes and the length of the path, we can choose two possible crossover points. The possible crossover points $C_{\text{P1}}$ and $C_{\text{P2}}$ are chosen by chance in the range of the active function nodes which are stored in $M_1$ and $M_2$. The possible crossover points may not be input or output nodes. The crossover procedure is done by performing the following steps:

1. **Define a general crossover point**
   A general crossover point $C_P$ is defined by choosing the smaller crossover point from $C_{\text{P1}}$ and $C_{\text{P2}}$. The reason for this is that the subgraphs of the parents which will be placed in front of or behind the crossover point of the offspring genome should be balanced. The representation of CGP allows active paths of an individual which can start in the middle or back of the graph. The subgraph which will be placed in front of the crossover point has to start at more leading active nodes. If $C_P$ is defined as the possible point $C_{\text{P1}}$, the subgraph of $P_1$ in front of $C_P$ will be placed in front of $C_P$ in the offspring genome. The subgraph behind $C_P$ of $P_2$ will be placed behind $C_P$ in the offspring genome.

2. **Copy the genetic material in front of the crossover point**
   The genetic material for the section in front of $C_P$ is copied from the parent $P_1$ to the offspring genome. This includes the function nodes from the start of the genome of $P_1$ until the function node given by $C_P$. Since the inactive nodes are also genetic material which can become active, we also copy these nodes for further genetic variation steps.

3. **Copy the genetic material behind the crossover point**

   The genetic material for the section behind the crossover point of the offspring genome is copied from the parent $P_2$ starting at the crossover point until the output. The resulting subgraph $S_2$ including the output is copied to the offspring genome behind the crossover point. However, the active nodes of the section behind the crossover point can alter the active path in front of the crossover point by referring to inactive nodes. Further steps are necessary to connect both sections.

4. **Connect both sections of the offspring genome**

   (a) Both sections are connected with a special step that we call *neighbourhood connect*. This step refers to the first active node of the section behind the crossover point which is connected to the last active node of the section in front of the crossover point. This is done by adjusting the connection gene of the first active node of the section behind the crossover point.

   (b) To ensure that active nodes of the section behind the crossover point do not refer to inactive nodes of the section in front of the crossover point, we perform a step which we call *random active connect*. All connection genes of the active nodes of the section behind the crossover point are adjusted to the active nodes of the section in front of the crossover point, previous active nodes of $S_2$ or input nodes. The nodes which are suitable for a random active connection are named as *permissible nodes*. The connection is done by changing the connection gene of a node which refers to an inactive node to a randomly chosen *permissible node*.

The crossover procedure produces a new genome which represents the offspring concerning the phenotypes of both parents. In the case that two children should be produced, the crossover procedure is performed twice with two different general crossover points. Since the representation of CGP provides connections to any of the previous function nodes of the graph, performing only the *neighbourhood connect* could result in a monotone data flow of the resulting phenotype.

Figure 1 exemplifies the crossover procedure. At the top of the figure, the arrays with the active nodes and crossover points are listed. Below this information, the genotypes and phenotypes of the parents and the offspring are shown, and the parts of the crossover are marked with dashed boxes.

### 3.1   Multiple Outputs

The proposed subgraph crossover primarily focuses on single output problems but we also investigate multiple output problems in this paper. On this kind of problems we deal with multiple active paths which share one genotype. The proposed step *neighbourhood connect* connects two nodes for one active path. This procedure becomes more complex if multiple active paths are involved. Therefore, on multiple output genotypes, we only connect the two parts of the parent genotypes by performing the *random active connect*. This procedure is also applied to all output nodes since they can refer to inactive nodes in the newly produced genotype.

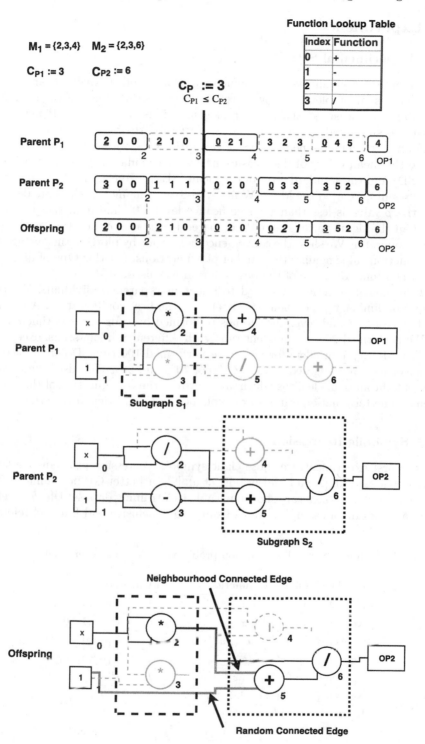

**Fig. 1.** The proposed subgraph crossover (An implementation for the Java-based Evolutionary Computation Research System (ECJ) is available at https://ls11-www.cs.uni-dortmund.de/staff/kalkreuth#resources)

## 4  Experiments

### 4.1  Experimental Setup

We performed experiments on symbolic regression, boolean functions and image operator design problems. To evaluate the search performance of the subgraph crossover, we measured the number of generations until the CGP algorithm terminates successfully (*generations-to-success*) and the best fitness value which was found after a predefined number of generations (*best-fitness-of-run*). In addition to the mean values of the measurements, we calculated the standard deviation (SD) and the standard error of the mean (SEM). To classify the significance of our results, we used the Mann-Whitney-U-Test. The mean values are denoted $a^\dagger$ if the $p$-value is less than the significance level 0.05 and $a^\ddagger$ if the $p$-value is less than the significance level 0.01 compared to the use of mutation as the sole genetic operator. We show the convergence behaviour by plotting the average fitness function value against the number of generations. For this type of diagram, the fitness function value of the best solution was used.

Tournament selection was used to select new parent individuals. We performed preliminary experiments to determine the best configuration. A tournament size of four and seven individuals performed best for our experiments.

We performed 200 independent runs with different random seeds, except for the Even-Parity-7 problem for which we performed 100 runs. Different rates of crossover were investigated, 0%, 20%, 50%, 70% and 90%. We also investigated different chromosome lengths which are shown in the configuration of the experiments. The termination criteria are explained in the particular experiments.

### 4.2  Symbolic Regression

For our first experiment, we chose eight symbolic regression problems from the work of Clegg et al. [2] and McDermott et al. [9] for better GP benchmarks. The functions of the problems are shown in Table 1. A training data set U[$a, b, c$] refers to $c$ uniform random samples drawn from $a$ to $b$ inclusive and E[$a, b, c$] refers to

**Table 1.** Symbolic regression problems of the first experiment

| Problem | Objective function | Vars | Training set |
|---------|--------------------|------|--------------|
| Koza-2 | $x^5 - 2x^3 + x$ | 1 | U$[-1, 1, 20]$ |
| Koza-3 | $x^6 - 2x^4 + x^2$ | 1 | U$[-1, 1, 20]$ |
| Nguyen-4 | $x^6 + x^5 + x^4 + x^3 + x^2 + x$ | 1 | U$[-1, 1, 20]$ |
| Nguyen-5 | $\sin(x^2)\cos(x) - 1$ | 1 | U$[-1, 1, 20]$ |
| Nguyen-6 | $\sin(x) + \sin(x + x^2)$ | 1 | U$[-1, 1, 20]$ |
| Nguyen-7 | $\ln(x + 1) + \ln(x^2 + 1)$ | 1 | U$[0, 2, 20]$ |
| Keijzer-6 | $\sum_i^x 1/i$ | 1 | E$[1, 50, 1]$ |
| Pagie-1 | $1/(1 + x^{-4}) + 1/(1 + y^{-4})$ | 2 | E$[-5, 5, 0.4]$ |

**Table 2.** Configuration of the first experiment

| Property | Koza-2,3/Nguyen-4,5,6 | Nguyen-7/Keijzer-6/Pagie-1 |
|---|---|---|
| Maximum node count | 10 | 20/30/30 |
| Maximum generation | - | 1000 |
| Number of inputs | 2 | 2 |
| Number of outputs | 1 | 1 |
| Population size | 50 | 50 |
| Function set | Koza | Koza/Keijzer/Koza |
| Mutation rate | 0.2 | 0.05/0.04/0.04 |
| Tournament selection size | 4 | 4 |
| Evaluation method | Generations-to-success | Best-fitness-of-run |

a grid of points evenly spaced with an interval of $c$, from $a$ to $b$ inclusive. We replaced the investigation of the problem Koza-1 ("quartic") by the problems Keijzer-6, Nguyen-7 and Pagie-1 which have been proposed as alternatives by White et al. [16] to this overused problem and have different reputations.

The fitness of the individuals was represented by a cost function value. The cost function was defined by the sum of the absolute difference between the real function values and the values of an evaluated individual. We defined the termination criteria for the experiment with a cost function value less or equal than 0.01 and a predefined number of generations. The configuration of the experiment is shown in Table 2. The maximum node count, population size, and the mutation rate were orientated on previous work [2]. The Koza function set consisted of eight mathematical functions $(+, -, *, /, \sin, \cos, \ln(|n|), e^n)$ and the Keijzer function set of five mathematical functions $(+, *, \frac{1}{n}, -n, \sqrt{n})$.

Table 3a and b show the results of the first experiment and it is clearly seen that the use of the subgraph crossover reduces the number of generations until the termination criterion triggers and results in a better fitness value. The standard deviations are also clearly reduced by the use of higher crossover rates.

Figure 2 shows the convergence behaviour. It can also clearly seen that the use of the subgraph crossover results in improved convergence curves for the presented number of generations.

## 4.3  Boolean Functions

To investigate the use of the subgraph crossover with one output in the boolean domain, we chose the three Even-Parity problems with $n = 5$, 6 and 7 boolean inputs. The goal was to find a program that produces the value of the boolean even parity depending on the $n$ independent inputs. The fitness was represented by the number of fitness cases for which the candidate solution failed to generate the correct value of the Even-Parity function.

Since former work by White et al. [16] outlined that this problem type was excessively used and investigated in the past, we also investigated multiple out-

**Table 3.** Results for the symbolic regression problems of the first experiment

(a) Koza 2,3 and Nguyen 4,5,6

| Problem | CR* | Mean Generations | SD | SEM |
|---|---|---|---|---|
| Koza-2 | 0 | 22917 | 30180 | ±2139 |
| | 20 | 15882[†] | 17852 | ±1259 |
| | 50 | 9781[‡] | 10554 | ±748 |
| | 70 | 8898[‡] | **9600** | **±680** |
| | 90 | **8642**[‡] | 10021 | ±708 |
| Koza-3 | 0 | 9795 | 14176 | ±1005 |
| | 20 | 5628[‡] | 10375 | ±735 |
| | 50 | 4405[‡] | 6576 | ±466 |
| | 70 | 3531[‡] | 5566 | ±394 |
| | 90 | **2346**[‡] | **2568** | **±182** |
| Nguyen-4 | 0 | 3270848 | 4660590 | ±330380 |
| | 20 | 2045079 | 2316987 | ±203213 |
| | 50 | 446490[‡] | 564625 | ±40025 |
| | 70 | 218063[‡] | 326772 | ±23164 |
| | 90 | **169837**[‡] | **234799** | **±16644** |
| Nguyen-5 | 0 | 20588 | 55131 | ±3908 |
| | 20 | 8316[†] | 14362 | ±1018 |
| | 50 | 5440[‡] | 10216 | ±724 |
| | 70 | 3936[‡] | 5836 | ±413 |
| | 90 | **2978**[‡] | **3550** | **±251** |
| Nguyen-6 | 0 | 116734 | 215400 | ±15269 |
| | 20 | 52711[‡] | 130260 | ±9233 |
| | 50 | 28399[‡] | 72211 | ±9233 |
| | 70 | 12480[‡] | 37103 | ±2630 |
| | 90 | **5303**[‡] | **12925** | **±916** |

(b) Nguyen-7, Keijzer-6, and Pagie-1

| Problem | CR* | Mean Best Fitness | SD | SEM |
|---|---|---|---|---|
| Nguyen-7 | 0 | 0.60 | 0.30 | ±0,021 |
| | 20 | 0.59 | 0.30 | ±0.021 |
| | 50 | 0.56 | 0.29 | ±0.020 |
| | 70 | 0.51 | **0.26** | **±0.018** |
| | 90 | **0.46**[†] | 0.27 | ±0,019 |
| Keijzer-6 | 0 | 3.90 | 1.68 | ±0,12 |
| | 20 | 3.47 | 1.53 | ±0.11 |
| | 50 | 2.96[†] | 1.31 | ±0.09 |
| | 70 | 2.93[†] | 1.03 | ±0.07 |
| | 90 | **2.81**[†] | **0.66** | **±0.06** |
| Pagie-1 | 0 | 128.07 | 53.77 | ±3.80 |
| | 20 | 116.82 | 48.74 | ±3.44 |
| | 50 | 109.74 | 42.84 | ±3.02 |
| | 70 | 106.80 | 46.27 | ±3.27 |
| | 90 | **91.20**[‡] | **42.07** | **±2.97** |

*Crossover rate in %

put problems as the digital adder, subtractor and multiplier. These sort of problems differ markedly from the parity problems, and the multiple output multiplier has been proposed as a suitable alternative. As a result, we receive a diverse set of problems in this domain.

To evaluate the fitness of the individuals on the multiple output problems, we defined the fitness value of an individual as the number of different bits to the corresponding truth table. When this number became zero, the algorithm terminated successfully. The configuration for the experiment is shown in Table 4.

Table 5a and b show the results of our second experiment which demonstrates a reduced demand of generations until the algorithm terminates successfully by the use of the subgraph crossover.

## 4.4  Image Operator Design

For our third experiment, we chose two image operator design problems from the field of evolvable hardware [12]. For these problems, we wanted to minimise the difference between a filtered image $I_f$ and a reference image $I_r$, which were based on an input image $I_i$. The images $I_i$ and $I_r$ were of size $N \times M$ pixels. We used

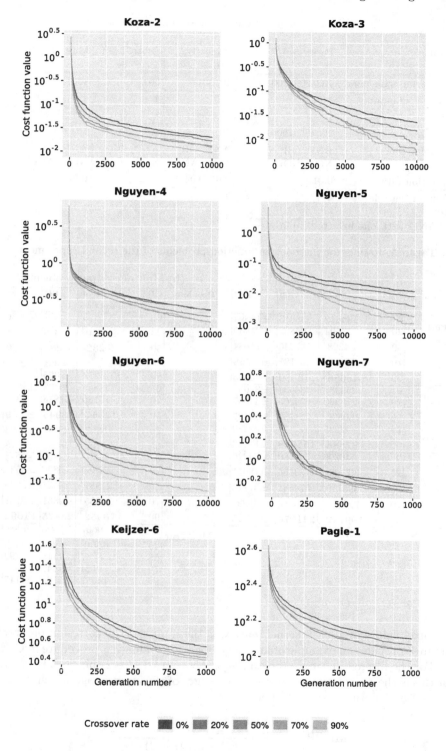

**Fig. 2.** Convergence curves for all problems of the first experiment

**Table 4.** Configuration for the second experiment

| Property | Parity-5/6/7 | Adder 1/2 Bit | Subtr. 2 Bit | Multipl. 2 Bit |
|---|---|---|---|---|
| Maximum node count | 100/200/200 | 10/30 | 30 | 10 |
| Number of inputs | 5/6/7 | 3/5 | 4 | 4 |
| Number of outputs | 1 | 2/3 | 3 | 4 |
| Population size | 100/200/200 | 50 | 50 | 50 |
| Function set | AND, OR, NAND NOT, NOR | AND, OR XOR, AND* | AND, OR XOR, AND* | AND, OR, XOR NOR, AND* |
| Mutation rate | 0.01 | 0.07/0.04 | 0.02 | 0.07 |
| Tournament selection size | 4/7/7 | 4 | 4 | 4 |

*AND with one inverted input

**Table 5.** Results for the boolean function problems of the second experiment

(a) Single output problems

| Problem | CR* | Mean Generations | SD | SEM |
|---|---|---|---|---|
| Parity-5 | 0 | 134872 | 238915 | ±16894 |
| | 20 | 155694 | 296802 | ±21039 |
| | 50 | 84298‡ | 167645 | ±11884 |
| | 70 | 52068‡ | 67121 | **±4758** |
| | 90 | **48688‡** | **73880** | ±5263 |
| Parity-6 | 0 | 256988 | 436000 | ±32055 |
| | 20 | 226899 | 629265 | ±44607 |
| | 50 | 160596‡ | 344737 | ±24437 |
| | 70 | 192225 | 412333 | ±29229 |
| | 90 | **92888‡** | **155648** | ±11033 |
| Parity-7 | 0 | 546206 | 738980 | ±74270 |
| | 20 | 362886† | 1183156 | ±118911 |
| | 50 | 396758† | 574841 | ±57773 |
| | 70 | 220611‡ | 318338 | ±31994 |
| | 90 | **199142‡** | **241070** | ±24228 |

*Crossover rate in %

(b) Multiple output problems

| Problem | CR* | Mean Generations | SD | SEM |
|---|---|---|---|---|
| Add.-1Bit | 0 | 373 | 507 | ±36 |
| | 20 | 240‡ | 320 | ±22 |
| | 50 | **232‡** | **287** | **±20** |
| | 70 | 245‡ | 302 | ±21 |
| | 90 | 277 | 325 | ±23 |
| Add.-2Bit | 0 | 23864 | 36463 | ±2578 |
| | 20 | 19180 | 31813 | 1356 |
| | 50 | **12854†** | **13945** | **±909** |
| | 70 | 15867† | 28187 | ±1993 |
| | 90 | 15102‡ | 28546 | ±2018 |
| Sub.-2Bit | 0 | 28501 | 43495 | ±3083 |
| | 20 | 27989 | 49836 | ±3532 |
| | 50 | 21581† | 29557 | ±2095 |
| | 70 | 21335† | 52351 | ±3711 |
| | 90 | **16952‡** | **23875** | **±1692** |
| Mul.-2Bit | 0 | 9607 | 12506 | ±886 |
| | 20 | **6350†** | 6263 | ±443 |
| | 50 | 8887 | 1118 | ±792 |
| | 70 | 9124 | 7441 | ±527 |
| | 90 | 14587 | 15766 | ±1117 |

the pixel values of a $3 \times 3$ kernel matrix as the inputs of the cartesian program. Figure 3 provides an example of an image operator evaluation procedure in CGP.

For the filtered image we took a size of $(N - 2) \times (M - 2)$ pixels. The reason for this is the convolution problem on the edges of the input image which is often treated by ignoring the edges. Let

$$\text{MDPP} := \frac{\sum_{i=1}^{N-2} \sum_{j=1}^{M-2} |I_f(i, j) - I_r(i, j)|}{(N - 2)(M - 2)}$$

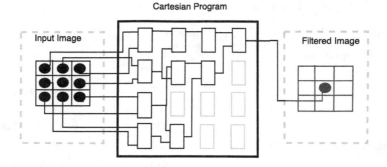

**Fig. 3.** Example of an image operator evaluation procedure in CGP

be the mean difference per pixel between the reference image $I_r$ and the filtered image $I_f$. If MDPP is zero, the images $I_f$ and $I_r$ are identical, except the pixels on the edges. As test image, we chose the famous Lena image. We added Gaussian Noise (Fig. 4) and Salt & Pepper noise (Fig. 5) to the Lena image and chose them as input images.

**Fig. 4.** Lena image with Gaussian noise

**Fig. 5.** Lena image with Salt & Pepper noise

**Fig. 6.** Lena image without noise

The goal of the resulting problem was to evolve a filter which reduces the noise in reference to the Lena image without noise as shown in Fig. 6. We defined the termination criterion with a MDPP of less or equal than 3. The configuration and function set for the experiment are shown in Table 6. The function set consisted of eight low-level image processing functions.

**Table 6.** Configuration of the third experiment

| Property | Configuration |
|---|---|
| Maximum node count | 20 |
| Number of inputs | 9 |
| Number of outputs | 1 |
| Population size | 100 |
| Function set | OR, AND, XOR, ADD, ADD*, MAX, MIN, AVG |
| Mutation rate | 0.05 |
| Tournament selection size | 4 |
| Image size | 64 × 64 pixels |

*ADD with saturation

**Table 7.** Results for the problems of the third experiment

| Problem | Crossover rate in % | Mean generations | SD | SEM |
|---|---|---|---|---|
| Salt & Pepper noise | 0 | 2155 | 2785 | ±197 |
| | 20 | 1582$^\dagger$ | 2225 | ±157 |
| | 50 | 1370$^\ddagger$ | 1749 | ±123 |
| | 70 | 1669$^\ddagger$ | 2192 | ±155 |
| | 90 | **1175$^\ddagger$** | **1419** | **±100** |
| Gaussian noise | 0 | 3816 | 4319 | ±306 |
| | 20 | 3142 | 3141 | ±222 |
| | 50 | 3092 | 2813 | ±199 |
| | 70 | 2715$^\dagger$ | 2528 | ±179 |
| | 90 | **2383$^\dagger$** | **2086** | **±147** |

Table 7 shows the results of our third experiment and it is clearly visible that the use of the subgraph crossover leads to a reduced number of generations until the algorithm terminated successfully, similar as in our previous experiments.

## 4.5 Crossover Comparison

We compared the arithmetic crossover [2] to the subgraph crossover. We measured the generations until the termination criteria triggered (*generation-to-termination*) and the best fitness of run, similar as in our previous experiments. The parameters for CGP were also similar. The crossover rate was set to 90%.

**Table 8.** Results of the crossover comparison for the symbolic regression problems

(a) Koza 2,3 & Nguyen 4,5,6

| Problem | Method | Mean Generations | SD | SEM |
|---|---|---|---|---|
| Koza-2 | Arithm. | 15801 | 23558 | ±1665 |
| | Subgr. | **8642$^\ddagger$** | 10021 | ±708 |
| Koza-3 | Arithm. | 2561 | 5085 | ±360 |
| | Subgr. | **2346$^\dagger$** | 2568 | ±182 |
| Nguyen-4 | Arithm. | 7314661 | 10699730 | ±758483 |
| | Subgr. | **169837$^\ddagger$** | 234799 | ±16644 |
| Nguyen-5 | Arithm. | 28437 | 78706 | ±5579 |
| | Subgr. | **2906$^\ddagger$** | 3413 | ±241 |
| Nguyen-6 | Arithm. | 343343 | 875241 | ±62044 |
| | Subgr. | **5459$^\ddagger$** | 11624 | ±821 |

(b) Nguyen-7, Keijzer-6, Pagie-1

| Problem | Method | Mean Best Fitness | SD | SEM |
|---|---|---|---|---|
| Nguyen-7 | Arithm. | 0.63 | **0.22** | **±0,016** |
| | Subgr. | **0.46$^\ddagger$** | 0.27 | ±0.019 |
| Keijzer-6 | Arithm. | 8.89 | 7.23 | ±0.51 |
| | Subgr. | **2.81$^\dagger$** | **0.66** | **±0.06** |
| Pagie-1 | Arithm. | 125.20 | 51.07 | ±3.61 |
| | Subgr. | **91.20$^\dagger$** | 42.07 | ±2.97 |

Tables 8, 9 and 10 show the results of the crossover technique comparison. Compared to the arithmetic crossover, the subgraph crossover performs significantly better on our tested problems.

**Table 9.** Results of the crossover comparison for the boolean function problems

(a) Single output problems

| Problem | Method | Mean Generations | SD | SEM |
|---|---|---|---|---|
| Parity-5 | Arithm. | 898477 | 259872 | ±18515 |
| | Subgr. | **48688**‡ | **73880** | **±5263** |
| Parity-6 | Arithm. | 942262 | 206752 | ±18418 |
| | Subgr. | **92888**‡ | **155648** | **±11033** |
| Parity-7 | Arithm. | 1825487 | 504094 | ±75145 |
| | Subgr. | **199142**‡ | **241070** | **±24228** |

(b) Multiple output problems

| Problem | Method | Mean Generations | SD | SEM |
|---|---|---|---|---|
| Add.-1Bit | Arithm. | 54225 | 124974 | 8859 |
| | Subgr. | **277**‡ | **325** | **±23** |
| Add.-2Bit | Arithm. | 99389 | 125828 | ±8919 |
| | Subgr. | **15102**‡ | **28546** | **±2018** |
| Sub.-2Bit | Arithm. | 637929 | 4301481 | ±30492 |
| | Subgr. | **16952**‡ | **23875** | **±1692** |
| Mul.-2Bit | Arithm. | 47444 | 141631 | ±10040 |
| | Subgr. | **14587**‡ | **15766** | **±1117** |

**Table 10.** Results of the crossover comparison for the image operator problems

| Problem | Method | Mean generations | SD | SEM |
|---|---|---|---|---|
| Salt & Pepper noise | Arithmetic | 5410 | 3463 | ±245 |
| | Subgraph | **1175**‡ | **1419** | **±100** |
| Gaussian noise | Arithmetic | 9762 | 0987 | ±69 |
| | Subgraph | **2383**‡ | 2086 | ±147 |

**Table 11.** Upper generation limits and success rates for the arithmetic crossover

| Problem | Upper generation count limit | Sucess rate |
|---|---|---|
| Nguyen-4 | $3 * 10^7$ | 0,83 |
| Parity-5 | $10^6$ | 0,16 |
| Parity-6 | $10^6$ | 0,08 |
| Parity-7 | $2 * 10^6$ | 0,10 |
| Sub.-2Bit | $10^6$ | 0,45 |
| Salt & Pepper noise | $10^4$ | 0,75 |
| Gaussian noise | $10^4$ | 0,06 |

On some of the tested problems, the arithmetic crossover caused occasional runs which took a huge number of generations. Consequently, we defined upper limits which are shown in Table 11 with the corresponding success rates.

## 5   Discussion

The primary concern of our experiments was to find significant contributions of the subgraph crossover technique to the search performance of CGP in different problem domains. One point which should be discussed are the crossover rates. On our tested single output problems, a higher crossover rate led to the best results. This is contrary to the majority of the tested multiple output problems in which small and medium rates of crossover showed more beneficial effects. Furthermore, a high crossover rate of 90% had a negative effect on the 2-Bit multiplier problem. For this behaviour, we have no general and provable explanation yet. A preliminary and hypothetical assumption could be that certain multiple output structures are very sensitive concerning genetic variation. Since multiple outputs lead to multiple semantics, significant variation steps which are caused by the subgraph crossover could be even more disruptive compared to single output structures. For more significant statements a more detailed investigation of the subgraph crossover is needed and should be based on the automated tuning of the crossover rate on a diverse set of problems.

Another point which should be discussed is the population size. For our experiments we utilised sizes which are oriented with former work on CGP [2] and sizes which are empirically determined. However, it should be investigated which population sizes perform most efficient in different problem domains when the crossover is used. Former work by Miller [11] outlined that very small population sizes can be very efficient for boolean function problems. Based on Miller's experiments we plan to investigate the use of the subgraph crossover with a $(\mu + \lambda)$ evolutionary algorithm and small population sizes.

Concerning the selection pressure, we would like to point that the size of the tournament in our experiment was based on empirical observations and should be taken carefully. Furthermore, we find that an automated tuning of this setting would be interesting and should be included into further work.

Our results open the question in which way the subgraph crossover improves the search performance of CGP. At this time we can only provide a preliminary and hypothetical assumption. We assume that the use of the subgraph crossover leads to more exploration of high fitness regions in the search space. When parent graphs of high fitness are recombined the number of diverse children with high fitness is maybe increased. However, for more proven statements a detailed investigation is needed and should include fitness landscape analyses.

Related to our comparison to the arithmetic crossover we would like to point that the parameters for the comparison were determined empirically. For a more significant comparison, an automated tuning of the CGP parameters for both techniques is necessary and should be included into future work.

Since the CGP representation allows inactive nodes and the connection of each function node to input nodes, the genotypes can become imbalanced in their structure. This behaviour makes it hard to generate two children when parts of the active paths are recombined. Therefore we think an advancement of the representation of CGP which creates more balanced and structured genotypes would be beneficial and can enable the generation of two children with the same crossover point.

## 6   Conclusion and Future Work

In this paper, we proposed a new subgraph crossover technique for CGP. In our experiments, the use of the subgraph crossover resulted in a significantly smaller number of generations until the CGP algorithm terminated successfully, a better fitness value, and a better convergence behaviour. The experiments on our test problems indicate that the subgraph crossover technique may be beneficial for the search performance of CGP. Moreover, we have demonstrated the potential of the subgraph crossover for a diverse set of problems including different functions and types of fitness. Our preliminary comparison to the arithmetic crossover technique indicates a significantly better performance for the subgraph crossover on our tested problems.

In the future, we will primarily focus on a detailed study with more detailed experiments and comparisons. This work will tackle the question how the subgraph crossover exactly contributes to the search performance of CGP. Furthermore, we will perform experiments with different population sizes, selection strategies and crossover rates in various test problem domains and on a real world application. Our work will also include comparisons to the original mutation-only CGP which is performed within a $(1 + \lambda)$ EA. Another part of our future work is devoted to the advancement of the representation of CGP to enable a subgraph crossover which allows the production of two children with the same crossover point.

# References

1. Cai, X., Smith, S.L., Tyrrell, A.M.: Positional independence and recombination in cartesian genetic programming. In: Collet, P., Tomassini, M., Ebner, M., Gustafson, S., Ekárt, A. (eds.) EuroGP 2006. LNCS, vol. 3905, pp. 351–360. Springer, Heidelberg (2006). doi:10.1007/11729976_32
2. Clegg, J., Walker, J.A., Miller, J.F.: A new crossover technique for cartesian genetic programming. In: Proceedings of the 9th Annual Conference on Genetic and Evolutionary Computation, GECCO 2007, vol. 2, pp. 1580–1587. ACM Press, London, 7–11 July 2007
3. Kaufmann, P., Platzner, M.: Advanced techniques for the creation and propagation of modules in cartesian genetic programming. In: Proceedings of the 10th Annual Conference on Genetic and Evolutionary Computation, GECCO 2008, pp. 1219–1226. ACM, Atlanta, GA, USA, 12–16 July 2008
4. Koza, J.: Genetic programming: a paradigm for genetically breeding populations of computer programs to solve problems. Technical report STAN-CS-90-1314, Department of Computer Science, Stanford University, June 1990
5. Koza, J.R.: Genetic Programming: On the Programming of Computers by Means of Natural Selection. MIT Press, Cambridge (1992)
6. Koza, J.R.: Genetic Programming II: Automatic Discovery of Reusable Programs. MIT Press, Cambridge (1994)
7. Luke, S., Spector, L.: A comparison of crossover and mutation in genetic programming. In: Proceedings of the Second Annual Conference on Genetic Programming 1997, pp. 240–248. Morgan Kaufmann, Stanford University, CA, USA, 13–16 July 1997
8. Luke, S., Spector, L.: A revised comparison of crossover and mutation in genetic programming. In: Proceedings of the Third Annual Conference on Genetic Programming 1998, pp. 208–213. Morgan Kaufmann, University of Wisconsin, Madison, Wisconsin, USA, 22–25 July 1998

9. McDermott, J., White, D.R., Luke, S., Manzoni, L., Castelli, M., Vanneschi, L., Jaskowski, W., Krawiec, K., Harper, R., De Jong, K., O'Reilly, U.M.: Genetic programming needs better benchmarks. In: Proceedings of the Fourteenth International Conference on Genetic and Evolutionary Computation Conference, GECCO 2012, pp. 791–798. ACM, Philadelphia, Pennsylvania, USA, 7–11 July 2012

10. Miller, J.F., Thomson, P.: Cartesian genetic programming. In: Poli, R., Banzhaf, W., Langdon, W.B., Miller, J., Nordin, P., Fogarty, T.C. (eds.) EuroGP 2000. LNCS, vol. 1802, pp. 121–132. Springer, Heidelberg (2000). doi:10.1007/978-3-540-46239-2_9

11. Miller, J.F.: An empirical study of the efficiency of learning Boolean functions using a cartesian genetic programming approach. In: Proceedings of the Genetic and Evolutionary Computation Conference, vol. 2, pp. 1135–1142. Morgan Kaufmann, Orlando, Florida, USA, 13–17 July 1999

12. Sekanina, L.: Image filter design with evolvable hardware. In: Cagnoni, S., Gottlieb, J., Hart, E., Middendorf, M., Raidl, G.R. (eds.) EvoWorkshops 2002. LNCS, vol. 2279, pp. 255–266. Springer, Heidelberg (2002). doi:10.1007/3-540-46004-7_26

13. Slaný, K., Sekanina, L.: Fitness landscape analysis and image filter evolution using functional-level CGP. In: Ebner, M., O'Neill, M., Ekárt, A., Vanneschi, L., Esparcia-Alcázar, A.I. (eds.) EuroGP 2007. LNCS, vol. 4445, pp. 311–320. Springer, Heidelberg (2007). doi:10.1007/978-3-540-71605-1_29

14. Walker, J.A., Miller, J.F.: Evolution and acquisition of modules in cartesian genetic programming. In: Keijzer, M., O'Reilly, U.-M., Lucas, S., Costa, E., Soule, T. (eds.) EuroGP 2004. LNCS, vol. 3003, pp. 187–197. Springer, Heidelberg (2004). doi:10.1007/978-3-540-24650-3_17

15. Walker, J.A., Miller, J.F., Cavill, R.: A multi-chromosome approach to standard and embedded cartesian genetic programming. In: Proceedings of the 8th Annual Conference on Genetic and Evolutionary Computation, GECCO 2006, vol. 1, pp. 903–910. ACM Press, Seattle, 8–12 July 2006

16. White, D.R., McDermott, J., Castelli, M., Manzoni, L., Goldman, B.W., Kronberger, G., Jaskowski, W., O'Reilly, U.M., Luke, S.: Better GP benchmarks: community survey results and proposals. Genet. Program. Evolvable Mach. **14**(1), 3–29 (2013)

17. White, D.R., Poulding, S.: A rigorous evaluation of crossover and mutation in genetic programming. In: Vanneschi, L., Gustafson, S., Moraglio, A., Falco, I., Ebner, M. (eds.) EuroGP 2009. LNCS, vol. 5481, pp. 220–231. Springer, Heidelberg (2009). doi:10.1007/978-3-642-01181-8_19

# A Comparative Study of Different Grammar-Based Genetic Programming Approaches

Nuno Lourenço[1](✉), Joaquim Ferrer[1], Francisco B. Pereira[1,2], and Ernesto Costa[1]

[1] CISUC, Department of Informatics Engineering, University of Coimbra, Polo II - Pinhal de Marrocos, 3030 Coimbra, Portugal
{naml,xico,ernesto}@dei.uc.pt, jferrer@student.dei.uc.pt
[2] Instituto Superior de Engenharia de Coimbra, Quinta da Nora, 3030-199 Coimbra, Portugal

**Abstract.** Grammars are useful formalisms to specify constraints, and not surprisingly, they have attracted the attention of Evolutionary Computation (EC) researchers to enforce problem restrictions. Context-Free-Grammar GP (CFG-GP) established the foundations for the application of grammars in Genetic Programming (GP), whilst Grammatical Evolution (GE) popularised the use of these approaches, becoming one of the most used GP variants. However, studies have shown that GE suffers from issues that have impact on its performance. To minimise these issues, several extensions have been proposed, which made the distinction between GE and CFG-GP less noticeable. Another direction was followed by Structured Grammatical Evolution (SGE) that maintains the separation between genotype and phenotype from GE, but overcomes most of its issues. Our goal is to perform a comparative study between CFG-GP, GE and SGE to examine their relative performance. The results show that in most of the selected benchmarks, CFG-GP and SGE have a similar performance, showing that SGE is a good alternative to GE.

**Keywords:** Genetic programming · Grammar-based genetic programming · Grammatical evolution

## 1 Introduction

Grammars are widely used by computer scientists to represent complex structures by specifying restrictions on general domains, thus limiting the number of expressions that one can use. They are commonly applied in the specification of a programming language syntax, to define type restrictions or to describe the interaction constraints between the different components of a system. So it does not come as a surprise that grammars have attracted the attention of EC researchers to specify problem restrictions and guide the evolutionary process, specifically in GP [3].

© Springer International Publishing AG 2017
J. McDermott et al. (Eds.): EuroGP 2017, LNCS 10196, pp. 311–325, 2017.
DOI: 10.1007/978-3-319-55696-3_20

Whigham pioneered the use of grammars in GP with his CFG-GP approach [16], while GE [8,13] mainstreamed the use of grammar-based approaches to the point where it has become one of the most used GP methods. The most distinctive factor between CFG-GP and GE has to do with the different representation of individual solutions in the search space. While the former relies on a derivation-tree based representation, the latter uses a variable length linear integer string and a grammar to map individuals from search space into problem space. This separation between genotype and phenotype is usually seen as an advantage of GE over other techniques, since it is possible to decouple the search engine from the problem we are tackling, simplifying its application to different domains.

Despite the popularity of GE, recent studies have shown that it suffers from high levels of redundancy and low locality [12]. A representation is said to be redundant when several different genotypes map in the same phenotype, whereas locality is concerned with measuring how variations at the genotype level reflect on differences at the phenotype level [11]. In a representation with high locality, a small modification on the genotype usually results in a small modification on the phenotype, thus harbouring conditions for an effective sampling of the search space. If this condition is not satisfied, the search performed by an Evolutionary Algoritm (EA) tends to resemble that of a random search [10].

Recently, Whigham et al. [15] compared the relative performance of CFG-GP and GE. Results presented in the aforementioned study confirmed the ineffectiveness of GE, as it was outperformed by CFG-GP in most of the problems. Also, it was reported that, in some situations, the performance of GE is comparable to that of a random search method.

SGE is a new representation that aims to overcome the limitations of standard GE [6]. Its most noteworthy characteristic is the one-to-one correspondence between genes and non-terminals of the grammar being used. In the aforementioned reference, Lourenço et al. analysed the properties of both SGE and GE, and concluded that the new representation, not only reduces the levels of redundancy, but also increases locality. These results justify the increased performance of SGE over the traditional GE representation. The goal of this work is to compare the enhanced SGE proposal with existing state-of-the art grammar-based representations. We extend the study of Whigham et al. by including SGE in the experimental framework. This analysis will help to gain insight on the strengths of the novel GE representation and identify the relative performance of the various flavours of grammar-based GP stand relatively to each other.

The remainder of this paper is organised as follows: Sect. 2 introduces the grammar-based approaches used in this study. We briefly describe how they work, and how they are connected with each other. Section 3 details the experimental framework used to compare the different grammar-based approaches, followed by the analysis and discussion of the results (Sect. 4). Finally, Sect. 5 gathers the main conclusions.

# 2 Grammar-Based Genetic Programming

## 2.1 Contex-Free-Grammar Genetic Programming (CFG-GP)

Context-Free-Grammars (CFGs) have been widely used to represent and control the search bias of EAs [7]. Formally, a CFG is a tuple $G = (N, T, S, P)$, where $N$ is a non-empty set of non-terminal symbols, $T$ is a non-empty set of terminal symbols, $S$ is an element of $N$ called the axiom, and $P$ is a set of production rules of the form $A:: = \alpha$, with $A \in N$ and $\alpha \in (N \cup T)^*$. $N$ and $T$ are disjoint. Each grammar $G$ defines a language $L(G)$ composed by all sequences of terminal symbols (the words) that can be derived from the axiom: $L(G) = \{w : S \overset{*}{\Rightarrow} w, w \in T^*\}$. An example of a CFG is presented in Fig. 1.

One of the first GP methods to use CFG to guide the evolutionary search was proposed by Whigham [16]. Each individual consists of a derivation tree from the axiom using the grammar. The leaves of the tree correspond to a specific word of the language defined by the grammar. The translation of the genotype (i.e. derivation tree) into the phenotype is performed by reading the derivation tree leaves (i.e. the terminals symbols) from left to right, starting at the left-most leaf. Consider the set of production rules defined in Fig. 1 to create polynomial expressions. One possible derivation tree for this grammar is depicted in Fig. 2. Looking at the tree, we can see the the left most leaf is the terminal $($. So starting from that node, we scan the tree from left to right, appending the symbols that are terminals to our expression. The resulting polynomial expression is: $(0.5/1)/x_1 * x_1$.

## 2.2 Grammatical Evolution (GE)

Grammatical Evolution is different from the CFG-GP for there is a separation between the genotype, i.e., a linear string of integers, and the phenotype, i.e., a program in the form of a tree expression. As a consequence, a mapping process is required to map the string into an executable program, using the productions rules of a CFG. The translation of the genotype into the phenotype is done by

```
<start> ::= <expr><op><expr>      (0)
         |   <expr>               (1)

<expr> ::= <term><op><term>       (0)
         | (<term><op><term>)     (1)

<op> ::= +    (0)
       | -    (1)
       | /    (2)
       | *    (3)

<term> ::= x₁   (0)
         | 0.5  (1)
```

Fig. 1. Example of a grammar in the Backus-Naur Form (BNF).

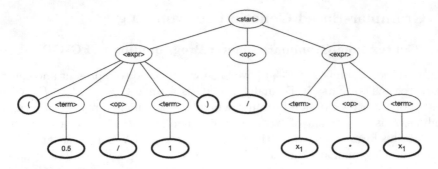

**Fig. 2.** CFG-GP individual genotype i.e., a derivation tree. The highlighted nodes represent the terminal symbols that correspond to the polynomial expression represented in the tree.

simulating a leftmost derivation from the axiom of the grammar. This process scans the linear sequence from left to right and each integer (*i.e.*, each codon) is used to determine the grammar rule that expands the leftmost non-terminal symbol of the current partial derivation tree. Consider de the set of production rules defined in Fig. 1, where there are two options to rewrite the left-hand side symbol <*start*>. In the beginning we have a sentential form equal to the axiom <*start*>. To rewrite the axiom one must choose which alternative will be used by taking the first integer of the genotype and dividing it by the number of options in which we can derive <*start*>. The remainder of that operation will indicate the option to be used. In the example above, assuming that the first integer is 23, it follows that $23\%2 = 1$ and the axiom is rewritten in <*expr*><*op*><*expr*>. Then the second integer is read, and the same method is used to the left most non-terminal of the derivation. The complete mapping of an individual is showed in Table 1. Sometimes the length of the genotype is not sufficient to complete the

**Table 1.** GE mapping procedure that translates the genotype of an individual into a polynomial expression (phenotype). Each row represents a derivation step. The used grammar is represented in Fig. 1.

| Derivation step | Integers left |
|---|---|
| <start> | $[23, 7, 55, 22, 3, 10, 30, 16, 203, 24]$ |
| <expr><op><expr> | $[7, 55, 22, 3, 10, 30, 16, 203, 24]$ |
| (<term><op><term>)<op><expr> | $[55, 22, 3, 10, 30, 16, 203, 24]$ |
| (0.5 <op><term>)<op><expr> | $[22, 3, 10, 30, 16, 203, 24]$ |
| (0.5/<term>)<op><expr> | $[3, 10, 30, 16, 203, 24]$ |
| (0.5/0.5)<op><expr> | $[10, 30, 16, 203, 24]$ |
| (0.5/1)/<expr> | $[30, 16, 203, 24]$ |
| (0.5/1)/<term><op><term> | $[15, 203, 24]$ |
| (0.5/1)/$x_1$ <op><term> | $[203, 24]$ |
| (0.5/1)/$x_1$ * <term> | $[24]$ |
| (0.5/1)/$x_1$ * $x_1$ | $[]$ |

mapping. In those cases the sequence is repeatedly reused in a process known as wrapping. If mapping exceeds a pre-determined number of wrappings, the process stops and the worst possible fitness value is assigned to the individual.

## 2.3  Structured Grammatical Evolution (SGE)

SGE is a recent genotypic representation for GE. In SGE each gene is linked to a specific non-terminal, and it is composed by a list of integers used to select the expansion option. The length of each list is determined by computing the maximum possible number of expansions of the corresponding non-terminal. This structure ensures that the modification of a gene does not affect the derivation options of other non-terminals, thus limiting the number of changes that can occur at the phenotypic level, which result in an higher locality. The values inside each list are bounded by the number of possible expansion options of the corresponding non-terminal. Therefore, mapping does not rely on the modulo rule, which reduces the redundancy associated with it.

As an example consider the grammar depicted in Fig. 1. The non-terminals set is $\{<start>, <expr>, <term>, <op>\}$. Then, the SGE genotype is composed by four genes, each one linked to a specific non-terminal. To determine the length of the gene's lists we calculate the maximum number of expansions of each non-terminal. The $<start>$ symbol is expanded only once, as it is the grammar axiom. The $<expr>$ symbol is expanded, at most, twice, because of the rule $<expr><op><expr>$. The computation of the list size for $<term>$ establishes a direct dependence between this non-terminal and $<expr>$: each time $<expr>$ is expanded, $<term>$ is expanded twice (in the two possible expansion options). As the grammar allows a maximum of two $<expr>$ expansions, it immediately follows that the list size for the $<term>$ gene is four. Following the same line of reasoning, the list size for the $<op>$ gene is 3. Thus, the list sizes for each gene are: $<start> : 1, <expr> : 2, <term> : 4, <op> : 3$. To complete the list inside each gene we take the number of derivation options $c_N$ of the corresponding non-terminal, and assign a random value from the interval $[0, c_N - 1]$ to every position. The $<start>$, $<expr>$ and $<term>$ symbols have $c_N = 2$, whereas $<op>$ has $c_N = 4$. Figure 3 shows an example of the genotype of a SGE individual. The complete mapping of this individual into a polynomial expression is depicted in Table 2.

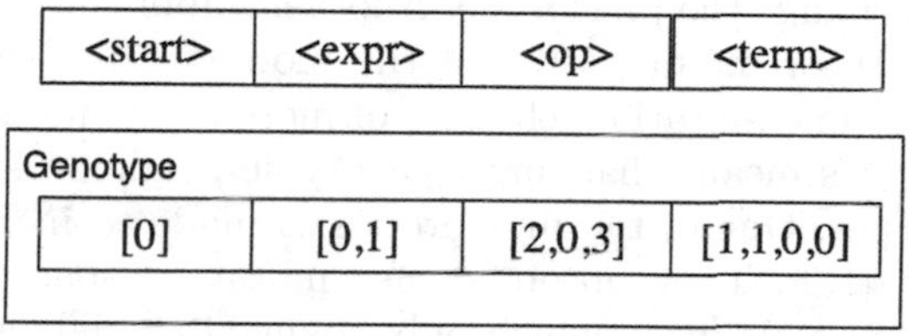

| <start> | <expr> | <op> | <term> |
|---------|--------|------|--------|

| Genotype | | | |
|----------|--------|---------|-----------|
| [0] | [0,1] | [2,0,3] | [1,1,0,0] |

**Fig. 3.** Example of a SGE genotype for the grammar showed in Fig. 1.

**Table 2.** SGE mapping procedure that converts a SGE individual into a polynomial expression. Each row represents a derivation step. The used grammar is represented in Fig. 1. The list of codons represents the integers needed for expanding <start>, <expr>, <op> and <value>, respectively.

| Derivation step | Integers left |
|---|---|
| <start> | $[[0], [0, 1], [2, 0, 3], [1, 1, 0, 0]]$ |
| <expr><op><expr> | $[[], [1, 0], [2, 0, 3], [1, 1, 0, 0]]$ |
| (<value><op><value>)<op><expr> | $[[], [0], [2, 0, 3], [1, 1, 0, 0]]$ |
| (0.5 <op><value>)<op><expr> | $[[], [0], [2, 0, 3], [1, 0, 0]]$ |
| (0.5/<value>)<op><expr> | $[[], [0], [0, 3], [1, 0, 0]]$ |
| (0.5/1)<op><expr> | $[[], [0], [0, 3], [0, 1]]$ |
| (0.5/1) + <expr> | $[[], [0], [3], [0, 1]]$ |
| (0.5/1) + <value><op><value> | $[[], [], [3], [0, 0]]$ |
| (0.5/1) + $x_1$ <op><value> | $[[], [], [3], [0]]$ |
| (0.5/1) + $x_1$ * <value> | $[[], [], [], [0]]$ |
| (0.5/1) + $x_1$ * $x_1$ | $[[], [], [], []]$ |

## 2.4  Related Work

In this section we briefly describe the main developments related with grammar-based evolutionary approaches. For an in depth review the reader might refer to [7].

GE is perhaps the most well known grammar-based GP variant. The separation between the search space and the problem space is seen as one of its advantages, which allows it to be easily used in different problem domains. These characteristics appeals to practitioners of various scientific domains, since they can easily use GE to solve their problems. However it is not exempted from criticisms. One of the first was about its initialisation procedure, which made it difficult to create populations with valid individuals [9]. To overcome the initialisation problem, GE adopted a method similar to the one proposed in GP and CFG-GP [14].

Another criticisms of GE are its high redundancy, and its low locality. Rothlauf *et al.* in [12] showed that in approximately 90% of the times a change in the genotype does not change the phenotype. Another important result of this work is concerned with the remaining 10% of the modifications. Specifically, when the genotype suffers one mutation, changes of more than one units occur at the phenotypic level. This means that many genotypic neighbours originate highly dissimilar phenotypes. One of the first proposals made to increase the locality is by Byrne et al. [1,2]. They proposed new mutation operators for GE that worked on the phenotypic level, which only changed the labels of the nodes in the derivation tree.

Another important development that aims at overcoming the locality and redundancy issues is the aforementioned SGE. This form of GE took a different approach on how to represent the solutions in the search space (i.e., the genotype), resulting in a EA that can properly sample the search space. This raises the question of how does the SGE performance compare to CFG-GP, which we try to answer in this work.

## 3   Experimental Framework

The framework used to compare the different grammar-based approaches is the one described in [15]. The CFG-GP system is based on the standard derivation-tree model [14], since this was one of the early grammar-based systems and worked as the foundation for later approaches. We use the *vannila* version of the GE implementation, with no improved search operators nor special initialisation mechanisms. The SGE version is identical to the one described in [6]. In addition to these methods we also consider a random search baseline model. This model iteratively generates individuals relying on the random initialisation method of GE, and records the best individual found in all the trials. To make it consistent with the GE, SGE, and CFG-GP methods, 51 groups of 1000 individuals are generated for each run. This is equivalent to generate 51 initial populations of 1000 individuals. The parameters used for each approach are presented in Table 3.

**Table 3.** Parameters used in the experimental analysis for each method.

| Parameter | Value | | |
|---|---|---|---|
| | SGE | GE | CFG-GP |
| Population size | 1000 | | |
| Generations | 50 | | |
| Selection method | Tournament with size 3 | | |
| Elitism | 10% | | |
| Crossover rate | 0.9 | | |
| Mutation rate | 0.05 | 0.05 | 0.1 |
| Crossover | Strctured | Single-point (variable length) | Subtree |
| Mutation | Pointwise (per codon) | Pointwise (per codon) | Subtree |
| Crossover node slection | Uniform | Uniform | Koza-style |
| Initialisation | Random | Random | Ramped-half & half |
| Min. initialisation depth | - | - | 2 |
| Max. initialisation depth | - | - | 6 |
| Max. tree depth | - | - | 17 |
| Initial codon length | - | 200 | - |
| Wrapping | - | None | - |
| Max. recursion level | 6 | - | - |

## 3.1   Problem Description

The problems that were considered in the experiments are the quartic symbolic regression, the Boston housing symbolic regression, the 5-bit even parity, the 11-Bit Boolean multiplexer and the santa fe ant trail. All the problems consider the minimisation of an error. For the regression problems, i.e., quartic and the Boston Housing, we used the Root Relative Squared Error (RRSE) which is 0 for a model with a perfect fit. For the 5-Bit Parity and the 11-Bit Boolean multiplexer we count the number of test cases that were incorrectly predicted. Finally, for the Santa Fe Ant trail we consider the number of food pellets left after the maximum number of steps has been achieved.

*Quartic Symbolic Regression.* The goal of this problem is to approximate the function defined by:

$$f(x) = x^4 + x^3 + x^2 + x + 1 \tag{1}$$

where x is sample in the interval $[-1, 1]$ with a step $s = 0.1$ The set of production rules considered for this problem is:

$$< start > :: = < expr >$$
$$< expr > :: = < expr >< op >< expr >$$
$$|(< expr >< op >< expr >)$$
$$| < pre\_op > (< expr >)$$
$$| < var >$$
$$< op > :: = + | - | * | /$$
$$< pre\_op > :: = sin|cos|exp|inv|log$$
$$< var > :: = x|1.0$$

where *inv* is $\frac{1}{f(x)}$. Moreover, we considered that division, and the logarithm functions are protected, i.e., $\frac{1}{0} = 1$ and $log(f(x)) = 0$ *iff* $f(x) \leq 0$.

*Boston Housing Problem.* This is a regression dataset from the UCI repository [5]. The dataset is composed by the housing prices from the suburbs of Boston, and the aim is to create a regression model that can predict the median house price, given a set of demographic features. The dataset is composed by 506 examples, each one composed by 13 features (12 continuous, 1 binary), and one continuous output variable in the range $[0, 50]$. This problem corresponds to a typical machine learning task and we need to measure the ability of the evolved models to work with unseen instances. Following the guidelines from [15], the dataset is divide in 90% of the examples, used as the training set to learn a model, and the remaining 10% as the test set, to assess the performance of the model. The set of production rules is extended from the quartic problem, with the inclusion of the additional descriptive variables used for predicting house price to the non-terminal <var>.

*5-Bit Parity.* The aim of this problem is to evolve a boolean function that takes binary string of length 5 as input and returns a value that indicates whether the number of 1 s in the string is even (0) or odd (1). The production set for this problem is:

$$< start > ::= < B >$$
$$< B > ::= < B > and < B >$$
$$| < B > or < B >$$
$$|not(< B > and < B >)$$
$$|not(< B > or < B >)$$
$$| < var >$$
$$< var > ::= b_0|b_1|b_2|b_3|b_4$$

where $b_0, b_1, b_2, b_3, b_4$ are the input bits.

*11-Bit Boolean Multiplexer.* The task of the 11-bit boolean multiplexer is to decode a 3-bit address and return the value of the corresponding data register (d0, d1, d2, d3, d4, d5, d6, d7). The the boolean 11-multiplexer is a function of 11 arguments: three, $s_0$ to $s_2$ which correspond to the addresses and eight data registers, $i_0$ to $i_7$. The production set for this problem is defined as:

$$< start > ::= < B >$$
$$< B > ::= < B > and < B >$$
$$| < B > or < B >$$
$$|not(< B >)$$
$$|not(< B > or < B >)$$
$$|(< B >)if(< B >)else(< B >)$$
$$| < var >$$
$$< var > ::= s_0|s_1|s_2|i_0|i_1|i_2|i_3|i_4|i_5|i_6|i_7$$

*Santa Fe Ant Trail.* The goal of the Santa Fe Ant Trail problem is to evolve a set of instructions for an artificial agent so that it can collect a certain number of food pellets in a limited number of steps (650). The trail consists of 89 food pellets distributed in a $32 \times 32$ toroidal grid. The agent starts in the top-left corner of the grid facing east, and it can turn left, right, move one square forward, and check if the square ahead contains food. The production set for this problem is:

$$<start>::=<code>$$
$$<code>::=<line>$$
$$|<code>$$
$$<line>$$
$$<line>::=if\ ant.sense\_food():$$
$$<line>$$
$$else:$$
$$<line>$$
$$|<op>$$
$$<op>::=ant.turn\_left()$$
$$|ant.turn\_right()$$
$$|ant.move\_forward()$$

## 4   Experimental Results

The experimental results described in this section are depicted in terms of the mean best fitness value obtained for each generation, over the course of 30 independent runs. For each approach we also present the 95% confidence interval about the mean. To compare all approaches we used the Kruskal-Wallis nonparametric test due to the characteristics of the data, and since the initial populations are different for each method. When statistical significant differences were found at the group level, we used the Mann-Whitney *post-hoc* test, with Bonferroni Correction, to perform the pairwise comparisons. We considered a global level of significance of $\alpha = 0.05$.

The results for the quartic problem are shown in Fig. 4. Looking at the performance of the CFG-GP, it is possible to see a steep descent in the error, and, after 15 generations, the method has already converged to solutions with error 0. The general behaviour of SGE is relatively similar, although convergence is slower. The error gradually decreases over the course of generations to values

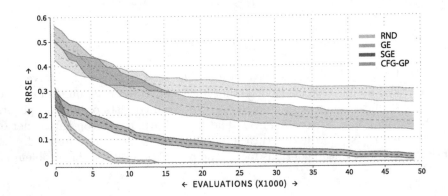

**Fig. 4.** Result for the quartic regression problem.

very close to 0. Although SGE does not reach a mean error of 0 in the time assigned to the optimisation, the line trend suggests that an extension of a few more generations would be enough to reach solutions with error 0. The behaviour of GE is noticeably different. For the first 10 generations its fitness is no better than the random model. After that time there is an improvement on the quality of the solutions, but its effectiveness is never comparable to neither CFG-GP, nor SGE. Moreover, looking at the confidence intervals around the mean, it is possible to see a substantial difference between the tested approaches. Whereas in SGE and CFG-GP the variance tends to decrease, in GE it remains stable throughout the run. This type of behaviour has been observed in GE [17], and it can been seen as an indication that GE is performing only slightly better than random search.

In terms of the statistical analysis, the Kruskal-Wallis test indicated that there are differences between the means. The *post-hoc* test revealed GE is marginally better than the random model at the end of the run, i.e., the p-value obtained is relatively close to the significance level. Concerning CFG-GP and SGE, at the end of the run, there are no statistically significant differences. However, when considering the solutions at generation 15, the CFG-GP is statistically better than SGE.

The results for the Boston Housing problem are depicted in Fig. 5. Looking at the plot lines from the training results (left panel), the CFG-GP and the SGE exhibit little differences amongst them. In spite of not finding statistical differences between CFG-GP and SGE, both approaches are statistically better than GE. In fact, looking at the plots from the GE and the random model there are almost no differences between them. This type of behaviour can be explained by the issues mentioned earlier affecting by GE, particularly the low locality [6,12].

The test results (right panel) are more important than the training ones when building predictive models. In this problem they reveal that both CFG-GP and SGE are building models that can work well on unseen data. Again,

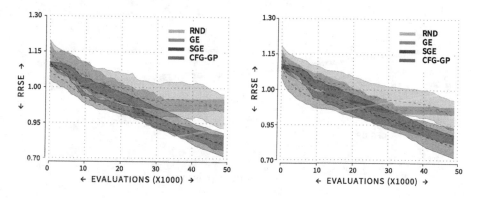

**Fig. 5.** Results for the Boston housing problem. The panel on the left shows the results for training while the panel on the right correspond to test results.

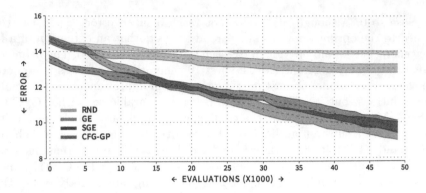

**Fig. 6.** Results for the 5-bit parity

there are no statistical significant differences between the solutions found by the two approaches. On the contrary, the behaviour exhibited by GE is very similar to the random search model, with no statistical differences between them. In addition, after 23 generations GE shows none or little improvement.

The results of the 5-bit even parity problem are depicted in Fig. 6. For the first 14 generations, SGE is better than the other three. Then, CFG-GP starts to exhibit an enhanced performance, decreasing the error faster. However, by the end of the evolutionary run, both models have similar results, with no statistical significant differences between them. Note that the variance of the SGE model is sightly smaller than the CFG-GP. Concerning the GE model, it keeps up with CFG-GP int the first 10 generations. However after this point, the rate of improvement decreases, resulting in a loss of performance, and by the end of the run, the GE results are only marginally better than the ones of the random model, with the comparison *p-value* being very close to the significance level.

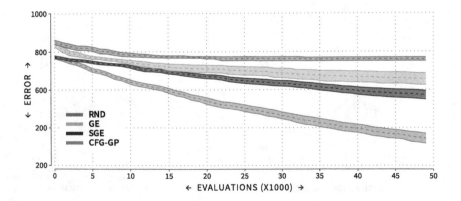

**Fig. 7.** Results for the 11-bit Boolean multiplexer

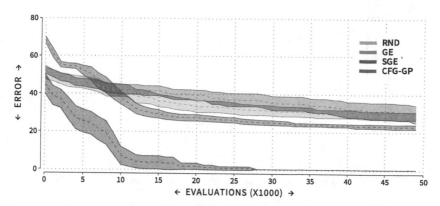

**Fig. 8.** Results for the Santa Fe Ant trail

The 11-Bit Boolean multiplexer results are shown in Fig. 7. The plot shows that the CFG-GP model clearly outperforms all the others. SGE is statistically better than GE, which in turn is statistically better than the random model. The difference between the results obtained by the CFG-GP and the SGE representations can be explained by the fact that latter has a fixed level of recursion. In spite of being possible to have good quality solutions of depth 5, it might be necessary for the evolutionary process to create a large tree (of depth 9 or 10) and then prune it. Whilst the CFG-GP method allows for this, the SGE method does not, due to the fixed limit on the recursion depth. To get similar solutions SGE would require more evaluations to ensure a proper exploration of the search space.

Finally the results for the Santa Fe ant problem are shown in Fig. 8. The results show that SGE rapidly decreases the error, and after 30 generations, has found solutions that can eat all the food pellets on the trail. On the contrary, the performance of CFG-GP is very poor. These results are confirmed by previous studies that show that the problem is hard for GP [4]. The difference between the two approaches can be explained by different tree shapes in the initial sample of each approach. In Fig. 9 we present a sampling of 10000 individuals randomly generated for CFG-GP with maximum initial depth of 6 and for SGE. The plot shows the number of non-terminal nodes presented in each tree. We relied on the number of non-terminal to avoid counting nodes that are part of the syntactic sugar needed to execute the program. We can see that SGE does a proper sampling of the search space, generating trees with different sizes. On the other hand, CFG-GP seems to be biased towards the generation of small trees with no more than 15 nodes. With this initial sampling, CFG-GP is not able to find good quality solutions within the time given. Concerning GE, its performance, on average, is no better than random search.

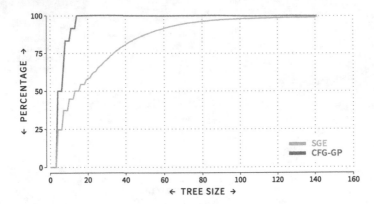

**Fig. 9.** Cumulative frequency of different tree sizes in a random sample with 10000 individuals for SGE and CFG-GP. The size of the tree is equal to the number of non-terminal nodes that composes it.

## 5    Conclusions

Grammars are very useful tools to represent constraints. Since the early 1990s, when GP was becoming a well established field, researchers have used grammars as a way of guiding the evolutionary process. One of these first proposals was CFG-GP proposed by Whigham. However, the evolutionary methods based on grammars hit the spotlight with GE. Nowadays GE is one of the most used and studied GP methods. It is known that it suffers from some issues that impact its overall performance. In order to overcome these issues, SGE is a recent representation to GE that has showed a good performance in some difficult problems.

In recent work, Whigham *et al.* [15] have performed a comparison between CFG-GP and GE, which has brought to light the issues with GE. The conclusions of this work was that the standard GE tended to perform worse than CFG-GP in the benchmarks used for comparison.

In this paper we used the same experimental framework detailed by Whigham *et al.* work, and compared the novel SGE representation with CFG-GP and GE. The results confirm that SGE is always better than GE in the benchmark suite used. When we compared the CFG-GP method with SGE we saw that no method is better than the other in regression problems, nor in the 5-Bit Parity problem. On the 11-Bit Boolean multiplexer, the CFG-GP is better than SGE. Finally, on the Santa Fe Ant trail is the other way around, i.e., the SGE model is better than the CFG-GP.

We provide some preliminary insights suggesting why a given representation outperforms the other on a specific scenario, but certainly more research is needed to understand which type of problems are appropriated for each approach.

In spite of being a recent grammar-based method SGE has evinced a good performance over different kinds of problems, even when compared with established well state-of-the-art methods, such as CFG-GP. Paraphrasing the title of the work presented in [15], SGE might be a relevant step towards the "Best of Both Worlds" of GE and CFG-GP.

# References

1. Byrne, J., O'Neill, M., Brabazon, A.: Structural and nodal mutation in grammatical evolution. In: Proceedings of the 11th Annual Conference on Genetic and Evolutionary Computation, New York, pp. 1881–1882 (2009)
2. Byrne, J., O'Neill, M., McDermott, J., Brabazon, A.: An analysis of the behaviour of mutation in grammatical evolution. In: Esparcia-Alcázar, A.I., Ekárt, A., Silva, S., Dignum, S., Uyar, A.Ş. (eds.) EuroGP 2010. LNCS, vol. 6021, pp. 14–25. Springer, Heidelberg (2010). doi:10.1007/978-3-642-12148-7_2
3. Koza, J.R.: Genetic Programming: On the Programming of Computers by Means of Natural Selection, vol. 1. MIT Press, Cambridge (1992)
4. Langdon, W.B., Poli, R.: Why ants are hard. In: Koza, J.R., Banzhaf, W., Chellapilla, K., Deb, K., Dorigo, M., Fogel, D.B., Garzon, M.H., Goldberg, D.E., Iba, H., Riolo, R. (eds.) Genetic Programming 1998: Proceedings of the Third Annual Conference, 22–25 July 1998, pp. 193–201. Morgan Kaufmann, University of Wisconsin, Madison, Wisconsin, USA (1998)
5. Lichman, M.: UCI machine learning repository (2013). http://archive.ics.uci.edu/ml
6. Lourenço, N., Pereira, F.B., Costa, E.: Unveiling the properties of structured grammatical evolution. Genet. Program. Evolvable Mach. **17**(3), 251–289 (2016)
7. Mckay, R.I., Hoai, N.X., Whigham, P.A., Shan, Y., ONeill, M.: Grammar-based genetic programming: a survey. Genet. Program. Evolvable Mach. **11**(3–4), 365–396 (2010)
8. O'Neill, M., Ryan, C.: Grammatical evolution. IEEE Trans. Evol. Comput. **5**(4), 349–358 (2001)
9. ONeill, M., Ryan, C.: Grammatical Evolution: Evolutionary Automatic Programming in an Arbitrary Language. Genetic Programming, vol. 4. Springer, New York (2003)
10. Rothlauf, F.: On the locality of representations. In: Cantú-Paz, E., et al. (eds.) GECCO 2003. LNCS, vol. 2724, pp. 1608–1609. Springer, Heidelberg (2003). doi:10.1007/3-540-45110-2_48
11. Rothlauf, F.: Representations for Genetic and Evolutionary Algorithms. Springer, Heidelberg (2006)
12. Rothlauf, F., Oetzel, M.: On the locality of grammatical evolution. In: Collet, P., Tomassini, M., Ebner, M., Gustafson, S., Ekárt, A. (eds.) EuroGP 2006. LNCS, vol. 3905, pp. 320–330. Springer, Heidelberg (2006). doi:10.1007/11729976_29
13. Ryan, C., Collins, J.J., Neill, M.O.: Grammatical evolution: evolving programs for an arbitrary language. In: Banzhaf, W., Poli, R., Schoenauer, M., Fogarty, T.C. (eds.) EuroGP 1998. LNCS, vol. 1391, pp. 83–96. Springer, Heidelberg (1998). doi:10.1007/BFb0055930
14. Whigham, P.A.: Inductive bias and genetic programming. In: First International Conference on Genetic Algorithms in Engineering Systems: Innovations and Applications, GALESIA (Conf. Publ. No. 414), pp. 461–466. IET (1995)
15. Whigham, P.A., Dick, G., Maclaurin, J., Owen, C.A.: Examining the best of both worlds of grammatical evolution. In: Proceedings of the 2015 Annual Conference on Genetic and Evolutionary Computation, pp. 1111–1118. ACM (2015)
16. Whigham, P.A., et al.: Grammatically-based genetic programming. In: Proceedings of the Workshop on Genetic Programming: From Theory to Real-World Applications, vol. 16, pp. 33–41 (1995)
17. White, B.C., Reif, D.M., Gilbert, J.C., Moore, J.H.: A statistical comparison of grammatical evolution strategies in the domain of human genetics. In: 2005 IEEE Congress on Evolutionary Computation, vol. 1, pp. 491–497. IEEE (2005)

# A Comparative Analysis of Dynamic Locality and Redundancy in Grammatical Evolution

Eric Medvet[(✉)]

Department of Engineering and Architecture, University of Trieste,
Trieste, Italy
emedvet@units.it

**Abstract.** The most salient feature of Grammatical Evolution (GE) is a procedure which maps genotypes to phenotypes using the grammar production rules; however, the search effectiveness of GE may be affected by low locality and high redundancy, which can prevent GE to comply with the basic principle that offspring should inherit some traits from their parents. Indeed, many studies previously investigated the locality and redundancy of GE as originally proposed in [31]. In this paper, we extend those results by considering redundancy and locality during the evolution, rather than statically, hence trying to understand if and how they are influenced by the selective pressure determined by the fitness. Moreover, we consider not only the original GE formulation, but three other variants proposed later (BGE, $\pi$GE, and SGE). We experimentally find that there is an interaction between locality/redundancy and other evolution-related measures, namely diversity and growth of individual size. In particular, the combined action of the crossover operator and the genotype-phenotype mapper makes SGE less redundant at the beginning of the evolution, but with very high redundancy after some generations, due to the low phenotype diversity.

**Keywords:** Genetic programming · Diversity · Genotype-phenotype mapping · Genetic operators

## 1 Introduction

Grammatical Evolution (GE) [19,31] is a form of grammar-based [14] Genetic Programming (GP) [9] which can evolve programs in an arbitrary language, provided that a context-free grammar (CFG) describing that language is available. The salient and distinguishing feature of GE is that each candidate solution, i.e., each individual, is represented as a bit string (*genotype*) which is transformed into the actual program by means of a mapping procedure. The mapping procedure consumes the genotype in order to select a specific sequence of production rules of the grammar to be used: the outcome (*phenotype*) is a tree whose leaves are non-terminal symbols of the grammar which constitute the program.

GE has been successfully used for approaching a wide range of applications, where the evolved artifacts spanned from actual computer programs, e.g., for

© Springer International Publishing AG 2017
J. McDermott et al. (Eds.): EuroGP 2017, LNCS 10196, pp. 326–342, 2017.
DOI: 10.1007/978-3-319-55696-3_21

computing the similarity among strings [1], to rules expressed according to a specific language, e.g., for market trading [23], or geometrical shapes complying with predefined constraints [17]. In all cases, GE allowed practitioners and researchers to exploit Evolutionary Algorithms (EAs) without requiring them to adapt some EA aspects to their specific application. In other words, given any grammar and an appropriate fitness function defining a problem, GE allows to (try to) solve that problem.

Despite its widespread adoption, however, GE has been criticized for being, among the many different EAs, a variant which suffers from two issues: low locality [29] and high redundancy [32]. In general, *locality* is the property of an evolutionary algorithm according to which small modifications to an individual correspond to small deviations in its behavior: if this property is not or poorly reflected by an EA variant, that variant is said to have low locality. In GE, locality declines in two ways which reflect the working principle of GE: (a) the degree to which small modifications in the genotype correspond to small modifications in the phenotype, and (b) the degree to which small modifications in the phenotype correspond to small modifications in the fitness. While the latter is mainly determined by the problem itself, i.e., by the language and the fitness function, the former depends on the genotype-phenotype mapping procedure and, in the context of the evolution, on the genetic operators which cause the modifications of the genotype.

Other than locality, which applies to every EA where an individual is assessed by means of a fitness function, redundancy is specific to those EAs where individual representation is someway decoupled, as in GE. In brief, *redundancy* in GE is the property according to which many different genotypes result in the same phenotype. As for the locality of GE mapping procedure, redundancy in GE is not related to the problem being solved, but to the mapping procedure, which in essence can be viewed as a function operating between two spaces (genotype and phenotype) whose injectivity determines the redundancy. Redundancy in GE may be harmful because, at least, it slows down the search and, in case it is not uniform, makes some regions of the search space very unlikely to be reached [32].

Low locality and high redundancy of GE are well known in the EA research community: the fact that they are issues or, rather, features of GE is itself subject of an ample debate. In this paper, we intend to advance the understanding of locality and redundancy in GE by means of two contributions. First, we perform a *comparative* analysis in which contenders are four among the most relevant and recent variants in the genotype-phenotype mapping procedure of GE: the standard original version [31], the Breadth-first variant [4], Position-independent Grammatical Evolution ($\pi$GE) [22], and Structured Grammatical Evolution (SGE) [11]. Second, we study locality and redundancy in a *dynamic* context, that is, we observe if and how the locality and redundancy of a given GE variant change during the evolution, as a result of the combined action of several factors, the foremost being—besides the mapping procedure—the operators, the genotype/phenotype size, and the selective pressure. The latter is clearly related to the fitness function and hence this kind of analysis cannot be done abstractly, i.e., by reasoning on one single problem and extending the findings to the general case. To this end, we experimented with different problems, including some benchmark problems following the guidelines for experimental assessment of GP approaches [13, 35], and a problem which we defined ad hoc in order to facilitate our analysis.

Our experiments show that there is a strong interaction between redundancy/locality and other factors, mainly individual size and population diversity. This interaction manifests after the initial phase of the evolution has ended and hence cannot be observed by performing a static experimentation, i.e., one in which the fitness function does not play any role.

## 2   Related Work

As sketched in the previous section, since the birth of GE in [31], many studies have been carried out about this technique. We here focus on two kinds of works: (a) those explicitly targeting locality/redundancy in GE and (b) those studying other aspects of GE that might impact on locality/redundancy.

The seminal study about locality in GE has been conducted by Rothlauf and Oetzel [29]. The authors focus on the GE genotype-phenotype mapping procedure and start from the long-established principle [3] that if similar solutions do not similarly solve the problem, the optimization performed by an EA tends to resemble a random search. In their study, they define locality in GE as the degree to which neighboring genotypes are also neighboring phenotypes, where two genotypes (phenotypes) are neighbors if their distance is minimal. The authors use the Hamming distance for genotypes and the tree edit distance for phenotypes. In our work, we used the edit distance in the former case, because, in the dynamic scenario, the length of the genotypes may change as a result of operators application, and the Hamming distance is not defined for sequences with different length; moreover, we also explored the use of edit distance (besides tree edit distance) for the phenotypes (see Sect. 4). In [29], the locality is assessed experimentally by applying one-bit flip mutations to a number of randomly generated genotypes mapped to phenotypes according to a simple grammar. As a side effect, [29] also measures the percentage of cases in which such distance is zero, a figure which represents an estimate of the redundancy. The result is twofold: in 90% of cases a mutation does not change the phenotype and, in the remaining 10% of cases, a significant fraction corresponds to mutations causing heavy changes in the phenotype.

A more recent study [33] by the same research group focused on the locality of search operators and considered both GE and GP. In this case, the key motivation of the study is to assess experimentally to which degree the search operators produce an offspring which is similar to its parent(s). In particular, mutation should guarantee that the distance between the offspring and its parent is bounded, whereas crossover should guarantee that the distance of the offspring from its farthest parent is lower than the distance between the two parents—this principle being introduced previously [10] and later denoted as "locality of search operators" [27] or "geometry of search operators" [15]. As in [29], the locality is assessed experimentally by measuring distances resulting from operator application during a random search—i.e., without applying the selective pressure given by the fitness, as we do in our present work. The result of the cited paper is that both GE and GP standard operators suffers from low locality: however, in GE locality is slightly greater.

A different point of view on how GE mutation and crossover affect phenotypes is given by [2]. The authors aim at investigating the destructive (or constructive) nature of those operators and, in particular, to verify if it is related to the position in which the operator actually applies: in facts, due to the mapping procedure of GE, modifications to the leftmost bits of a genotype are expected to cause changes close to the root of the tree in the phenotype. Differently than aforementioned works, in [2] the focus is not on locality as defined before, but rather on whether an operator application results in an increased (constructive) or decreased (destructive) fitness: however, the authors do not study if and how that figure varies during the evolution, as we do in the present study with locality. The result is that genotype modifications which occur at leftmost positions are in general more destructive on the phenotype—as expected—but, at the same time, may result in the largest increments in the fitness.

The locality of different genotype representations, here intended as the way in which integer are codified by bits slices (codons) in the genotype, is studied in [6,7]. The authors compare experimentally the binary and grey code representations and their combined effect on locality with 4 different mutation operators. They find that the standard combination is not improved by other options, neither in locality nor in effectiveness: moreover, this result induces the authors to wonder if locality is indeed beneficial to GE performance.

Since the locality issue in GE has been often attributed to the genotype-phenotype procedure, it is not surprising that improvements to that procedure have been proposed which try to tackle the issue. This trend has also been studied recently in [34], where the authors noted how GE, which was born as a purely grammar-agnostic approach, slowly incorporated information from the grammar into crossover, mutation, and individual initialisation, blurring the distinction between genotype and phenotype.

A first comparative analysis of different genotype-phenotype mappings has been conducted in [4], including the standard GE mapping, its breadth-first variant, and $\pi$GE. The experimental campaign showed an advantage of the latter, but the authors themselves acknowledged that more research was needed to fully comprehend the reasons of their findings.

Recently, and later than the previously cited works, a novel variant of GE, called Structural GE (SGE), has been introduced [11] (see Sect. 3.4). SGE mapping procedure has been designed with high locality and low redundancy as first-class goals and its experimental evaluation on a set of benchmark problems showed that it outperformed the standard GE. The good performance of SGE has been later analysed in [12], where several experiments are described, including some concerning locality and redundancy. The authors find that SGE has a better locality than GE and argue that, together with lower redundancy, this is among the key factors for the superiority of their approach. In this present study, we included SGE in our comparative analysis.

Reduncancy in GE has been the focus of fewer works than locality—more works, instead, considered redundancy in other EAs or in general [28]. In a very recent study [32], redundancy in GE has been characterized beyond its simple quantification. The author analyzed the entire solution space resulting from two simple grammars and a fixed genotype length. She concludes that the standard GE mapping procedure is strongly non-uniformly redundant, i.e., some phenotypes are mapped from many genotypes whereas other phenotypes are mapped by few genotypes.

## 3   GE Variants

We compared four variants of GE in our analysis: standard GE, breadth-first GE, $\pi$GE, and SGE. Three of them differ only in the mapping procedure, whereas the fourth and most recent (SGE) also bases on different versions of the genetic operators. We did not include in the analysis few other GE variants we were aware of (such as, e.g. [8,30]), because they introduced minor changes over standard GE or did not result in relevant improvements. We present the 4 variants in the following sections, detailing in particular the genotype-phenotype mapping procedure.

We recall that, for all GE variants, the mapping procedure maps a genotype $g$ into a phenotype $p$ using the production rules of a CFG $\mathcal{G} = (N, T, s_0, R)$, where $N$ is the non-empty set of non-terminal symbols, $T$ is the non-empty set of terminal symbols, $s_0 \in N$ is the starting symbol (or axiom), and $R$ is the set of production rules. The phenotype $p$ can be either viewed (a) as a tree whose leaves are elements of $T$ that, visited from left to right, are string of the language $\mathcal{L}(\mathcal{G})$ defined by $\mathcal{G}$; or, more simply, as (b) a string of the language $\mathcal{L}(\mathcal{G})$.

### 3.1   Standard GE

In standard GE [31], the genotype is a variable-length bit string: it can be read as a sequence of integers by grouping $n$ consecutive bits, each integer being called *codon*. The value of $n$ is a parameter of the mapping procedure and should be large enough so that $2^n$ is greater or equal than the maximum number of options in the rules of the grammar. Conventionally, it is set to $n = 8$, but in some applications (e.g., [1]) it has been set to the lowest value which met the above criterion.

The mapping procedure of GE works as follows. Initially, the phenotype is set to $p = s_0$, i.e., to the grammar starting symbol, the counter $i$ is set to 0, and the counter $w$ is set to 0. Then, the following steps are iterated. (1) Expand the leftmost non-terminal $s$ in $p$ by using the $j$-th option (zero-based indexing) in the production rule $r_s$ for $s$ in $\mathcal{G}$. The value of $j$ is set to the remainder of the division between the value $g_i$ of the $i$-th codon (zero-based indexing) and the number $|r_s|$ of options in $r_s$, i.e., $j = g_i \mod |r_s|$. (2) Increment $i$; if $i$ exceeds the number of codons, i.e., if $i > \frac{|g|}{n}$, then set $i$ to 0 and increment $w$—the latter operation is called wrapping and $w$ represents the number of wraps performed during the mapping. (3) If $w$ exceeds a predefined threshold $n_w$, then abort the mapping, i.e., produce a *null phenotype* and set its fitness to the worst possible value. (4) If $p$ contains at least one non-terminal to be expanded, return to step 1, otherwise end.

With wrapping the genotype may be reused: each codon may hence concur in defining the expansion of different non-terminals—a feature which has been called polymorphism [18]. Wrapping has been included in order to make GE work with any CFG, including recursive grammars which, in turn, correspond to languages containing non-finite strings. On the other hand, the upper bound $n_w$ to the number of wrapping operations is imposed to avoid an endless mapping. Clearly, the choice of the size $|g|$ of the genotype plays a crucial role in this respect and cannot be determi ned easily. In some application, such as in [1], $|g|$ is chosen such that one or more target phenotypes can be generated without resorting to wrapping.

It is straightforward to relate the role of the rule option choice driven by a module operation ($j = g_i \mod |r_s|$) with the redundancy. For instance, in a genotype for which the mapping does not result in any wrapping, all the values of a codon for which the remainder of the division remains the same correspond to the same phenotype. Moreover, if the genotype $g$ is such that only the first $k$ codons are used in the mapping, then any genotypes $g'$ that differs from $g$ in codons whose index is larger than $k$ will correspond to the same phenotype of $g$.

Evolution in GE is commonly driven by GA, but other approaches have been proposed [24], such as Particle Swarm Optimization (PSA) [21] and Differential Evolution (DE) [16]. In this work, we consider a classical GA configuration in which the genetic operators are the bit flip mutation (i.e., each bit in the phenotype may be flipped according to a predefined probability $p_{\text{mut}}$) and the one-point crossover. Among the many options for the two operators (mutation and crossover), we chose the two mentioned because they can result in genotype modifications of varying impact, which is functional to our analysis of locality. Concerning crossover, it has been shown that one-point crossover is not significantly worse than other options [5, 20].

## 3.2    Breadth-First GE

Breadth-first GE (we here denote it with BGE) has been introduced in [4] and only slightly differs from standard GE. In the iterative mapping procedure (at step 1), instead of choosing the leftmost non-terminal in $p$, the least deep (closest to the tree root) non-terminal is chosen.

Despite experiments in [4] show that BGE is not significantly better or worse than GE, we include it in our analysis because its underlying mapping mechanism might suggest that codon positions have a different role in determining the tree:

while in standard GE first codons concur in building the "left" portion of the tree, up to the leaves, in BGE first codons concur in building the "higher" portion of the tree (i.e., the root and its immediate descendant). This difference could in turn result in a different impact on locality.

All the other approach components of BGE—namely the operators—are the same of standard GE. The same considerations concerning redundancy in standard GE apply also to BGE.

### 3.3  πGE

A significant modification has been proposed to the standard GE mapping procedure by O. Neill et al. in [22], named Position-independent GE (πGE). The salient feature of πGE is that it decouples the choice of the non-terminal to be expanded from the choice of the specific expansion (i.e., option in the rule). The main idea behind this design choice is, according to authors, to break the codon dependency chain of the mapping in order to allow the arising, in the genotype, of short codon sequences which should act as building blocks, as subtrees do in GP. Despite the fact that the cited work did not provide any strong evidence of the actual arising of such building blocks, the experimental campaign showed that πGE indeed delivered better results than GE in the majority of the problems analysed in [22]. That result was further confirmed later in [4].

Differently than in GE, in πGE each codon consists of a pair $g_i^{\text{nont}}, g_i^{\text{rule}}$ integers, each of $n$ bits—conventionally, $n$ is set to 8. In the mapping procedure (at step 1), the non-terminal of $p$ to be expanded is not chosen statically (i.e., the leftmost in GE or the closest to the root in BGE) but is instead chosen as follows: let be $n_s$ the number of non-terminals in $p$, then the $j^{\text{nont}}$-th one is chosen, where $j^{\text{nont}} = g_i^{\text{nont}} \mod n_s$. Then, the rule option to be used is determined, as in standard GE, with $j^{\text{rule}} = g_i^{\text{rule}} \mod |r_s|$. In other words, in πGE it is the genotype itself which encodes the positions of the phenotype which are to be expanded and those positions evolve independently from corresponding contents.

As for BGE, operators and implications on redundancy do not change in πGE w.r.t. standard GE.

### 3.4  SGE

The most recent GE variant which we consider in our comparison is Structural GE (SGE), which has been first described in [11] and then more deeply analyzed in [12]. The key idea behind SGE is to structure the mapping such that each codon corresponds to exactly one expansion of a given non-terminal. The aim of this idea is, according to the authors, to increase the locality, since "the modification of a single gene [codon] does not affect the derivation options of other non-terminals". Moreover, SGE introduce two other relevant variations to the standard GE: (i) it redefines the genotype structure which results in a remarkable reduction in redundancy and (ii) it applies only to non-recursive grammars which results in the guarantee of mapping any genotype to a non-null phenotype (differently from GE, BGE, and $\pi$GE). The latter point may appear as a strong limitation in the approach applicability, since for most problems of practical interest the grammar is recursive. However, the authors of [11] state that the limitation should not severely affect SGE practicality and recall that in many other EAs there is an upper bound to the size of an individual (e.g., in GP). They also provide the sketch of a procedure for transforming any possibly recursive grammar $G$ into a non-recursive grammar $G'$ by imposing a maximum tree depth: it is yet not fully clear how that parameter should be set in problems where the tree depth of an acceptable solution is not known in advance and, from another point of view, if and how its value affects SGE effectiveness in searching for a solution.

Differently than in previously presented variants, in SGE the genotype $g$ is a fixed-length integer string. The string is itself composed of $|N|$ substrings, i.e., one substring $g_s$ for each non-terminal $s \in N$ of the grammar $G$. The length of (i.e., number of codons in) each substring $g_s$ is determined by the maximum number of expansions which can be applied to the corresponding non-terminal $s$ according to the non-recursive grammar $G'$. Moreover, the domain of each codon is set to $\{0, \ldots, |r_s| - 1\}$, where $r_s$ is the production rule for $s$. It follows that the redundancy in SGE should be reduced w.r.t. GE, BGE, and $\pi$GE, since the modulo is not needed: the only redundancy which remains is the one related to the possibility that some codons are not used in the mapping because of the small size of the phenotype.

The mapping procedure of SGE works as follows. Initially, the phenotype is set to $p = s_0$, and a counter $i_s$ for each non-terminal $s \in N$ is set to 0. Then, the following steps are iterated. (1) Expand the leftmost non-terminal $s$ in $p$ by using the $g_{s,i_s}$-th option (zero-based indexing) of the rule $r_s$, $g_{s,i_s}$ being the value of the $i_s$-th codon (zero-based indexing) in $g_s$. (2) Increment $i_s$. (3) If $p$ contains at least one non-terminal to be expanded, return to step 1, otherwise end. It can be noted that, in step 1, there is no need to use the modulo on the codon value, since its domain is $\{0, \ldots, |r_s| - 1\}$.

The operators in SGE are peculiar to that approach. The mutation is a probabilistic (according to a parameter $p_{mut}$) codon mutation which is guaranteed to set a random new codon value which stays in the proper domain. The crossover works as follows: given two genotypes $g^1$ and $g^2$, a subset $N' \subseteq N$ of non-terminals is randomly chosen; then, for each $s \in N'$, the two corresponding genotype substrings $g_s^1, g_s^2$ are swapped.

## 4   Experiments and Results

Our aim is to study if and how locality and redundancy vary during the evolution across different GE variants; as a secondary goal, we are interested in studying the tendency to generate null phenotypes. We here define the *indexes* by means of which we measured these concepts; coping with a dynamic scenario, we cast definitions considering measures to be collected at each generation. We denote with $d_g(i_1, i_2)$ and $d_p(i_1, i_2)$ the genotype and phenotype distance among two individuals $i_1, i_2$, respectively.

- *Invalidity.* The percentage of not-*valid* individuals being generated, i.e., those in which the phenotype is null.
- *Redundancy.* The percentage of valid individuals $i$ being generated in which the genotype of $i$ is different from both parent genotypes and the phenotype of $i$ is equal to at least one of the two parent phenotypes.
- *Locality.* The correlation between the genotype distance from the closest parent $\min(d_g(i, i_1), d_g(i, i_2))$ and the phenotype distance from the closest parent $\min(d_p(i, i_1), d_p(i, i_2))$, considering valid individuals for which both distances are not zero—we used Pearson correlation. The rationale for this index is to generalize the high level definition of locality as a measure of the degree to which small modifications in the genotype correspond to small modification in the genotype. From another point of view, our definition extends the one introduced in [26] (and later used in [12]) which, basically, corresponds to the mean of $\min(d_p(i, i_1), d_p(i, i_2))$, measured separately for mutation and crossover and called Mutation Innovation (MI) and Crossover Innovation (CI), respectively.

All of the 4 indexes can be measured w.r.t. a specific operator or globally: we present the results for invalidity globally, and for each operator for locality and redundancy.

We considered 4 problems: 3 are established benchmark problems (Harmonic regression, Polynomial regression, and Santa-Fe trail—see, e.g., [12] for the corresponding grammars) whereas the last (Text) is a problem that we designed specifically for this analysis.

- *Harmonic.* This is a symbolic regression problem in which the goal is to approximate the function $f(x) = \sum_i^x \frac{1}{i}$ in the $x$ values $\{1, \ldots, 50\}$. The fitness is given by the sum of absolute errors, to be minimized.
- *Polynomial.* This is another symbolic regression problem in which the goal is to approximate the function $f(x) = x^4 + x^3 + x^2 + x$ in the $x$ values $\{-1, -0.9, \ldots, 0.9, 1\}$. The fitness is given by the sum of absolute errors, to be minimized.
- *Santa-Fe.* This is a classic path-finding problem which consists in finding a program able to guide an artificial ant to reach 89 statically placed food items in a $32 \times 32$ grid given a maximum number of steps. The fitness is given by the number of missed food items, to be minimized.
- *Text.* The goal of this problem is to generate a string which matches a statically defined target, which was `Hello world!` in our experimentation. The fitness is given by the edit distance between the string encoded by the individual and the target string. The rationale for including this problem is to (try to) neutralize the influence of the phenotype to fitness mapping: in facts, the phenotype distance between individuals corresponds (when using the edit distance for phenotypes) by definition to their fitness distance. Figure 1 shows the grammar for this problem.

For each problem and each GE variant, we performed 30 different runs, with the settings shown in Table 1—for SGE, we experimented with two values for the max tree depth. In each run, after each generation, we measured the 3

```
       <text> ::= <sentence> ␣ <text> | <sentence>
   <sentence> ::= <Word> ␣ <sentence> | <word> ␣ <sentence> | <word> <punct>
       <word> ::= <letter> <word> | <letter>
       <Word> ::= <Letter> <word>
     <letter> ::= <vowel> | <consonant>
      <vowel> ::= a | e | i | o | u
  <consonant> ::= b | c | d | ... | z
     <Letter> ::= <Vowel> | <Consonant>
      <Vowel> ::= A | E | I | O | U
  <Consonant> ::= B | C | D | ... | Z
      <punct> ::= ! | ? | .
```

**Fig. 1.** The CFG for the text problem.

indexes defined above. For the locality, we used as $d_g$ the edit distance (for GE, BGE, and $\pi$GE) and the Hamming distance (for SGE) and as $d_p$ both the edit distance (between phenotypes viewed as strings of $\mathcal{L}(\mathcal{G})$) and the tree edit distance (between phenotypes viewed as trees and computed using the method of [25]) for all the variants. For brevity, we here present results only for the tree edit distance, but in essence they do not change when considering the edit distance, which is also much faster to compute.

**Table 1.** Settings for the experimental analysis.

| | GE, BGE, $\pi$GE | SGE |
|---|---|---|
| Population | 500 | 500 |
| Pop. initialization | Random | Random |
| Generations | 50 | 50 |
| Crossover rate | 0.8 | 0.8 |
| Crossover operator | One-point | SGE crossover |
| Mutation rate | 0.2 | 0.2 |
| Mutation operator | Bit flip with $p_{mut} = 0.01$ | SGE mutation with $p_{mut} = 0.02$ |
| Selection | Tournament with size 5 | Tournament with size 5 |
| Replacement | $m + n$ strategy with $m = n$ | $m + n$ strategy with $m = n$ |
| Initial genotype size | 1024 | n.a. |
| Max wraps | 5 | n.a. |
| Max tree depth | n.a | 6, 10 |

## 4.1   Results and Discussion

Figure 2 presents the results in form of several plots. There is one column of plots for each problem and one row of plots for each index. In each plot, the $x$ axis represents the generation, the $y$ axis represents the index: there is one curve for

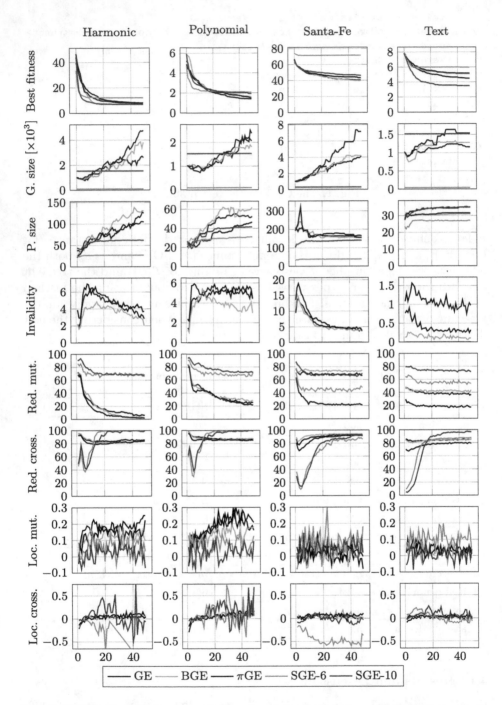

**Fig. 2.** Results: one plot row per index, one plot column per problem, one curve per GE variant.

each GE variant. The curves are obtained by averaging the measures collected at each generation across the 30 runs. The figure is the result of the evaluation of an overall $\approx 14700000$ operator applications.

**Context: Fitness and Individual Size.** In order to provide some context for the investigation about invalidity, locality, and redundancy, the first 3 plot rows of Fig. 2 show the fitness, genotype size, and phenotype size of the best individual. The plots about best fitness show that (in our configuration, which, we remark, has been chosen to facilitate our analysis rather than for obtaining good fitness values) GE, BGE, and $\pi$GE exhibit similar behavior. Concerning SGE, there is a difference between the two values of maximum depth: SGE-10 is significantly better ($p \leq 0.001$ with the Mann-Whitney test) than all the other variants on the Text problem, whereas SGE-6 is significantly worse than all the others on the Santa-Fe problem. This difference suggests that the choice of the maximum depth may be crucial.

The plots about the genotype size of the best individual show that for GE, BGE, and $\pi$GE, the index increases during the evolution, roughly linearly, for all but the Text problem—for SGE, the genotype size is fixed by design. For the Text problem, the genotype size tends to remain constant after some tens of generations and is, in general, smaller than for the other problems at the end of the evolution. Concerning phenotype size, in GE, BGE, and $\pi$GE it increases for the two symbolic regression problems, whereas it stabilizes after few generations for the other two problems: this can be explained by the fact that, for Santa-Fe and Text, large phenotype sizes may negatively affect the fitness (more instructions, longer text), whereas more complex expressions may not result in less accurate regression than simple expressions. Differently, with SGE the phenotype early reaches a stable size which is, for all but the Text problem, lower than for the other GE variants, being always lower for SGE-6 than for SGE-10. We analyzed the runs in order to better understand this behavior and we found that SGE suffers (in our experimentation) of a low phenotype diversity, measuread ratio between the number of unique phenotypes and the population size. A detailed analysis about diversity in different variants of GE is beyond the scope of this paper, but we speculate that low diversity likely limits the performance of SGE: indeed, it can be seen in Fig. 2 that the fitness curves for SGE become and remain flat after few generations in all the problems for which other variants evolve a larger phenotype. More in general, we think that diversity, redundancy, and locality interact

**Invalidity.** The 4th row of plots in Fig. 2 shows the invalidity, i.e., the tendency to generate null phenotypes. We recall that, by design, it never happens in SGE that a genotype is not mapped to a valid phenotype. Concerning the other three mappers, it can be seen that for three on four problems (all but Text), invalidity tends to set around 5%, slowly decreasing during the evolution: at the beginning of the evolution, for Santa-Fe, GE, BGE, and $\pi$GE exhibit a much larger invalidity ($\approx 15\%$). In the Text problem, the figure is much lower ($\leq 1.5\%$) and the differences among the three mappers are sharper, BGE being the one with the lowest values. We conjecture that there is some dependency between the invalidity and the complexity of the problem grammar: in particular, in the Text grammar, the production rules with the largest number of options are those including only non-terminals, differently from the other problems.

**Redundancy.** The bottommost 4 rows of plots in the figure show the most salient results, i.e., redundancy and locality indexes measured separately for the two operators (mutation and crossover). The redundancy for mutation is very high at the beginning of the evolution for all but the Text problem—between 60% and 90%: this result is somewhat consistent with the findings of [29] which, however, are limited to a static scenario (our generation zero). In regression problems and for GE, BGE, and $\pi$GE, redundancy decreases during the evolution, reaching very low ($\leq 10\%$) or low ($\leq 30\%$) values—however, no significant differences are noticeable among the three variants. Notably, $\pi$GE exhibits a lower redundancy both in Santa-Fe and Text, where GE and BGE behave similarly. Mutation redundancy curves related to SGE show instead a diverse shape: the decreasing is much less important in regression problems and absolute values are in general higher. Moreover, SGE-6 exhibits in general a lower mutation redundancy than SGE-10, which is indeed consistent with the fact that the genotype is much smaller for the former—there are hence fewer codons which do not take part in the mapping.

We think that the findings about mutation redundancy can be explained mainly in terms of the phenotype size. When, at the beginning of the evolution, the phenotypes are small, the redundancy is high for all the GE variants because a large part of the genotype is not expressed (i.e., does not play any role in the mapping procedure); as far as the phenotypes grow, the mutation redundancy settles down to lower values. Since in SGE the phenotype size early reaches and maintains rather small values, redundancy cannot be strongly reduced. From another point of view, we think that mutation redundancy is only marginally related to the modulo and is instead more determined by unexpressed codons—a condition which occurs in all variants.

Concerning crossover, it can be seen that there are two clearly different phenomena. For GE, BGE, and $\pi$GE, redundancy tends to remain around 80%–90% in all cases (with $\pi$GE being always slightly less redundant). We think that this behavior is motivated by how one-point crossover works: the rightmost portions of the two parents are swapped in the offspring; if, however, in the mapping procedure the new genotype portion is never used, the child phenotype is equal to the first parent phenotype. For SGE, instead, the redundancy is initially always lower than for the other GE variants (initial values are consistent with those found in [12]), but quickly increases reaching very high values, close to 100% in almost all cases: moreover, no significant differences exist between SGE-6 and SGE-10. We justify the low initial values by the different working principle of SGE crossover and SGE mapping. With respect to the subsequent very high redundancy values, we think they are a consequence of the very low phenotype diversity in the population after some tens of generations. That is, no matter how good the operator or the mapping procedure are in avoiding redundancy by design if they work on identical or almost identical parents.

**Locality.** Finally, the last two rows of plots in Fig. 2 show the locality for the two operators. Looking at these plots, no clear conclusions can be drawn concerning locality. However, a different behavior of GE, BGE, and $\pi$GE, on one side, and SGE, on the other, can be observed.

Concerning GE, BGE, and $\pi$GE, the locality is always rather low, with a correlation which, in the best case, scores 0.3. Moreover, no significant trend can be observed during the evolution: only for the Polynomial problem some general tendency to increase locality may be observed during the first half of the evolution.

For SGE, the locality greatly varies during the evolution, sometimes being greater than for the other three variants, and often being negative—i.e., the smaller $d_g$, the larger $d_p$—which may appear surprising. We analyzed the collected measures in details and found that the values for $d_p$ and $d_g$ were much smaller for SGE than for the other variants, due to the low diversity in the population (shorter distances), and, in general, assumed few different values, hence causing a very unstable trend of the curve. We motivate this finding as follows: if the population does not provide the crossover with parents which are sufficiently different, the operator fails in generating a new individual which is somewhat midway between the two parents. From a more general point of view, we think that despite the fact that in principle SGE should exhibit a greater locality, the actual presence of such property may be frustrated in actual conditions by other factors, in particular by low diversity.

## 5   Concluding Remarks and Future Work

We considered 4 different variants of GE, including the most recent SGE, and performed an experimental campaign aimed at understanding to which degree locality and redundancy properties are exhibited by those variants during the evolution, rather than statically. To this end, we considered 4 problems, including 3 benchmark problems commonly used in GP and GE approaches assessment, and measured redundancy (by means of an established definition) and locality (for which we propose a definition which generalizes previous definitions) after each generation in 30 independent runs for each problem/variant.

According to our experimental results, previous findings about redundancy and locality in a static scenario are partially confirmed. However, we also found that, considering the dynamic scenario occurring during the evolution, there is a strong interaction between individual size and redundancy, which may positively affect the latter in variable size GE variants (GE, BGE, $\pi$GE). Moreover, locality appears to be influenced by the diversity in the population, which may hamper, in particular, the by-design ability of SGE mapping procedure and operators to favor an high locality.

**Acknowledgements.** The author is grateful to Alberto Bartoli and Fabio Daolio for their insightful comments.

# References

1. Bartoli, A., De Lorenzo, A., Medvet, E., Tarlao, F.: Syntactical similarity learning by means of grammatical evolution. In: Handl, J., Hart, E., Lewis, P.R., López-Ibáñez, M., Ochoa, G., Paechter, B. (eds.) PPSN 2016. LNCS, vol. 9921, pp. 260–269. Springer, Heidelberg (2016). doi:10.1007/978-3-319-45823-6_24
2. Castle, T., Johnson, C.G.: Positional effect of crossover and mutation in grammatical evolution. In: Esparcia-Alcázar, A.I., Ekárt, A., Silva, S., Dignum, S., Uyar, A.Ş. (eds.) EuroGP 2010. LNCS, vol. 6021, pp. 26–37. Springer, Heidelberg (2010). doi:10.1007/978-3-642-12148-7_3
3. Doran, J.E., Michie, D.: Experiments with the graph Traverser program. In: Proceedings of the Royal Society of London A: Mathematical, Physical and Engineering Sciences, vol. 294, pp. 235–259. The Royal Society (1966)
4. Fagan, D., O'Neill, M., Galván-López, E., Brabazon, A., McGarraghy, S.: An analysis of genotype-phenotype maps in grammatical evolution. In: Esparcia-Alcázar, A.I., Ekárt, A., Silva, S., Dignum, S., Uyar, A.Ş. (eds.) EuroGP 2010. LNCS, vol. 6021, pp. 62–73. Springer, Heidelberg (2010). doi:10.1007/978-3-642-12148-7_6
5. Harper, R., Blair, A.: A structure preserving crossover in grammatical evolution. In: 2005 IEEE Congress on Evolutionary Computation, vol. 3, pp. 2537–2544. IEEE (2005)
6. Hugosson, J., Hemberg, E., Brabazon, A., ONeill, M.: An investigation of the mutation operator using different representations in grammatical evolution. In: Proceedings of 2nd International Symposium Advances in Artificial Intelligence and Applications, vol. 2, pp. 409–419 (2007)
7. Hugosson, J., Hemberg, E., Brabazon, A., ONeill, M.: Genotype representations in grammatical evolution. Appl. Soft Comput. **10**(1), 36–43 (2010)
8. Keijzer, M., O'Neill, M., Ryan, C., Cattolico, M.: Grammatical evolution rules: the mod and the bucket rule. In: Foster, J.A., Lutton, E., Miller, J., Ryan, C., Tettamanzi, A. (eds.) EuroGP 2002. LNCS, vol. 2278, pp. 123–130. Springer, Heidelberg (2002). doi:10.1007/3-540-45984-7_12
9. Koza, J.R.: Genetic Programming: On the Programming of Computers by Means of Natural Selection, vol. 1. MIT press, Cambridge (1992)
10. Liepins, G.E., Vose, M.D.: Representational issues in genetic optimization. J. Exp. Theoret. Artif. Intell. **2**(2), 101–115 (1990)
11. Lourenço, N., Pereira, F.B., Costa, E.: SGE: a structured representation for grammatical evolution. In: Bonnevay, S., Legrand, P., Monmarché, N., Lutton, E., Schoenauer, M. (eds.) EA 2015. LNCS, vol. 9554, pp. 136–148. Springer, Heidelberg (2016). doi:10.1007/978-3-319-31471-6_11
12. Lourenço, N., Pereira, F.B., Costa, E.: Unveiling the properties of structured grammatical evolution. Genet. Program. Evol. Mach. **17**, 251–289 (2016)
13. McDermott, J., White, D.R., Luke, S., Manzoni, L., Castelli, M., Vanneschi, L., Jaskowski, W., Krawiec, K., Harper, R., De Jong, K., et al.: Genetic programming needs better benchmarks. In: Proceedings of the 14th Annual conference on Genetic and Evolutionary Computation, pp. 791–798. ACM (2012)
14. Mckay, R.I., Hoai, N.X., Whigham, P.A., Shan, Y., O'Neill, M.: Grammar-based genetic programming: a survey. Genet. Program. Evol. Mach. **11**(3–4), 365–396 (2010)

15. Moraglio, A., Poli, R.: Topological interpretation of crossover. In: Deb, K. (ed.) GECCO 2004. LNCS, vol. 3102, pp. 1377–1388. Springer, Heidelberg (2004). doi:10. 1007/978-3-540-24854-5_131
16. O'Neill, M., Brabazon, A.: Grammatical differential evolution. In: IC-AI, pp. 231–236 (2006)
17. O'Neill, M., McDermott, J., Swafford, J.M., Byrne, J., Hemberg, E., Brabazon, A., Shotton, E., McNally, C., Hemberg, M.: Evolutionary design using grammatical evolution and shape grammars: designing a shelter. Int. J. Des. Eng. 3(1), 4–24 (2010)
18. O'Neill, M., Ryan, C.: Automatic generation of programs with grammatical evolution. In: Proceedings of AICS, pp. 72–78 (1999)
19. O'Neill, M., Ryan, C.: Grammatical evolution. IEEE Trans. Evol. Comput. 5(4), 349–358 (2001)
20. O'Neill, M., Ryan, C., Keijzer, M., Cattolico, M.: Crossover in grammatical evolution. Genet. Program. Evol. Mach. 4(1), 67–93 (2003)
21. O'Neill, M., Brabazon, A.: Grammatical swarm the generation of programs by social programming. Nat. Comput. 5(4), 443–462 (2006)
22. O'Neill, M., Brabazon, A., Nicolau, M., Garraghy, S.M., Keenan, P.: πgrammatical evolution. In: Deb, K. (ed.) GECCO 2004. LNCS, vol. 3103, pp. 617–629. Springer, Heidelberg (2004). doi:10.1007/978-3-540-24855-2_70
23. O'Neill, M., Brabazon, A., Ryan, C., Collins, J.J.: Evolving market index trading rules using grammatical evolution. In: Boers, E.J.W. (ed.) EvoWorkshops 2001. LNCS, vol. 2037, pp. 343–352. Springer, Heidelberg (2001). doi:10.1007/ 3-540-45365-2_36
24. O'Sullivan, J., Ryan, C.: An investigation into the use of different search strategies with grammatical evolution. In: Foster, J.A., Lutton, E., Miller, J., Ryan, C., Tettamanzi, A. (eds.) EuroGP 2002. LNCS, vol. 2278, pp. 268–277. Springer, Heidelberg (2002). doi:10.1007/3-540-45984-7_26
25. Pawlik, M., Augsten, N.: Efficient computation of the tree edit distance. ACM Trans. Database Syst. (TODS) 40(1), 3 (2015)
26. Raidl, G.R., Gottlieb, J.: Empirical analysis of locality, heritability and heuristic bias in evolutionary algorithms: a case study for the multidimensional knapsack problem. Evol. Comput. 13(4), 441–475 (2005)
27. Rothlauf, F.: Design of Modern Heuristics: Principles and Application. Springer Science Business Media, Berlin (2011)
28. Rothlauf, F., Goldberg, D.E.: Redundant representations in evolutionary computation. Evol. Comput. 11(4), 381–415 (2003)
29. Rothlauf, F., Oetzel, M.: On the locality of grammatical evolution. In: Collet, P., Tomassini, M., Ebner, M., Gustafson, S., Ekárt, A. (eds.) EuroGP 2006. LNCS, vol 3905, pp. 320–330. Springer, Heidelberg (2006). doi:10.1007/11729976_29
30. Ryan, C., Azad, A., Sheahan, A., O'Neill, M.: No coercion and no prohibition, a position independent encoding scheme for evolutionary algorithms – the chorus system. In: Foster, J.A., Lutton, E., Miller, J., Ryan, C., Tettamanzi, A. (eds.) EuroGP 2002. LNCS, vol. 2278, pp. 131–141. Springer, Heidelberg (2002). doi:10. 1007/3-540-45984-7_13
31. Ryan, C., Collins, J.J., Neill, M.O.: Grammatical evolution: evolving programs for an arbitrary language. In: Banzhaf, W., Poli, R., Schoenauer, M., Fogarty, T.C. (eds.) EuroGP 1998. LNCS, vol. 1391, pp. 83–96. Springer, Heidelberg (1998). doi:10.1007/BFb0055930

32. Thorhauer, A.: On the non-uniform redundancy in grammatical evolution. In: Handl, J., Hart, E., Lewis, P.R., López-Ibáñez, M., Ochoa, G., Paechter, B. (eds.) PPSN 2016. LNCS, vol. 9921, pp. 292–302. Springer, Heidelberg (2016). doi:10.1007/978-3-319-45823-6_27

33. Thorhauer, A., Rothlauf, F.: On the locality of standard search operators in grammatical evolution. In: Bartz-Beielstein, T., Branke, J., Filipič, B., Smith, J. (eds.) PPSN 2014. LNCS, vol. 8672, pp. 465–475. Springer, Heidelberg (2014). doi:10.1007/978-3-319-10762-2_46

34. Whigham, P.A., Dick, G., Maclaurin, J., Owen, C.A.: Examining the best of both worlds of grammatical evolution. In: Proceedings of the 2015 Annual Conference on Genetic and Evolutionary Computation, pp. 1111–1118. ACM (2015)

35. White, D.R., Mcdermott, J., Castelli, M., Manzoni, L., Goldman, B.W., Kronberger, G., Jaśkowski, W., O'Reilly, U.M., Luke, S.: Better GP benchmarks: community survey results and proposals. Genet. Program. Evol. Mach. 14(1), 3–29 (2013)

# On Evolutionary Approximation of Sigmoid Function for HW/SW Embedded Systems

Milos Minarik[✉] and Lukas Sekanina

Faculty of Information Technology, IT4Innovations Centre of Excellence,
Brno University of Technology, Brno, Czech Republic
{iminarikm,sekanina}@fit.vutbr.cz

**Abstract.** Providing machine learning capabilities on low cost electronic devices is a challenging goal especially in the context of the Internet of Things paradigm. In order to deliver high performance machine intelligence on low power devices, suitable hardware accelerators have to be introduced. In this paper, we developed a method enabling to evolve a hardware implementation together with a corresponding software controller for key components of smart embedded systems. The proposed approach is based on a multi-objective design space exploration conducted by means of extended linear genetic programming. The approach was evaluated in the task of approximate sigmoid function design which is an important component of hardware implementations of neural networks. During these experiments, we automatically re-discovered some approximate sigmoid functions known from the literature. The method was implemented as an extension of an existing platform supporting concurrent evolution of hardware and software of embedded systems.

**Keywords:** Sigmoid · Linear genetic programming · HW/SW co-design

## 1 Introduction

There are many applications in which it is too expensive or impractical to employ a general purpose processor programmed to perform a given task. For example, in small electronic subsystems such as sensors, it is often impossible to perform basic signal processing on a processor because of its relatively high cost. On the other hand, a general-purpose processor could be acceptable in terms of cost, but it is insufficient in delivering expected computing power with a given power budget. This is clearly the case of complex machine learning algorithms such as deep neural networks (DNN) which are currently ported to low power electronic devices. Hence a boom of new hardware implementations of DNN is currently observed in which the requested performance is achieved by using multiple processing units and smart memory access optimized for energy efficiency [11].

As the target processing unit can show both combinational and sequential behavior, its implementation is based on (i) a data processing part composed of functional modules and registers, and (ii) a (micro)program stored in a memory.

© Springer International Publishing AG 2017
J. McDermott et al. (Eds.): EuroGP 2017, LNCS 10196, pp. 343–358, 2017.
DOI: 10.1007/978-3-319-55696-3_22

The circuits of the data processing part can be configured, for example, in terms of the number of registers and their bit width, the number of modules, functions supported by each module, and interconnection options. The program then defines a sequence of operations over the preselected resources. The resulting HW/SW system can be programmed and configured to minimize power consumption, area or delay in a multi-objective optimization scenario. As this optimization task is difficult, a framework was developed which allows the designer to automatically evolve a control program together with the most suitable data processing circuits [10].

The objective of this paper is to extend the framework [10] in order to support the evolutionary design and optimization of elementary processing elements that are typical for recent neural networks implemented on a chip and demonstrate its performance in comparison with human-created designs. It is expected that the improved designs will lead to significant power reduction.

A clear disadvantage of current framework is the inability to effectively use the subsets of modules provided. If the framework were able to use arbitrary combination of modules, all these combinations would have to be specified in the instruction set. If there is a large number of modules, the instruction set can be quite extensive. This significantly increases the probability of disruptive mutation. Therefore it could be beneficial to introduce a mechanism enabling a better control of module utilization at the microinstruction level.

In order to optimize this kind of HW/SW systems, linear genetic programming (LGP) is used. The chromosome then contains two parts: (i) microinstructions to be executed and (ii) definition of the processing element (circuits of the data processing datapath). Resulting Pareto fronts then typically represent various design alternatives, where some solutions show better performance using more hardware resources and other solutions show better cost using fewer hardware components but more complicated program.

The proposed solution is evaluated in the task of sigmoid function approximation which is typically used as an activation function in the artificial neurons. Evolved approximate functions are compared with approximate sigmoid functions available in the literature. We show that a rich spectrum of sigmoid approximations can be obtained using the proposed approach.

The rest of the paper is organized as follows. Section 2 summarizes relevant state of the art and introduces the evolvable HW/SW platform. Section 3 is devoted to the extension of the framework enabling the deactivation of modules at microinstruction level. Experimental results are presented in Sect. 4. Conclusions are given in Sect. 5.

## 2  Previous Work

In this section similar approaches from the GP literature will be briefly reviewed. The proposed approach can be classified as a combination of genetic programming and evolvable hardware. Although we believe our approach is new and unique, there are some features similar to conventional evolutionary algorithms based HW/SW co–design [4,5,14], co-evolution of programs and cellular MOVE processors [15]. The approach having the most features in common with the proposed solution is genetic parallel programming (GPP) [3]. GPP evolves efficient parallel programs by mapping a problem on parallel resources (ALUs), whereas the proposed method is more hardware oriented and allows optimizations at the level of the underlying digital circuits. Therefore we consider it more suitable for embedded systems where area, speed or power consumption is critical. The rest of this section contains a brief description of some basic terms (regarding the framework proposed in [10]) that will be used in the following sections.

## 2.1   Hardware

The HW part is composed of a configurable datapath which is controlled by a microprogram. The structure of the HW part can be seen in Fig. 1. Some parts are fixed and are not affected by the evolution process. These parts are drawn in gray. The other parts are subjects to the evolution. There is a set of registers connected to modules' inputs via multiplexers that are configurable by the microprogram. The outputs of those modules are then connected back to the registers using a set of decoders.

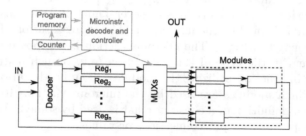

**Fig. 1.** HW architecture

**Registers.** The number of registers is given by the initial specification of the architecture and remains the same throughout the whole evolution. However the bit width of these registers can be affected by the evolution and it can range from 0 to the maximum width specified by the user. When the bit width of the register is set to 0, it is considered unused as it cannot influence the program execution. Therefore it is possible to let the evolution optimize the number of registers even if their number is constant.

**Modules.** The modules can be thought of as black boxes transforming the inputs to the outputs using an arbitrary function. Formally the module can be described as a 6–tuple $M = <n_i, n_o, a, p, d, f_t>$, where $n_i$ is the number of module inputs, $n_o$ the number of outputs, $a$ is the area occupied by the module and $p$ is its power consumption. Function $d$ defines the processing delay of the module. Provided that **D** is the user chosen data type (integer or floating point type), function $d$ can be thought of as a projection $\mathbf{D}^{n_i} \to \mathbb{N}$ as it can use the values of inputs to asses the processing delay. This is useful in the case of modules realizing internally different functions based on the inputs provided. Finally function $f_t$ is the output function transforming the inputs to the outputs. It can be described as $\mathbf{D}^{n_i} \times \mathbf{Q} \to \mathbf{D}^{n_o}$ where $\mathbf{Q}$ is the set of internal module states. The module can therefore retain some internal state during the program execution. However it is crucial that this state is reset between independent runs of the program.

**Architecture.** The HW part is described by following components:

| | |
|---|---|
| $\mathbf{i}$ | the number of inputs |
| $\mathbf{o}$ | the number of outputs |
| $\mathbf{R} = \{r_1, r_2, \ldots r_r\}$ | a set of registers |
| $\mathbf{w} : \mathbf{R} \to \mathbb{N}$ | a function defining the widths of the registers |
| $\mathbf{A} = \{M_1, M_2, \ldots M_m\}$ | a set of available modules |
| $\mathbf{u} : \mathbf{A} \to \{0,1\}$ | a function specifying module utilization |

### 2.2  Software

Each program is composed of instructions $i_1, i_2, \ldots, i_s$, where $s$ is the program size. Each of the instructions can consist of several microinstructions that get executed in the order defined by the instruction. The representation of the microinstruction is depicted in Fig. 2. The microinstruction is composed of the header specifying primarily the type of the instruction (e.g. branch instruction, reset instruction, or instruction utilizing the modules). The header also contains the information, which modules are used by the microinstruction. Right after the header there is a definition of a constant (that is used by some instructions) and definitions of input and output connections or values.

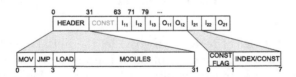

**Fig. 2.** Microinstruction format

### 2.3  Encoding and Search Method

The individuals are represented by chromosomes composed of integers. The first part of the chromosome describes the software part, where the program is encoded in LGP-like style [2]. The second part of the chromosome describes the hardware part. It contains the description of register bit widths, the usage of modules and the $\mu$ permutation encoding the order of modules. The $\mu$ permutation is the inversion sequence describing how many values precede the value at particular position while being greater than that value. The main advantage of this encoding is its straightforward use with genetic operators, because it stays valid even after recombination or mutation. Therefore the software part stays valid even when the order of modules is changed, so there is no need to validate or fix the software part as would be the case if a direct encoding of modules order was used. The details can be found in [17].

The initial population is generated randomly by default. After the generation of initial population the evolution is started. It utilizes two–point crossover operator which performs crossover at the level of instructions in the software part and at the level of modules in the hardware part. A mutation operator implements several modifications of the chromosome. It can change the order of modules, their usage, bit width of the registers, instructions order and inputs, outputs and parameters of microinstructions. The selection is performed by a tournament method with the base of two.

The fitness of an individual is composed of four components: functionality fitness, speed fitness, area fitness and power consumption fitness. Due to the fact that there are multiple components of the fitness, the NSGA-II algorithm is utilized as it supports non-dominated sorting of candidate solutions and multi-objective optimization. However, due to the configurable design of the framework it is possible to change the algorithm easily or to select just a subset of predefined objectives.

## 3   Proposed Extension: Microinstruction-Level Modules Deactivation and a New Mutation Operator

Throughout previous experiments with the framework the disadvantage regarding the strategy in which modules are used was found. As already stated, the header of a microinstruction includes the information about the modules used. This information had to be hardcoded in the instruction set and could not be changed in any way by the EA. Therefore if the architecture should be able to perform various instructions utilizing different combinations of modules, all such instructions would have to be specified in the instruction set. For example, if the architecture employs 8 modules, there should be $\binom{8}{1} + \binom{8}{2} + \binom{8}{3} + \binom{8}{4} + \binom{8}{5} + \binom{8}{6} + \binom{8}{7} = 254$ instructions operating with the modules in the instruction set. The number of instructions can be easily handled as the instruction set is generated automatically. However, there are other problems imposed by the excessive number of instructions.

Let us analyze the following situation. The architecture contains 8 modules where two of them perform addition and another one performs multiplication. The framework discovered a candidate solution providing the expected outputs. This solution is depicted by black parts of Fig. 3. It is obvious that the output of the multiplier module is not used in the first instruction. Presuming the delay of the multiplier is longer than the delay of the adder, this solution is sub-optimal, as the first instruction takes longer than it has to. If the multiplier is disabled at the architecture level, it will be skipped during the execution. This will lead to

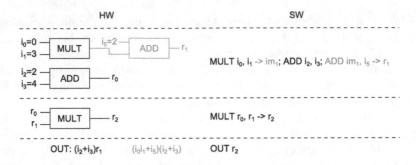

**Fig. 3.** Candidate solution. The instructions are separated by the dashed lines.

shorter delay of the first instruction and thus a better speed fitness. However, as the multiplier is disabled at the architecture level, it will be skipped also during the second instruction execution, therefore the $r_2$ register will not be set and the output of the third instruction will be wrong. The only possible way of achieving better speed fitness is the mutation that replaces the first instruction with the instruction performing only the addition. The mutation operator is implemented in such way it randomly chooses new instruction and generates random input connections for the modules. To achieve the desired effect, the mutation would need to choose the right one of 254 possible instructions and generate the same input connections $(i_2, i_3)$ and output connection $(r_0)$, what is quite unlikely.

To address this issue a modification of the SW part of the chromosome is proposed. This modification adds another property to microinstructions encoded in the SW part of the chromosome (the format of the instructions is not changed). It can be thought of as a bit string defining which modules are utilized by the microinstruction. This property is used during the microinstruction execution and if the module is not utilized by the microinstruction, it is skipped and its outputs are not available as the inputs for subsequent modules. Regarding this change, additional mutation operator was introduced that randomly flips the bits in aforementioned bit string of a particular microinstruction. The situation described in the previous paragraph can therefore be simply solved by deactivating the multiplier in the first microinstruction using this new mutation operator.

The downside of this approach is that the inputs of subsequent modules that were previously connected to the outputs of the deactivated module have to be connected to other points. Let's presume the candidate solution in Fig. 3 contains also the gray part. Presuming the $i_0$ input holds at zero value for all input samples, the multiplication in the first instruction is not needed. If the multiplier is disabled at the microinstruction level, the bottom input of the gray adder in the first instruction has to be connected somewhere else. If it gets connected to $i_0$ or to any of the uninitialized registers, the outputs will remain valid while the speed fitness increases. However, if it gets connected for example to $i_1$, the outputs will be wrong and the functional fitness will decrease.

On the other hand the proposed modification introduces some other advantages. During the experiments with the modified framework it was found that the deactivation of a module at the microinstruction level does not only lead to better speed, area and power consumption fitness values, but also prevents the deactivated modules from spoiling the registers by their unneeded outputs. This side effect is important especially with respect to the architectures utilizing just a few registers. For example, if the output of the multiplier (in the first instruction in Fig. 3) was connected to $r_0$, it would overwrite the value stored by the adder and the outputs would not be correct.

Another advantage of the proposed extension is simpler generation of the instruction set. The simplest approach is to specify just the instruction utilizing all the modules and let the evolution disable the unneeded modules at the microinstruction level. There is also a possibility to divide modules to separate groups. For example, one instruction can utilize all the modules performing Boolean operations and another instruction can utilize the modules performing arithmetic operations as there is usually no point in combining Boolean and arithmetic modules in the scope of one instruction.

## 4 Experiments

The proposed framework will be evaluated in the task of sigmoid function approximation. In order to reduce a bias of the method, only the inputs and expected outputs will be provided in the training set.

### 4.1 Problem Description

The sigmoid function is defined as

$$y = \frac{1}{1 + e^{-x}}$$

and has the derivative

$$\frac{dy}{dx} = y(1 - y)$$

The existence of the first derivative is crucial for artificial neural network training algorithms. Moreover, the sigmoid function is symmetrical with the point of symmetry at $(0, 0.5)$. Therefore, its value for negative values of $x$ can be computed as $y(-x) = 1 - y(x)$.

A straightforward implementation of the sigmoid function is very resource demanding. Hence there is a need for its approximation. Most implementations of such approximations can be divided into three groups: piecewise linear approximations, piecewise second-order approximations and purely combinational approximations. There are also other approaches, e.g. lookup tables or recursive interpolation, but as stated in [16] they are outperformed by the three aforementioned approaches in terms of precision, area or speed, so they will not be further discussed in this paper.

The main goal of the experiments is to find out, whether the proposed framework is capable of finding some of these solutions on its own. As little information as possible was exposed to the framework so the solutions found can be considered as new designs discovered by the evolution. Some decisions were made regarding the inputs and outputs representation. Although the framework is capable of working with floating point numbers, the HW implementation of floating-point arithmetics is more resource demanding compared with fixed-point arithmetics [12]. Hence fixed-point representation of inputs and outputs was chosen. In terms of bit width, we decided to use the representation with 6 fractional bits, as according to [16], this precision should be sufficient for implementing a reliable network forward operation.

## 4.2   Experiment 1: Using the Arithmetic Operations

The first experiment was based on the premise that the sigmoid function could be approximated on some interval by another function using less HW resources, but with required precision.

**Experiment Setup.** The modules allowed for use by the framework were 1 input module, 2 multipliers, 2 ALU modules and 2 bit shifters. The ALU modules can implement various basic arithmetical operations based on the function selection input. Namely these operations include addition, subtraction, incrementing and decrementing by one (in terms of the chosen fixed-point representation, where the increment is equal to 0.015625 for 6 bit inputs).

The training set was composed of 32 evenly distributed samples from the interval <0; 4>. The testing set was the whole set of possible inputs (i.e. 256 values). The parameters of the evolution are summarized in Table 1. The values of population size, crossover and mutation probabilities were chosen empirically based on several hundreds of runs with different values of these parameters. The maximum program length was chosen to be 10 instructions as during the aforementioned runs the output usually took place among the first ten instructions. The maximum logical time was inferred from the delays of the modules available and the maximum program length. The functionality fitness was defined as

$$f_o = \sum_{i=1}^{n_e} \frac{100}{n_s} \frac{1}{1 + (e_i - o_i)^2},$$

**Table 1.** EA parameters used for the first experiment

| Parameter | Value |
| --- | --- |
| Population size | 50 |
| Max. generation count | 200,000 |
| Crossover probability | 0.1 |
| Mutation probability | 0.7 |
| Max. logical time | 300 |
| Max. program length | 10 |

where $e_i$ is $i^{th}$ item from the sequence of expected outputs $(e_1, e_2, \ldots, e_n)$, $o_i$ is the $i^{th}$ output generated by the framework and $n_s$ is the number of samples. The functionality fitness could therefore range from 0 to 100. Other parts of the fitness (speed, area and power consumption) were left to default (see [9]).

**Results.** 100 independent runs were carried out and the solutions found were then examined to asses their quality. The framework was able to find the solution in 19% of runs. The computational effort needed to find the solution was calculated according to [6]. To the best of our knowledge, no study has investigated the same problem. Therefore the sextic polynomial symbolic regression problem was chosen for comparison as it is a problem of comparable complexity and it has already been tested with one of the older versions of the proposed framework. Table 2 shows the computational efforts needed to find the solution by various approaches. Note there was no successful run in the sigmoid approximation experiment when conducted on the original framework [10], while it succeeded in the sextic polynomial regression experiment with the computational effort comparable to other methods. The sigmoid approximation could be therefore considered more complex problem than the sextic polynomial regression.

**Table 2.** Comparison of the computational effort with sextic polynomial symbolic regression

| Problem | Method | Computational effort |
|---------|--------|---------------------|
| Sextic polynomial | GPP $\mathcal{M}_{1,2}$ [7] | 5,310,000 |
| | GP [6] | 1,440,000 |
| | Original framework [9] | 990,000 |
| | GPP $\mathcal{M}_{8,8}$ [7] | 540,000 |
| Sigmoid approximation | Proposed framework | 7,520,000 |
| | Original framework | No solution |

Afterwards the solutions found were examined. One of the solutions found is depicted in Fig. 4. It implements the formula

$$y = 1 - 2^{-1}(1 - 2^{-2}x)^2,$$

which is the expression realizing piecewise second-order approximation proposed by Zhang et al. [18]. Minor differences are present due to the fact the solution found by the framework implements the approximation only for the interval $<0; 4>$. Moreover, the approximation of Zhang et al. utilizes the nature of chosen binary encoding to replace the addition/subtraction by exclusive-or.

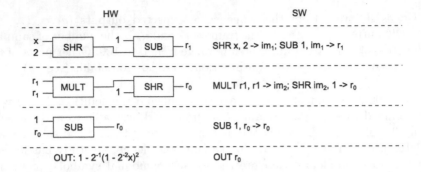

Fig. 4. Example of evolved solution in Experiment 1

Multiple variations of the correct solution were found. Most of them were computing the approximated value in the expanded form as

$$y = -2^{-5}x^2 + 2^{-2}x + 2^{-1}.$$

Some of those solutions were found to be sub-optimal in terms of area and speed, as they used multiplication by constant instead of a simple bit shift. Apart from that some solutions were found that even utilized both multipliers in parallel and therefore achieved lower area fitness but the highest speed fitness of all the solutions found. Figure 5 shows the non-dominated solutions discovered. The tradeoff between the speed and area fitness is clearly seen.

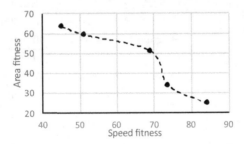

Fig. 5. Nondominated solutions for Experiment 1. The values were computed according to [9] and scaled from the $<0; 1>$ interval to $<0; 100>$ interval, where 0 is the worst fitness and 100 is the best (i.e. no modules used or zero execution time).

## 4.3   Experiment 2: No Multiplication

The framework was able to find a solution utilizing the multiplier module. However, multiplication is quite expensive in terms of the area used. Therefore, the next step was to find a solution that would not need the multiplier. One of such solutions (PLAN approximation) was proposed in [1]. This solution approximates the sigmoid by 4 linear segments (see Fig. 6).

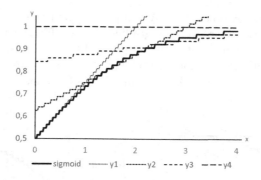

**Fig. 6.** PLAN approximation [1]

Each of the segments is used for some part of the interval. These segments can be described by the equations and appropriate intervals presented in Table 3. As can be seen, the multiplications by coefficients can be replaced by bit shifts. The HW implementation is, however, quite complex as it uses direct transformation of the inputs to outputs. Finding such a complex system at once would probably be nearly impossible. So the goal of our experiment was just to find a piecewise linear approximation of the sigmoid function.

**Table 3.** PLAN approximation of the sigmoid function [1]

| Function | Interval |
|---|---|
| $y_1 = 0.25x + 0.5$ | $0 \leq x < 1$ |
| $y_2 = 0.125x + 0.625$ | $1 \leq x < 2.375$ |
| $y_3 = 0.03125x + 0.84375$ | $2.375 \leq x < 5$ |
| $y_4 = 1$ | $5 \leq x$ |

**Experiment Setup.** The setup of the experiment remained almost the same as in previous experiment except the multipliers being removed. In the first experiment, only the first output for each sample was processed. However in the case of linear approximation, it could be beneficial to process more outputs and choose the one best approximating the sigmoid function for a given sample. This should enable the framework to evolve a program computing and outputting multiple linear approximations.

**Results.** 200 independent runs were performed and the results were examined. In 2.5% of runs a suitable solution was found. Outputs of one of the most precise solutions are depicted in Fig. 7. The solution differs from the original PLAN approximation. This is mainly due to the fact that the PLAN approximation is not bound only to interval <0; 4> as the evolved solution is. The evolved solution utilizes this restriction to approximate the last segment of the interval by constant 0.96875, whereas in PLAN approximation the constant segment is used for inputs $x \geq 5$. Moreover the gradient of the third segment differs from the PLAN approximation. That is, again, probably due to the interval restriction as the evolved solution does not have to approximate values between 4 and 5, where the gradient of the PLAN approximation is feasible.

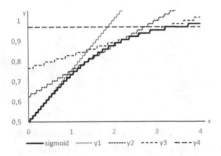

**Fig. 7.** Linear approximation A

**Fig. 8.** Linear approximation B

**Table 4.** Description of segments – solution A

| Function | Interval |
|---|---|
| $y_1 = 0.25x + 0.5$ | $0 \leq x < 1$ |
| $y_2 = 0.125x + 0.625$ | $1 \leq x < 2.5$ |
| $y_3 = 0.0625x + 0.765625$ | $2.5 \leq x < 3$ |
| $y_4 = 0.96875$ | $3 \leq x$ |

**Table 5.** Description of segments – solution B

| Function | Interval |
|---|---|
| $y_1 = 0.25x + 0.5$ | $0 \leq x < 1$ |
| $y_2 = 0.125x + 0.625$ | $1 \leq x < 2.5$ |
| $y_3 = 0.0625x + 0.75$ | $2.5 \leq x$ |

Table 4 shows the description of the linear functions obtained by the analysis of SW part of the evolved solution. The intervals are given by the fitness function (i.e. the segment with the closest value is taken). In the case there were more possible points, the boundary of the interval was chosen to be the number with the least fractional bits for better readability of the table. Another solution is shown in Fig. 8 and the description of the linear segments is given in Table 5. It is similar to solution A, but uses just three linear segments. Moreover, it allows to design the HW implementation of such variation of a PLAN approximation, in which the offsets of the segments of solution B are more feasible.

Finally, the original PLAN approximation and the two proposed solutions were compared in terms of the average and maximum error. The results are listed in Table 6. The maximum error is the same but the average error is smaller for both evolved solutions. Therefore those solutions could be considered superior to original PLAN approximation in the interval <0; 4>. The framework therefore succeeded in evolving the solution suited specifically for this restricted interval and as [16] states, this interval is sufficient in many applications. The solution can therefore be considered as an improvement of an existing solution.

**Table 6.** Comparison of average $E_{avg}$ and maximum $E_{max}$ error

| Approximation | $E_{max}$ | $E_{avg}$ |
|---|---|---|
| Plan approximation | 3.13% | 0.92% |
| Solution A | 3.13% | 0.61% |
| Solution B | 3.13% | 0.83% |

## 4.4   Experiment 3: Combinational Approximation

The last approach is a purely combinational approximation. It is based on the fact that when both the input and output have a bit width restricted to only a few bits, it is possible to perform a direct bit-level mapping. More formally, the bit-level mapping can be expressed as a sum-of-products (SOP). The SOP can be minimized using a procedure such as Quine–McCluskey [8] and the result can be implemented by AND, OR and NOT gates.

**Encoding.** The bit widths should be as small as possible while still maintaining the required precision. As the previous experiments were carried out at the $<0;4>$ interval, the inputs were chosen to have 2 integral bits, so the $<0;4)$ interval is covered. The output is restricted to interval $<0.5;1>$, so the outputs were chosen to have 0 integral and 6 fractional bits to provide the same precision as the previous experiments. The number of input fractional bits was decided to be 3 as, according to [16], it should provide a sufficient precision for neural network operation.

**Experiment Setup.** The input was chosen to have 2 integral and 3 fractional bits, therefore the modules allowed to use by the framework were 5 input modules (one for each bit) providing the bit value and its complement and 20 Boolean modules. Boolean modules can implement bitwise AND, OR, NAND or NOR. These modules were chosen to have 5 inputs. Four of them are used for actual inputs and the last one is used for Boolean operation selection. Four inputs were chosen as it is a typical number of binary inputs for a function that can be realized by a single look-up table in field programmable gate array (FPGA). The number of 20 Boolean modules is quite high compared to experiments performed with the previous version of the framework. It should, however, assure the evolution wouldn't be too limited by available resources and confirm that the proposed modules deactivation works as expected.

As only Boolean operations were used, the inputs were merged into 32 bit integers (according to [13]). The complete set of all the input combinations was chosen as the input set, as the goal of the experiment was to find a solution giving the correct outputs for all the input combinations. As there are only 5 input bits, the evaluation of whole training set could be done in one program execution. The outputs are processed in the same manner. The problem is that there are 6 outputs and it would not be possible to tell, which one should correspond to a particular expected output. Therefore the evolution was performed separately for individual outputs.

**Modification of Boolean Modules Evaluation.** After performing several runs, it was observed, that the candidate solutions tend to output 0 or $-1$ (all bits set). All possible input combinations are processed at once, which has an interesting side-effect. As it is known that none of the 32bit inputs nor the output is 0 or $-1$, these values can be ignored as they only spoil the computation. Thus the modification was made that the Boolean module ignores 0 values on its inputs when operating as AND/NAND and $-1$ values when operating as OR/NOR. This change reportedly lowered the computational effort by approx. two orders of magnitude.

**Results.** As the first fractional bit is known to be 1 for the whole positive domain, the first runs were performed for the second fractional bit of the output. The solution was found and compared to the solution sig_236 proposed in [16] (see Tables 7 and 8). The expressions in the tables use the notation from [16], where the input is of form $x_4 x_3 . x_2 x_1 x_0$. Redundant input bits in expressions in Table 7 are caused by redundant connections of corresponding module. This can happen as such connection influences neither the result correctness nor the speed or area fitness. Such connections could be easily identified and removed manually or some area would have to be assigned to the connections, so the framework would remove them while trying to minimize the area used. Another difference is the presence of $\overline{x_3} \vee \overline{x_0}$ instead of $x_3 \wedge x_0$. According to the rules of Boolean algebra these expressions are equivalent, so it's not a difference in terms of functionality.

**Table 7.** Logic equations of the solution found for the second bit

| Term | Expression |
| --- | --- |
| $im_1$ | $NOR(\overline{x_3}, \overline{x_0})$ |
| $im_2$ | $AND(x_3, x_3, x_1)$ |
| $im_3$ | $AND(x_3, x_2, x_2)$ |
| $im_4$ | $OR(im_2, x_4)$ |
| $Output$ | $OR(im_4, im_3, im_1)$ |

**Table 8.** Logic equations for the second bit of sig_236p [16]

| Term | Expression |
| --- | --- |
| $p_2$ | $AND(x_4)$ |
| $p_4$ | $AND(x_4, \overline{x_3}, \overline{x_2}, x_0)$ |
| $p_{17}$ | $AND(x_3, x_0)$ |
| $p_{19}$ | $AND(x_3, x_1)$ |
| $p_{22}$ | $AND(x_3, x_2)$ |
| $Output$ | $OR(p_2, p_4, p_{17}, p_{19}, p_{22})$ |

The most important difference, however, is the absence of $x_4 \wedge \overline{x_3} \wedge \overline{x_2} \wedge x_0$. After the examination it was concluded that the absence of this expression is correct, because it gets minimized due to $x_4$ input as $x_4 \vee (x_4 \wedge \overline{x_3} \wedge \overline{x_2} \wedge x_0)$ minimizes to $x_4$, which is included. Presence of such expression in sig_236 could possibly be a mistake or some side-effect of the synthesis. This could happen e.g. in the case when some gates are shared by multiple outputs. In such case, it would be in accordance with aforementioned disadvantage of separate outputs evolution. However no evidence has been found to prove this. Afterwards the solutions were successfully found for all subsequent output bits, therefore the evolution succeeded in reinventing the sig_236p approximation presented in [16].

# 5   Conclusions

In this paper the framework for development of small application-specific digital embedded architectures was extended with the possibility to deactivate the modules at microinstruction level. The proposed extension was evaluated by evolving several distinct solutions of sigmoid function approximation. The extended framework was able to evolve variations of two sequential and one combinational well-known sigmoid function approximation algorithms. These results together with the results in [9] and [10] have shown that the framework can be used to find solutions for a wide variety of problems without the need of modifying the underlying algorithms. In most of the experiments, it was sufficient to specify the inputs and expected outputs, choose the modules available and define the functionality component of the fitness function.

The future research regarding this platform will deal with improving overall efficiency of the method. Another possibility would be to assess the extended framework on other complex problems, such as the evolution of convolutional kernels for DNNs.

**Acknowledgments.** This work was supported by the Czech science foundation project GA16-17538S.

# References

1. Amin, H., Curtis, K.M., Hayes-Gill, B.R.: Piecewise linear approximation applied to nonlinear function of a neural network. IEE Proc. Circ. Dev. Syst. **144**(6), 313–317 (1997)
2. Brameier, M., Banzhaf, W.: Linear Genetic Programming. Springer, Berlin (2007)
3. Cheang, S.M., Leung, K.S., Lee, K.H.: Genetic parallel programming: design and implementation. Evol. Comput. **14**(2), 129–156 (2006)
4. Deniziak, S., Gorski, A.: Hardware/software co-synthesis of distributed embedded systems using genetic programming. In: Hornby, G.S., Sekanina, L., Haddow, P.C. (eds.) ICES 2008. LNCS, vol. 5216, pp. 83–93. Springer, Heidelberg (2008). doi:10.1007/978-3-540-85857-7_8
5. Dick, R.P., Jha, N.K.: MOGAC: a multiobjective genetic algorithm for hardware-software cosynthesis of distributed embedded systems. IEEE Trans. CAD Integr. Circ. Syst. **17**(10), 920–935 (1998)
6. Koza, J.R.: Genetic Programming II: Automatic Discovery of Reusable Programs. MIT Press, Cambridge (1994)
7. Leung, K.S., Lee, K.H., Cheang, S.M.: Parallel programs are more evolvable than sequential programs. In: Ryan, C., Soule, T., Keijzer, M., Tsang, E., Poli, R., Costa, E. (eds.) EuroGP 2003. LNCS, vol. 2610, pp. 107–118. Springer, Heidelberg (2003). doi:10.1007/3-540-36599-0_10
8. McCluskey, E.J.: Minimization of Boolean functions. Bell Syst. Tech. J. **35**(6), 1417–1444 (1956)
9. Minarik, M., Sekanina, L.: Concurrent evolution of hardware and software for application-specific microprogrammed systems. In: 2013 IEEE International Conference on Evolvable Systems (ICES), pp. 43–50 (2013). Proceedings of the 2013 IEEE Symposium Series on Computational Intelligence (SSCI). IEEE Computational Intelligence Society

10. Minarik, M., Sekanina, L.: Exploring the search space of hardware/software embedded systems by means of GP. In: Nicolau, M., Krawiec, K., Heywood, M.I., Castelli, M., García-Sánchez, P., Merelo, J.J., Rivas Santos, V.M., Sim, K. (eds.) EuroGP 2014. LNCS, vol. 8599, pp. 112–123. Springer, Heidelberg (2014). doi:10.1007/978-3-662-44303-3_10

11. Misra, J., Saha, I.: Artificial neural networks in hardware: a survey of two decades of progress. Neurocomputing **74**(1–3), 239–255 (2010)

12. Parhami, B.: Computer Arithmetic: Algorithms and Hardware Designs. Oxford University Press, Oxford (2000)

13. Poli, R., Langdon, W.B.: Sub-machine-code genetic programming. In: Advances in Genetic Programming, pp. 301–323. MIT Press, Cambridge (1999)

14. Shang, L., Dick, R.P., Jha, N.K.: SLOPES: hardware-software cosynthesis of low-power real-time distributed embedded systems with dynamically reconfigurable FPGAs. IEEE Trans. CAD Integr. Circ. Syst. **26**(3), 508–526 (2007)

15. Tempesti, G., Mudry, P.A., Zufferey, G.: Hardware/software coevolution of genome programs and cellular processors. In: First NASA/ESA Conference on Adaptive Hardware and Systems (AHS 2006), pp. 129–136. IEEE Computer Society (2006)

16. Tommiska, M.T.: Efficient digital implementation of the sigmoid function for reprogrammable logic. IEE Proc. Comput. Digital Tech. **150**(6), 403–411 (2003)

17. Üçoluk, G.: Genetic algorithm solution of the TSP avoiding special crossover and mutation. In: Sixth Turkish AI and NN Symposium (TAINN VI), Ankara, pp. 57–62 (1997)

18. Zhang, M., Vassiliadis, S., Delgado-Frias, J.G.: Sigmoid generators for neural computing using piecewise approximations. IEEE Trans. Comput. **45**(9), 1045–1049 (1996)

# Author Index

Printed in the United States
By Bookmasters